GREAT YARMOUTH

Great

EARNING

CESSION NO. .... **055670** ..........................

ASS NO. ............ 621 TIM ....................

TE ............ 04/09/03 ..............................

GYC LIBRARY

055670

CH00802942

**Newnes
Workshop
Engineer's
Pocket Book**

# Newnes
# Workshop
# Engineer's
# Pocket Book

Roger Timings

OXFORD   AUCKLAND   BOSTON
JOHANNESBURG   MELBOURNE   NEW DELHI

GREAT YARMOUTH COLLEGE

Newnes
An imprint of Butterworth-Heinemann
Linacre House, Jordan Hill, Oxford OX2 8DP
225 Wildwood Avenue, Woburn, MA 01801-2041
A division of Reed Educational and Professional Publishing Ltd

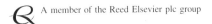 A member of the Reed Elsevier plc group

First published 2000

© Roger Timings 2000

All rights reserved. No part of this publication may be
reproduced in any material form (including photocopying
or storing in any medium by electronic means and whether
or not transiently or incidentally to some other use of
this publication) without the written permission of the
copyright holder except in accordance with the provisions
of the Copyright, Designs and Patents Act 1988 or under
the terms of a licence issued by the Copyright Licensing
Agency Ltd, 90 Tottenham Court Rd, London, England W1P 9HE.
Applications for the copyright holder's written permission
to reproduce any part of this publication should be
addressed to the publishers.

**British Library Cataloguing in Publication Data**
A catalogue record for this book is available from the British Library.

ISBN 0 7506 4719 1

Typeset by Laser Words, Madras, India
Printed in Great Britain by Antony Rowe Ltd, Reading, Berkshire

FOR EVERY TITLE THAT WE PUBLISH, BUTTERWORTH-HEINEMANN
WILL PAY FOR BTCV TO PLANT AND CARE FOR A TREE.

# Contents

## Part 3  Cutting Tools (HSS) and Abrasive Wheels

## Part 4    Miscellaneous

# Preface

This pocket book has been prepared as an aid to practising workshop engineers. The tables have been selected to provide a quick day-to-day reference for useful workshop information. For this reason many of the highly prescriptive British and ISO Standards, necessary for design engineers and managers, have been abridged and simplified in this book. However, wherever this has been done, the reference code for the full Standard is included should this be required.

For easy reference this book is divided into four parts, namely:

1. workshop calculations and conversion tables;
2. threaded fastenings;
3. cutting tools;
4. miscellaneous.

Within these parts, the material has been assembled in a logical sequence for easy reference, and a comprehensive list of contents has been provided which leads the reader directly to the item required. There is also a comprehensive alphabetical index.

Currently, many revisions of the British Standards is taking place. These revisions range from relatively minor amendments to complete withdrawal and replacement. This is necessary to reflect technological changes and to ensure harmonization with international (ISO) requirements. The currency and validity of any Standard can be identified as set out in the following notes which preface the catalogue issued by the British Standards Institute (BSI).

### How to use the BSI catalogue basic details of entries

The list of BSI publications in their catalogue is arranged in numerical order, within each series. The series can be identified from the alphabetical characters which precede the number of the Standard.

For example: BS AU = automobile series, or BSEN = European Standards adopted as British Standards.

Current publications can be identified by the use of bold type for the number of the publication and its title. The revision of any publication automatically supersedes all previous editions of the publication. Only the current editions are listed.

Withdrawn publications can be identified by the use of light type for the number of the publication and its title, and the word 'withdrawn' in parentheses.

### Definitions of entries in the catalogue

An entry appears in the catalogue as in the example below. The various elements are labelled A–J. The key explains each element.

**A**   **B**   **C**                                    **D**

**BS 1361** : 1971 (1986)                    ≠ IEC 269-1

### Specification for cartridge fuses for a.c. circuits in domestic and similar premises.

Requirements, ratings and tests for fuse links, fuse bases and carriers. Dimensions and time/current zones for fuse links. Type I-rate 240 V and 5 A to 45 A for replacement by domestic consumers; Type II-rated 415 V and 60 A, 80 A or 100 A for use by the supply authority in the incoming unit of domestic and similar premises.

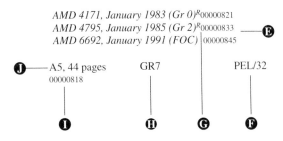

*AMD 4171, January 1983 (Gr 0)*[R]00000821
*AMD 4795, January 1985 (Gr 2)*[R]00000833 ——**E**
*AMD 6692, January 1991 (FOC)* 00000845

**J**—— A5, 44 pages      GR7              PEL/32
00000818

**I**          **H**      **G**      **F**

| | | |
|---|---|---|
| **A** | **BS 1361** | Product identifier. |
| **B** | 1971 | Original publication date. |
| **C** | (1986) | Confirmed in 1986, indicating the continuing currency of the Standard without full revision. |
| **D** | $\equiv$ | An identical Standard: a BSI publication identical in every detail with a corresponding European and/or international Standard. |
| or | $=$ | A (technically) equivalent Standard: a BSI publication in all technical respects the same as a corresponding European and/or international Standard, though the wording and presentation may differ quite extensively. |
| or | $\neq$ | A related but not equivalent Standard: a BSI publication that covers subject matters similar to that covered by a European and/or international Standard. The content however is short of complete identity or technical equivalence. |
| **E** | 00000833 | Unique product code for the amendment. |
| **F** | PEL/32 | BSI Technical Committee responsible for this publication. |
| **G** | *R* | Amendment incorporated in the reprinted text. No 'R' means the amendment is not part of the text. |
| **H** | Gr7 | The group price: refer to the flap on the inside back cover of the catalogue. |

**❶** 00000818      Unique product code.

**❶** A5, 44 pages    Most new and revised Standards are published in A4 size. Sizes other than A4 are listed.

## Amendments

All separate amendments to date of despatch are included with any main publication ordered. Prices are available on application. With the next reprint of the publication the amendment is incorporated into the text which then carries a statement drawing attention to this and includes an indication in the margin at the appropriate places on the amended pages.

## Review

The policy of BSI is for every Standard to be reviewed by the technical committee responsible not more than five years after publication, to establish whether it is still current and, if it is not, to identify and set in hand appropriate action. Circumstances may lead to an earlier review.

When reviewing a Standard, a committee has four options available:

**Withdrawal:** indicating that the Standard is no longer current.

**Declaration of obsolescence:** indicting by amendment that the Standard is not recommended for use in new equipment, but needs to be retained to provide for the servicing of existing equipment that is expected to have a long working life.

**Revision:** involving the procedure for new projects.

**Confirmation:** indicating the continuing currency of the Standard without full revision. Following confirmation of a publication, stock copies are overstamped with the month and year of confirmation.

The latest issue of Standards should always be used in new product designs and equipment. However many products are still being manufactured to obsolescent and obsolete Standards to satisfy a still buoyant demand. This is not only for maintenance purposes but also for current manufacture where market forces have not yet demanded an update in design. This is particularly true of screwed fasteners. For this reason the traditional screw thread tables have been retained and stand alongside the new BS EN requirements.

This pocket book is not a textbook but is a compilation of useful data. The author is indebted to the British Standards Institution for their cooperation in providing up-to-date data in so many technical areas. Unfortunately, limitations of space have allowed only abstracts to be included from the wealth of material provided. The tables in this pocket book should be adequate for day-to-day workshop use. However, where additional information is required, the reader is strongly recommended to consult the complete Standard, industrial manuals or catalogues after an initial perusal of the tables of data found in this book. To this end, an appendix is provided listing the names and addresses of the libraries and institutions where the complete Standards may be consulted or purchased. Many industrial manuals are available free of charge to bona fide users.

Within the restraints of commercial viability, it is still the intention of the author and the publisher to update this book from time to time. Therefore, the author would appreciate (via the publishers) suggestions from the users of this book for additions and/or deletions to be taken into account when producing new editions.

*Roger Timings*

# Part 1

# Conversion Tables and Workshop Calculations

## 1.1 Conversion table: fractions to decimals

| Fraction | Decimal | Fraction | Decimal | Fraction | Decimal |
|---|---|---|---|---|---|
| 1/64 | 0.015 625 | 11/32 | 0.343 750 | 43/64 | 0.671 875 |
| 1/32 | 0.031 250 | 23/64 | 0.359 375 | 11/16 | 0.687 500 |
| 3/64 | 0.046 875 | 3/8 | 0.375 000 | 45/64 | 0.703 125 |
| 1/16 | 0.062 500 | 25/64 | 0.390 625 | 23/32 | 0.718 750 |
| 5/64 | 0.078 125 | 13/32 | 0.406 250 | 47/64 | 0.734 375 |
| 3/32 | 0.093 750 | 27/64 | 0.421 875 | 3/4 | 0.750 000 |
| 7/64 | 0.109 375 | 7/16 | 0.437 500 | 49/64 | 0.765 625 |
| 1/8 | 0.125 000 | 29/64 | 0.453 125 | 25/32 | 0.781 250 |
| 9/64 | 0.140 625 | 15/32 | 0.468 750 | 51/64 | 0.796 875 |
| 5/32 | 0.156 250 | 31/64 | 0.484 375 | 13/16 | 0.812 500 |
| 11/64 | 0.171 875 | 1/2 | 0.500 000 | 53/64 | 0.828 125 |
| 3/16 | 0.187 500 | 33/64 | 0.515 625 | 27/32 | 0.843 750 |
| 13/64 | 0.203 125 | 17/32 | 0.531 250 | 55/64 | 0.859 375 |
| 7/32 | 0.218 750 | 35/64 | 0.546 875 | 7/8 | 0.875 000 |
| 15/64 | 0.234 375 | 9/16 | 0.562 500 | 57/64 | 0.890 625 |
| 1/4 | 0.250 000 | 37/64 | 0.578 125 | 29/32 | 0.906 250 |
| 17/64 | 0.265 625 | 19/32 | 0.593 750 | 59/64 | 0.921 875 |
| 9/32 | 0.281 250 | 39/64 | 0.609 375 | 15/16 | 0.937 500 |
| 19/64 | 0.296 875 | 5/8 | 0.625 000 | 61/64 | 0.953 125 |
| 5/16 | 0.312 500 | 41/64 | 0.640 625 | 31/32 | 0.968 750 |
| 21/64 | 0.328 125 | 21/32 | 0.656 250 | 63/64 | 0.984 375 |

## 1.2 Conversion table: millimetres to inches

| mm | in | mm | in | mm | in |
|---|---|---|---|---|---|
| 0.01 | 0.000 394 | 36 | 1.417 323 | 89 | 3.503 937 |
| 0.02 | 0.000 787 | 37 | 1.456 693 | 90 | 3.543 307 |
| 0.03 | 0.001 181 | 38 | 1.496 063 | 91 | 3.582 677 |
| 0.04 | 0.001 575 | 39 | 1.535 433 | 92 | 3.622 047 |
| 0.05 | 0.001 969 | 40 | 1.574 803 | 93 | 3.661 417 |
| 0.06 | 0.002 362 | 41 | 1.614 173 | 94 | 3.700 788 |
| 0.07 | 0.002 756 | 42 | 1.653 543 | 95 | 3.740 158 |
| 0.08 | 0.003 150 | 43 | 1.692 913 | 96 | 3.779 528 |
| 0.09 | 0.003 543 | 44 | 1.732 283 | 97 | 3.818 898 |
| 0.10 | 0.003 937 | 45 | 1.771 654 | 98 | 3.858 268 |
| 0.20 | 0.007 874 | 46 | 1.811 024 | 99 | 3.897 638 |
| 0.30 | 0.011 810 | 47 | 1.850 394 | 100 | 3.937 008 |
| 0.40 | 0.015 748 | 48 | 1.889 764 | 200 | 7.874 016 |
| 0.50 | 0.019 685 | 49 | 1.929 134 | 300 | 11.811 02 |
| 0.60 | 0.023 622 | 50 | 1.968 504 | 400 | 15.748 03 |
| 0.70 | 0.027 559 | 51 | 2.007 874 | 500 | 19.685 04 |
| 0.80 | 0.031 496 | 52 | 2.047 244 | 600 | 23.622 05 |
| 0.90 | 0.035 433 | 53 | 2.086 614 | 700 | 27.559 06 |
| 1 | 0.039 370 | 54 | 2.125 984 | 800 | 31.496 06 |
| 2 | 0.078 740 | 55 | 2.165 354 | 900 | 35.433 07 |

*(continued)*

**1.2** (*continued*)

| mm | in | mm | in | mm | in |
|----|-----|----|-----|----|-----|
| 3 | 0.118 110 | 56 | 2.204 725 | 1000 | 39.370 08 |
| 4 | 0.157 480 | 57 | 2.244 095 | 1100 | 43.307 09 |
| 5 | 0.196 850 | 58 | 2.283 465 | 1200 | 47.244 09 |
| 6 | 0.236 221 | 59 | 2.322 835 | 1300 | 51.181 10 |
| 7 | 0.275 591 | 60 | 2.362 205 | 1400 | 55.118 11 |
| 8 | 0.314 961 | 61 | 2.401 575 | 1500 | 59.055 12 |
| 9 | 0.354 331 | 62 | 2.440 945 | 1600 | 62.992 13 |
| 10 | 0.393 701 | 63 | 2.480 315 | 1700 | 66.929 14 |
| 11 | 0.433 071 | 64 | 2.519 685 | 1800 | 70.866 14 |
| 12 | 0.472 441 | 65 | 2.559 055 | 1900 | 74.803 15 |
| 13 | 0.511 811 | 66 | 2.598 425 | 2000 | 78.740 16 |
| 14 | 0.551 181 | 67 | 2.637 795 | 2100 | 82.677 17 |
| 15 | 0.590 551 | 68 | 2.677 166 | 2200 | 86.614 17 |
| 16 | 0.629 921 | 69 | 2.716 536 | 2300 | 90.551 19 |
| 17 | 0.669 291 | 70 | 2.755 906 | 2400 | 94.488 19 |
| 18 | 0.708 661 | 71 | 2.795 276 | 2500 | 98.425 2 |
| 19 | 0.748 032 | 72 | 2.834 646 | 2600 | 102.362 2 |
| 20 | 0.787 402 | 73 | 2.874 016 | 2700 | 106.299 2 |
| 21 | 0.826 772 | 74 | 2.913 386 | 2800 | 110.236 2 |
| 22 | 0.866 142 | 75 | 2.952 756 | 2900 | 114.173 2 |
| 23 | 0.905 512 | 76 | 2.992 126 | 3000 | 118.110 2 |
| 24 | 0.944 882 | 77 | 3.031 496 | 3100 | 122.047 2 |
| 25 | 0.984 252 | 78 | 3.070 866 | 3200 | 125.984 3 |
| 26 | 1.023 622 | 79 | 3.110 236 | 3300 | 129.921 3 |
| 27 | 1.062 992 | 80 | 3.149 606 | 3400 | 133.858 3 |
| 28 | 1.102 362 | 81 | 3.188 977 | 3500 | 137.795 3 |
| 29 | 1.141 732 | 82 | 3.228 347 | 3600 | 141.732 3 |
| 30 | 1.181 102 | 83 | 3.267 717 | 3700 | 145.669 3 |
| 31 | 1.220 472 | 84 | 3.307 087 | 3800 | 149.606 3 |
| 32 | 1.259 843 | 85 | 3.346 457 | 3900 | 153.543 3 |
| 33 | 1.299 213 | 86 | 3.385 827 | 4000 | 157.480 3 |
| 34 | 1.338 583 | 87 | 3.425 197 | 4100 | 161.417 3 |
| 35 | 1.377 953 | 88 | 3.464 567 | 4200 | 165.354 3 |

## 1.3 Conversion table: minutes of arc to degrees

| min. | degree | min. | degree | min. | degree |
|------|--------|------|--------|------|--------|
| 0.1 | 0.001 667 | 14 | 0.233 333 | 38 | 0.633 333 |
| 0.2 | 0.003 333 | 15 | 0.250 000 | 39 | 0.650 000 |
| 0.25 | 0.004 167 | 16 | 0.266 667 | 40 | 0.666 667 |
| 0.3 | 0.005 000 | 17 | 0.283 333 | 41 | 0.683 333 |

| | | | | | |
|---|---|---|---|---|---|
| 0.4 | 0.006 667 | 18 | 0.300 000 | 42 | 0.700 000 |
| 0.5 | 0.008 333 | 19 | 0.316 667 | 43 | 0.716 667 |
| 0.6 | 0.010 000 | 20 | 0.333 333 | 44 | 0.733 333 |
| 0.7 | 0.011 667 | 21 | 0.350 000 | 45 | 0.750 000 |
| 0.75 | 0.012 500 | 22 | 0.366 667 | 46 | 0.766 667 |
| 0.8 | 0.013 333 | 23 | 0.383 333 | 47 | 0.783 333 |
| 0.9 | 0.015 000 | 24 | 0.400 000 | 48 | 0.800 000 |
| 1 | 0.016 667 | 25 | 0.416 667 | 49 | 0.816 667 |
| 2 | 0.033 333 | 26 | 0.433 333 | 50 | 0.833 333 |
| 3 | 0.050 000 | 27 | 0.450 000 | 51 | 0.850 000 |
| 4 | 0.066 667 | 28 | 0.466 667 | 52 | 0.866 667 |
| 5 | 0.083 333 | 29 | 0.483 333 | 53 | 0.883 333 |
| 6 | 0.100 000 | 30 | 0.500 000 | 54 | 0.900 000 |
| 7 | 0.116 667 | 31 | 0.516 667 | 55 | 0.916 667 |
| 8 | 0.133 333 | 32 | 0.533 333 | 56 | 0.933 333 |
| 9 | 0.150 000 | 33 | 0.550 000 | 57 | 0.950 000 |
| 10 | 0.166 667 | 34 | 0.566 667 | 58 | 0.966 667 |
| 11 | 0.183 333 | 35 | 0.583 333 | 59 | 0.983 333 |
| 12 | 0.200 000 | 36 | 0.600 000 | 60 | 1.000 000 |
| 13 | 0.216 667 | 37 | 0.616 667 | | |

## 1.4 Circles: areas and circumferences

| Dia. | Area | Cir. | Dia. | Area | Cir. | Dia. | Area | Cir. |
|---|---|---|---|---|---|---|---|---|
| 1 | 0.785 4 | 3.142 | 34 | 907.92 | 106.8 | 67 | 3525.7 | 210.5 |
| 2 | 3.141 6 | 6.283 | 35 | 962.11 | 110.0 | 68 | 3631.7 | 213.6 |
| 3 | 7.068 6 | 9.425 | 36 | 1017.9 | 113.1 | 69 | 3739.3 | 216.8 |
| 4 | 12.566 | 12.57 | 37 | 1075.2 | 116.2 | 70 | 3848.5 | 219.9 |
| 5 | 19.635 | 15.71 | 38 | 1134.1 | 119.4 | 71 | 3959.2 | 223.1 |
| 6 | 28.274 | 18.85 | 39 | 1194.6 | 122.5 | 72 | 4071.5 | 226.2 |
| 7 | 38.485 | 21.99 | 40 | 1256.6 | 125.7 | 73 | 4185.4 | 229.3 |
| 8 | 50.265 | 25.13 | 41 | 1320.3 | 128.8 | 74 | 4300.8 | 232.5 |
| 9 | 63.617 | 28.27 | 42 | 1385.4 | 131.9 | 75 | 4417.9 | 235.6 |
| 10 | 78.540 | 31.42 | 43 | 1452.2 | 135.1 | 76 | 4536.5 | 238.8 |
| 11 | 95.033 | 34.56 | 44 | 1520.5 | 138.2 | 77 | 4656.6 | 241.9 |
| 12 | 113.10 | 37.70 | 45 | 1590.4 | 141.4 | 78 | 4778.4 | 245.0 |
| 13 | 132.73 | 40.84 | 46 | 1661.9 | 144.5 | 79 | 4901.7 | 248.2 |
| 14 | 153.94 | 43.98 | 47 | 1734.9 | 147.7 | 80 | 5026.5 | 251.3 |
| 15 | 176.71 | 47.12 | 48 | 1809.6 | 150.8 | 81 | 5153.0 | 254.5 |
| 16 | 201.06 | 50.27 | 49 | 1885.7 | 153.9 | 82 | 5381.0 | 257.6 |
| 17 | 226.98 | 53.41 | 50 | 1963.5 | 157.1 | 83 | 5410.6 | 260.8 |
| 18 | 254.47 | 56.55 | 51 | 2042.8 | 160.2 | 84 | 5541.8 | 263.9 |
| 19 | 283.53 | 59.69 | 52 | 2123.7 | 163.4 | 85 | 5674.5 | 267.0 |
| 20 | 314.16 | 62.83 | 53 | 2206.2 | 166.5 | 86 | 5808.8 | 270.2 |
| 21 | 346.36 | 65.97 | 54 | 2290.2 | 169.6 | 87 | 5944.7 | 273.3 |
| 22 | 380.13 | 69.11 | 55 | 2375.8 | 172.8 | 88 | 6082.1 | 276.5 |
| 23 | 415.48 | 72.26 | 56 | 2463.0 | 175.9 | 89 | 6221.1 | 279.6 |
| 24 | 452.39 | 75.40 | 57 | 2551.8 | 179.1 | 90 | 6361.7 | 282.7 |
| 25 | 490.87 | 78.54 | 58 | 2642.1 | 182.2 | 91 | 6503.9 | 285.9 |
| 26 | 530.93 | 81.68 | 59 | 2734.0 | 185.4 | 92 | 6647.6 | 289.0 |

(*continued*)

| Dia. | Area | Cir. | Dia. | Area | Cir. | Dia. | Area | Cir. |
|------|--------|-------|------|--------|-------|------|--------|-------|
| 27 | 572.56 | 84.82 | 60 | 2827.4 | 188.4 | 93 | 6792.9 | 292.2 |
| 28 | 616.75 | 87.96 | 61 | 2922.5 | 191.6 | 94 | 6939.8 | 295.3 |
| 29 | 660.52 | 91.11 | 62 | 3019.1 | 194.8 | 95 | 7088.2 | 298.5 |
| 30 | 706.86 | 94.25 | 63 | 3117.2 | 197.9 | 96 | 7238.2 | 301.6 |
| 31 | 754.77 | 97.39 | 64 | 3217.0 | 201.1 | 97 | 7389.8 | 304.7 |
| 32 | 804.25 | 100.5 | 65 | 3318.3 | 204.2 | 98 | 7543.0 | 307.9 |
| 33 | 855.30 | 103.7 | 66 | 3421.2 | 207.3 | 99 | 7697.7 | 311.0 |

area of a circle $= \pi r^2$ or $\pi \dfrac{d^2}{4}$

circumference of a circle $= 2\pi r$ or $\pi d$

where: $r$ = radius of the circle

$d$ = diameter of the circle

## 1.5 Twist drills: nearest equivalent sizes

| Drill designation | | | | Size | |
|--------|----------|--------|--------|---------|-------|
| Number | Fraction | Letter | Metric | Inches | mm |
| 80 | — | — | — | 0.0135 | 0.343 |
| — | — | — | 0.35 | 0.0138 | 0.350 |
| 79 | — | — | — | 0.0145 | 0.368 |
| — | 1/64 | — | — | 0.0156 | 0.396 |
| — | — | — | 0.40 | 0.0158 | 0.400 |
| 78 | — | — | — | 0.0160 | 0.406 |
| — | — | — | 0.45 | 0.0177 | 0.450 |
| 77 | — | — | — | 0.0180 | 0.457 |
| — | — | — | 0.50 | 0.0197 | 0.500 |
| 76 | — | — | — | 0.0200 | 0.508 |
| — | — | — | 0.52 | 0.0205 | 0.520 |
| 75 | — | — | — | 0.0210 | 0.533 |
| — | — | — | 0.55 | 0.0217 | 0.550 |
| 74 | — | — | — | 0.0225 | 0.572 |
| — | — | — | 0.58 | 0.0228 | 0.580 |
| — | — | — | 0.60 | 0.0236 | 0.600 |
| 73 | — | — | — | 0.0240 | 0.610 |
| — | — | — | 0.62 | 0.0244 | 0.620 |
| 72 | — | — | — | 0.0250 | 0.635 |
| — | — | — | 0.65 | 0.0256 | 0.650 |
| 71 | — | — | — | 0.0260 | 0.660 |
| — | — | — | 0.70 | 0.0276 | 0.700 |
| 70 | — | — | — | 0.0280 | 0.711 |
| 69 | — | — | — | 0.0292 | 0.742 |
| — | — | — | 0.75 | 0.0295 | 0.750 |

| | | | | | |
|---|---|---|---|---|---|
| 68 | — | — | — | 0.031 0 | 0.787 |
| — | 1/32 | — | — | 0.031 2 | 0.792 |
| — | — | — | 0.80 | 0.031 5 | 0.800 |
| 67 | — | — | — | 0.032 0 | 0.813 |
| 66 | — | — | — | 0.033 0 | 0.838 |
| — | — | — | 0.85 | 0.033 5 | 0.850 |
| 65 | — | — | — | 0.035 0 | 0.889 |
| — | — | — | 0.90 | 0.035 4 | 0.900 |
| 64 | — | — | — | 0.036 0 | 0.914 |
| 63 | — | — | — | 0.037 0 | 0.940 |
| — | — | — | 0.95 | 0.037 4 | 0.950 |
| 62 | — | — | — | 0.038 0 | 0.965 |
| 61 | — | — | — | 0.039 0 | 0.991 |
| — | — | — | 1.00 | 0.039 4 | 1.000 |
| 60 | — | — | — | 0.040 0 | 1.016 |
| 59 | — | — | — | 0.041 0 | 1.041 |
| — | — | — | 1.05 | 0.041 3 | 1.050 |
| 58 | — | — | — | 0.042 0 | 1.069 |
| 57 | — | — | — | 0.043 0 | 1.092 |
| — | — | — | 1.10 | 0.043 3 | 1.100 |
| — | — | — | 1.15 | 0.045 3 | 1.150 |
| 56 | — | — | — | 0.046 5 | 1.181 |
| — | 3/64 | — | — | 0.046 9 | 1.191 |
| — | — | — | 1.20 | 0.047 2 | 1.200 |
| — | — | — | 1.25 | 0.049 2 | 1.250 |
| — | — | — | 1.30 | 0.051 2 | 1.300 |
| 55 | — | — | — | 0.052 0 | 1.321 |
| — | — | — | 1.35 | 0.053 1 | 1.350 |
| 54 | — | — | — | 0.055 0 | 1.397 |
| — | — | — | 1.40 | 0.055 1 | 1.400 |
| — | — | — | 1.45 | 0.057 1 | 1.450 |
| — | — | — | 1.50 | 0.059 1 | 1.500 |
| 53 | — | — | — | 0.059 5 | 1.511 |
| — | — | — | 1.55 | 0.061 0 | 1.550 |
| — | 1/16 | — | — | 0.062 5 | 1.587 |
| — | — | — | 1.60 | 0.063 0 | 1.600 |
| 52 | — | — | — | 0.063 5 | 1.613 |
| — | — | — | 1.65 | 0.065 0 | 1.650 |
| — | — | — | 1.70 | 0.066 9 | 1.700 |
| 51 | — | — | — | 0.067 0 | 1.702 |
| — | — | — | 1.75 | 0.068 9 | 1.750 |
| 50 | — | — | — | 0.070 0 | 1.778 |
| — | — | — | 1.80 | 0.070 9 | 1.800 |
| — | — | — | 1.85 | 0.072 8 | 1.850 |
| 49 | — | — | — | 0.073 0 | 1.854 |

*(continued)*

7

| Drill designation | | | | Size | |
|---|---|---|---|---|---|
| Number | Fraction | Letter | Metric | Inches | mm |
| — | — | — | 1.90 | 0.0748 | 1.900 |
| 48 | — | — | — | 0.0760 | 1.930 |
| — | — | — | 1.95 | 0.0768 | 1.950 |
| — | 5/64 | — | — | 0.0781 | 1.984 |
| 47 | — | — | — | 0.0785 | 1.994 |
| — | — | — | 2.00 | 0.0787 | 2.000 |
| — | — | — | 2.05 | 0.0807 | 2.050 |
| 46 | — | — | — | 0.0810 | 2.057 |
| 45 | — | — | — | 0.0820 | 2.083 |
| — | — | — | 2.10 | 0.0827 | 2.100 |
| — | — | — | 2.15 | 0.0846 | 2.150 |
| 44 | — | — | — | 0.0860 | 2.184 |
| — | — | — | 2.20 | 0.0866 | 2.200 |
| — | — | — | 2.25 | 0.0886 | 2.250 |
| 43 | — | — | — | 0.0890 | 2.261 |
| — | — | — | 2.30 | 0.0906 | 2.300 |
| — | — | — | 2.35 | 0.0925 | 2.350 |
| 42 | — | — | — | 0.0935 | 2.375 |
| — | 3/32 | — | — | 0.0937 | 2.380 |
| — | — | — | 2.40 | 0.0945 | 2.400 |
| 41 | — | — | — | 0.0960 | 2.438 |
| — | — | — | 2.45 | 0.0965 | 2.450 |
| 40 | — | — | — | 0.0980 | 2.489 |
| — | — | — | 2.50 | 0.0984 | 2.500 |
| 39 | — | — | — | 0.0995 | 2.527 |
| 38 | — | — | — | 0.1015 | 2.578 |
| — | — | — | 2.60 | 0.1024 | 2.600 |
| 37 | — | — | — | 0.1040 | 2.642 |
| — | — | — | 2.70 | 0.1063 | 2.700 |
| 36 | — | — | — | 0.1065 | 2.705 |
| — | — | — | 2.75 | 0.1083 | 2.750 |
| — | 7/64 | — | — | 0.1094 | 2.779 |
| 35 | — | — | — | 0.1100 | 2.794 |
| — | — | — | 2.80 | 0.1102 | 2.800 |
| 34 | — | — | — | 0.1110 | 2.819 |
| 33 | — | — | — | 0.1130 | 2.870 |
| — | — | — | 2.90 | 0.1142 | 2.900 |
| 32 | — | — | — | 0.1160 | 2.946 |
| — | — | — | 3.00 | 0.1181 | 3.000 |
| 31 | — | — | — | 0.1200 | 3.048 |

| | | | | | |
|---|---|---|---|---|---|
| — | — | — | 3.10 | 0.1220 | 3.100 |
| — | 1/8 | — | — | 0.1250 | 3.175 |
| — | — | — | 3.20 | 0.1260 | 3.200 |
| — | — | — | 3.25 | 0.1280 | 3.250 |
| 30 | — | — | — | 0.1285 | 3.264 |
| — | — | — | 3.30 | 0.1299 | 3.300 |
| — | — | — | 3.40 | 0.1339 | 3.400 |
| 29 | — | — | — | 0.1360 | 3.454 |
| — | — | — | 3.50 | 0.1378 | 3.500 |
| 28 | — | — | — | 0.1405 | 3.569 |
| — | 9/64 | — | — | 0.1406 | 3.571 |
| — | — | — | 3.60 | 0.1417 | 3.600 |
| 27 | — | — | — | 0.1440 | 3.658 |
| — | — | — | 3.70 | 0.1457 | 3.700 |
| 26 | — | — | — | 0.1470 | 3.734 |
| — | — | — | 3.75 | 0.1476 | 3.750 |
| 25 | — | — | — | 0.1495 | 3.797 |
| — | — | — | 3.80 | 0.1496 | 3.800 |
| 24 | — | — | — | 0.1520 | 3.861 |
| — | — | — | 3.90 | 0.1535 | 3.900 |
| 23 | — | — | — | 0.1540 | 3.912 |
| — | 5/32 | — | — | 0.1562 | 3.967 |
| 22 | — | — | — | 0.1570 | 3.998 |
| — | — | — | 4.00 | 0.1575 | 4.000 |
| 21 | — | — | — | 0.1590 | 4.039 |
| 20 | — | — | — | 0.1610 | 4.089 |
| — | — | — | 4.10 | 0.1614 | 4.100 |
| — | — | — | 4.20 | 0.1654 | 4.200 |
| 19 | — | — | — | 0.1660 | 4.216 |
| — | — | — | 4.25 | 0.1673 | 4.250 |
| — | — | — | 4.30 | 0.1639 | 4.300 |
| 18 | — | — | — | 0.1695 | 4.305 |
| — | 11/64 | — | — | 0.1719 | 4.366 |
| 17 | — | — | — | 0.1730 | 4.394 |
| — | — | — | 4.40 | 0.1732 | 4.400 |
| 16 | — | — | — | 0.1770 | 4.496 |
| — | — | — | 4.50 | 0.1772 | 4.500 |
| 15 | — | — | — | 0.1800 | 4.572 |
| — | — | — | 4.60 | 0.1811 | 4.600 |
| 14 | — | — | — | 0.1820 | 4.623 |
| 13 | — | — | 4.70 | 0.1850 | 4.700 |
| — | — | — | 4.75 | 0.1870 | 4.750 |
| — | 3/16 | — | — | 0.1875 | 4.762 |

*(continued)*

| Drill designation | | | | Size | |
|---|---|---|---|---|---|
| Number | Fraction | Letter | Metric | Inches | mm |
| 12 | — | — | 4.80 | 0.1890 | 4.800 |
| 11 | — | — | — | 0.1910 | 4.851 |
| — | — | — | 4.90 | 0.1929 | 4.900 |
| 10 | — | — | — | 0.1935 | 4.915 |
| 9 | — | — | — | 0.1960 | 4.978 |
| — | — | — | 5.00 | 0.1968 | 5.000 |
| — | — | — | 5.05 | 0.1990 | 5.050 |
| 8 | — | — | — | 0.2000 | 5.080 |
| — | — | — | 5.10 | 0.2008 | 5.100 |
| 7 | — | — | — | 0.2010 | 5.105 |
| — | 13/64 | — | — | 0.2031 | 5.159 |
| 6 | — | — | — | 0.2040 | 5.182 |
| — | — | — | 5.20 | 0.2047 | 5.200 |
| 5 | — | — | — | 0.2055 | 5.220 |
| — | — | — | 5.25 | 0.2067 | 5.250 |
| — | — | — | 5.30 | 0.2087 | 5.300 |
| 4 | — | — | — | 0.2090 | 5.309 |
| — | — | — | 5.40 | 0.2126 | 5.400 |
| 3 | — | — | — | 0.2130 | 5.410 |
| — | — | — | 5.50 | 0.2165 | 5.500 |
| — | 7/32 | — | — | 0.2187 | 5.555 |
| — | — | — | 5.60 | 0.2205 | 5.600 |
| 2 | — | — | — | 0.2210 | 5.613 |
| — | — | — | 5.70 | 0.2244 | 5.700 |
| — | — | — | 5.75 | 0.2264 | 5.750 |
| 1 | — | — | — | 0.2280 | 5.791 |
| — | — | — | 5.80 | 0.2283 | 5.800 |
| — | — | — | 5.90 | 0.2323 | 5.900 |
| — | — | A | — | 0.2340 | 5.944 |
| — | 15/64 | — | — | 0.2344 | 5.954 |
| — | — | — | 6.00 | 0.2362 | 6.000 |
| — | — | B | — | 0.2380 | 6.045 |
| — | — | — | 6.10 | 0.2402 | 6.100 |
| — | — | C | — | 0.2420 | 6.147 |
| — | — | — | 6.20 | 0.2441 | 6.200 |
| — | — | D | — | 0.2460 | 6.248 |
| — | — | — | 6.25 | 0.2461 | 6.250 |
| — | — | — | 6.30 | 0.2480 | 6.300 |
| — | 1/4 | E | — | 0.2500 | 6.350 |
| — | — | — | 6.40 | 0.2520 | 6.400 |

| | | | | | |
|---|---|---|---|---|---|
| — | — | — | 6.50 | 0.2559 | 6.500 |
| — | — | F | — | 0.2570 | 6.528 |
| — | — | — | 6.60 | 0.2598 | 6.600 |
| — | — | G | — | 0.2610 | 6.629 |
| — | — | — | 6.70 | 0.2638 | 6.700 |
| — | 17/64 | — | — | 0.2656 | 6.746 |
| — | — | — | 6.75 | 0.2657 | 6.750 |
| — | — | H | — | 0.2660 | 6.756 |
| — | — | — | 6.80 | 0.2677 | 6.800 |
| — | — | — | 6.90 | 0.2717 | 6.900 |
| — | — | I | — | 0.2720 | 6.909 |
| — | — | — | 7.00 | 0.2756 | 7.000 |
| — | — | J | — | 0.2770 | 7.036 |
| — | — | — | 7.10 | 0.2795 | 7.100 |
| — | — | K | — | 0.2810 | 7.137 |
| — | 9/32 | — | — | 0.2812 | 7.142 |
| — | — | — | 7.20 | 0.2834 | 7.200 |
| — | — | — | 7.25 | 0.2854 | 7.250 |
| — | — | — | 7.30 | 0.2874 | 7.300 |
| — | — | L | — | 0.2900 | 7.366 |
| — | — | — | 7.40 | 0.2913 | 7.400 |
| — | — | M | — | 0.2950 | 7.493 |
| — | — | — | 7.50 | 0.2953 | 7.500 |
| — | 19/64 | — | — | 0.2969 | 7.541 |
| — | — | — | 7.60 | 0.2992 | 7.600 |
| — | — | N | — | 0.3020 | 7.671 |
| — | — | — | 7.70 | 0.3031 | 7.700 |
| — | — | — | 7.75 | 0.3051 | 7.750 |
| — | — | — | 7.80 | 0.3071 | 7.800 |
| — | — | — | 7.90 | 0.3110 | 7.900 |
| — | 5/16 | — | — | 0.3125 | 7.937 |
| — | — | — | 8.00 | 0.3150 | 8.000 |
| — | — | O | — | 0.3160 | 8.026 |
| — | — | — | 8.10 | 0.3189 | 8.100 |
| — | — | — | 8.20 | 0.3228 | 8.200 |
| — | — | P | — | 0.3230 | 8.204 |
| — | — | — | 8.25 | 0.3248 | 8.250 |
| — | — | — | 8.30 | 0.3268 | 8.300 |
| — | 21/64 | — | — | 0.3281 | 8.334 |
| — | — | — | 8.40 | 0.3307 | 8.400 |
| — | — | Q | — | 0.3320 | 8.433 |
| — | — | — | 8.50 | 0.3346 | 8.500 |
| — | — | — | 8.60 | 0.3386 | 8.600 |
| — | — | R | — | 0.3390 | 8.611 |

*(continued)*

| Drill designation | | | | Size | |
|---|---|---|---|---|---|
| Number | Fraction | Letter | Metric | Inches | mm |
| — | — | — | 8.70 | 0.3425 | 8.700 |
| — | 11/32 | — | — | 0.3437 | 8.730 |
| — | — | — | 8.75 | 0.3445 | 8.750 |
| — | — | — | 8.80 | 0.3465 | 8.800 |
| — | — | S | — | 0.3480 | 8.839 |
| — | — | — | 8.90 | 0.3504 | 8.900 |
| — | — | — | 9.00 | 0.3543 | 9.000 |
| — | — | T | — | 0.3580 | 9.093 |
| — | — | — | 9.10 | 0.3583 | 9.100 |
| — | 23/64 | — | — | 0.3594 | 9.129 |
| — | — | — | 9.20 | 0.3622 | 9.200 |
| — | — | — | 9.25 | 0.3642 | 9.250 |
| — | — | — | 9.30 | 0.3661 | 9.300 |
| — | — | U | — | 0.3680 | 9.347 |
| — | — | — | 9.40 | 0.3701 | 9.400 |
| — | — | — | 9.50 | 0.3740 | 9.500 |
| — | 3/8 | — | — | 0.3750 | 9.525 |
| — | — | V | — | 0.3770 | 9.576 |
| — | — | — | 9.60 | 0.3780 | 9.600 |
| — | — | — | 9.70 | 0.3819 | 9.700 |
| — | — | — | 9.75 | 0.3839 | 9.750 |
| — | — | — | 9.80 | 0.3858 | 9.800 |
| — | — | W | — | 0.3860 | 9.804 |
| — | — | — | 9.90 | 0.3898 | 9.900 |
| — | 25/64 | — | — | 0.3906 | 9.921 |
| — | — | — | 10.00 | 0.3937 | 10.000 |
| — | — | X | — | 0.3970 | 10.084 |
| — | — | — | 10.10 | 0.3976 | 10.100 |
| — | — | — | 10.25 | 0.4035 | 10.250 |
| — | — | Y | — | 0.4040 | 10.262 |
| — | 13/32 | — | — | 0.4062 | 10.317 |
| — | — | Z | — | 0.4130 | 10.490 |
| — | — | — | 10.50 | 0.4134 | 10.500 |
| — | 27/64 | — | — | 0.4219 | 10.716 |
| — | — | — | 10.75 | 0.4232 | 10.750 |
| — | — | — | 11.00 | 0.4331 | 11.000 |
| — | 7/16 | — | — | 0.4375 | 11.112 |
| — | — | — | 11.25 | 0.4429 | 11.250 |
| — | — | — | 11.50 | 0.4528 | 11.500 |

| | | | | | |
|---|---|---|---|---|---|
| — | 29/64 | — | — | 0.4531 | 11.509 |
| — | — | — | 11.75 | 0.4626 | 11.750 |
| — | 15/32 | — | — | 0.4687 | 11.905 |
| — | — | — | 12.00 | 0.4724 | 12.000 |
| — | — | — | 12.25 | 0.4823 | 12.250 |
| — | 31/64 | — | — | 0.4844 | 12.304 |
| — | — | — | 12.50 | 0.4921 | 12.500 |
| — | 1/2 | — | — | 0.5000 | 12.700 |
| — | — | — | 12.75 | 0.5020 | 12.750 |

## 1.6 Wire gauge equivalents

Standard Wire Gauge (SWG) for sheet metal, wire and rods.

| SWG No. | Size, in | Size, mm | SWG No. | Size, in | Size, mm |
|---|---|---|---|---|---|
| 1 | 0.300 | 7.62 | 16 | 0.064 | 1.62 |
| 2 | 0.276 | 7.06 | 17 | 0.056 | 1.42 |
| 3 | 0.252 | 6.40 | 18 | 0.048 | 1.22 |
| 4 | 0.232 | 5.89 | 19 | 0.040 | 1.02 |
| 5 | 0.212 | 5.38 | 20 | 0.036 | 0.91 |
| 6 | 0.192 | 4.88 | 21 | 0.032 | 0.81 |
| 7 | 0.176 | 4.46 | 22 | 0.028 | 0.71 |
| 8 | 0.160 | 4.06 | 23 | 0.024 | 0.61 |
| 9 | 0.144 | 3.66 | 24 | 0.022 | 0.56 |
| 10 | 0.128 | 3.24 | 25 | 0.020 | 0.51 |
| 11 | 0.116 | 2.94 | 26 | 0.018 | 0.46 |
| 12 | 0.104 | 2.64 | 27 | 0.016 | 0.41 |
| 13 | 0.092 | 2.34 | 28 | 0.0148 | 0.376 |
| 14 | 0.080 | 2.03 | 29 | 0.0136 | 0.345 |
| 15 | 0.072 | 1.83 | 30 | 0.012 | 0.304 |

## 1.7 Mensuration of plane figures

**Square**

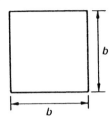

$$\text{area} = b^2$$
$$\text{length of diagonal} = \sqrt{(2)} \times b$$

Rectangle

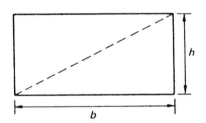

$$\text{area} = b \times h$$
$$\text{length of diagonal} = \sqrt{(b^2 + h^2)}$$

Parallelogram

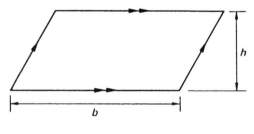

$$\text{area} = b \times h$$

Trapezium

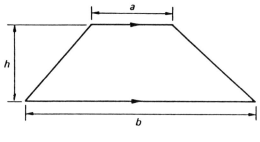

$$\text{area} = \tfrac{1}{2} \times (a + b) \times h$$

Triangle

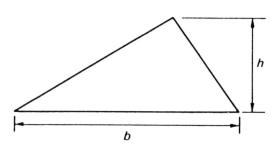

$$\text{area} = \tfrac{1}{2} \times b \times h$$

Circle

$$\text{area} = \pi \times r^2$$
$$\text{perimeter} = 2 \times \pi \times r$$

Sector of circle

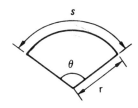

$$\text{area} = \tfrac{1}{2} \times r^2 \times \theta$$
$$\text{arc length } s = r \times \theta$$
$$(\theta \text{ is in radians})$$

Ellipse

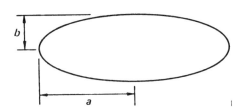

$$\text{area} = \pi \times a \times b$$
$$\text{perimeter} = \pi \times (a + b)$$

Irregular plane

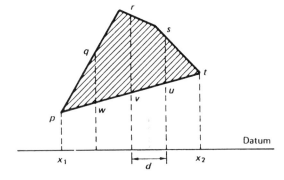

Several methods are used to find the shaded area, such as the mid-ordinate rule, the trapezoidal rule and

Simpon's rule. As an example of these, Simpson's rule is as shown. Divide $x_1 x_2$ into an even number of equal parts of width $d$. Let $p$, $q$, $r$, ... be the lengths of vertical lines measured from some datum, and let $A$ be the approximate area of the irregular plane, shown shaded. Then

$$A = \frac{d}{3}[(p + t) + 4(q + s) + 2r]$$
$$- \frac{d}{3}[(p + t) + 4(u + w) + 2v]$$

In general, the statement of Simpson's rule is

approximate area = $(d/3) \times$ [(first + last) + 4 $\times$ (sum of evens) + 2 $\times$ (sum of odds)]

where first, last, evens, odds refer to ordinate lengths and $d$ is the width of the equal parts of the datum line.

## 1.8 Mensuration of solids

### Rectangular prism

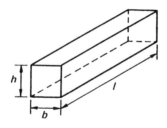

volume = $bhl$

total surface area = $2(bh + hl + lb)$

Cylinder

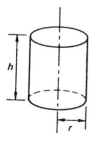

volume = $\pi r^2 h$

total surface area = $2\pi r(r + h)$

Cone

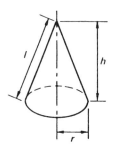

volume $= (1/3)\pi r^2 h$

total surface area $= \pi r(l + r)$

Frustrum of cone

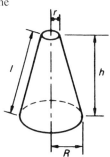

volume $= (1/3)\pi h(R^2 + Rr + r^2)$

total surface area $= \pi l(R + r) + \pi(R^2 + r^2)$

Sphere

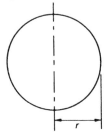

volume $= (4/3)\pi r^3$

total surface area $= 4\pi r^2$

18

Zone of sphere

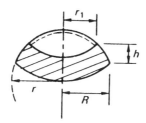

$$\text{volume} = (\pi h/6)(h^2 + 3R^2 + 3r_1^2)$$

$$\text{total surface area} = 2\pi rh + \pi(R^2 + r_1^2)$$

where $r$ is the radius of the sphere

Pyramid

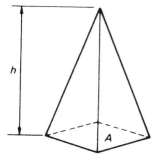

$\text{volume} = (1/3)Ah$

where $A$ is the area of the base and $h$ is the perpendicular height

Regular solids

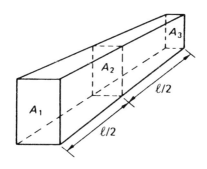

The volume of any regular solid can be found by using the prismoidal rule. Three parallel planes of areas $A_1$, $A_3$, $A_2$, are considered to be at the ends and at the centre of the solid respectively. Then

$$\text{volume} = (l/6)(A_1 + 4A_2 + A_3)$$

where $l$ is the length of the solid.

Irregular solids

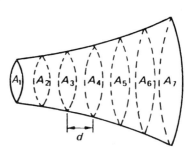

Various methods can be used to determine volumes of irregular solids; one of these is by applying the principles of Simpson's rule (see Section 1.7). The solid is considered to be divided into an even number of sections by equally spaced, parallel planes, distance $d$ apart and having areas of $A_1, A_2, A_3, \ldots$. Assuming, say, seven such planes, then approximate volume $= (d/3)[(A_1 + A_7) + 4(A_2 + A_4 + A_6) + 2(A_3 + A_5)]$.

## 1.9 Taper systems (metric units)

### 1.9.1 Self-holding tapers

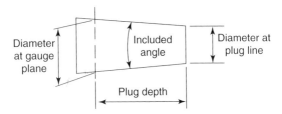

| Taper designation | Taper ratio | Included angle of taper | Diameter at gauge plane | Plug depth | Diameter at plug line |
|---|---|---|---|---|---|
| No. 4 metric 5% | 1:20 = 0.05 | 2°51.85' | 4.0 | 23.0 | 2.9 |
| No. 6 metric 5% | 1:20 = 0.05 | 2°51.85' | 6.0 | 32.0 | 4.4 |
| No. 0 Morse | 1:19.212 = 0.052 05 | 2°52.54' | 9.045 | 50.0 | 6.4 |
| No. 1 Morse | 1:20.047 = 0.049 88 | 2°51.45' | 12.065 | 53.5 | 9.4 |
| No. 2 Morse | 1:20.020 = 0.049 95 | 2°51.68' | 17.780 | 64.0 | 14.6 |
| No. 3 Morse | 1:19.922 = 0.050 20 | 2°52.52' | 23.825 | 81.0 | 19.8 |
| No. 4 Morse | 1:19.254 = 0.051 94 | 2°58.51' | 31.267 | 102.5 | 25.9 |
| No. 5 Morse | 1:19.002 = 0.052 63 | 3°00.87' | 44.399 | 129.5 | 37.6 |
| No. 6 Morse | 1:19.180 = 0.052 14 | 2°69.19' | 63.348 | 182.0 | 53.9 |
| No. 1 B & S | 1:23.904 = 0.041 83 | 2°23.79' | 6.076 | 23.813 | 5.080 |
| No. 2 B & S | 1:23.904 = 0.041 83 | 2°23.79' | 7.612 | 30.163 | 6.350 |
| No. 3 B & S | 1:23.904 = 0.041 83 | 2°23.79' | 9.530 | 38.100 | 7.938 |
| No. 80 metric 5% | 1:20 = 0.05 | 2°51.85' | 80.0 | 196.0 | 70.2 |
| No. 100 metric 5% | 1:20 = 0.05 | 2°51.85' | 100.0 | 232.0 | 88.4 |
| No. 120 metric 5% | 1:20 = 0.05 | 2°51.85' | 120.0 | 268.0 | 106.6 |
| No. 160 metric 5% | 1:20 = 0.05 | 2°51.85' | 160.0 | 340.0 | 143.0 |
| No. 200 metric 5% | 1:20 = 0.05 | 2°51.85' | 200.0 | 412.0 | 179.4 |

*Note:* B & S = Brown and Sharpe taper system.

21

### 1.9.2 Quick-release tapers (milling machine tapers for spindle nozes)

| Taper designation | Taper ratio | Included angle of taper | Diameter at gauge plane | Plug depth | Diameter at plug line |
|---|---|---|---|---|---|
| No. 30 MMT | 7:24 | 16°35.68′ | 31.75 | 47.625 | 17.859 |
| No. 40 MMT | 7:24 | 16°35.68′ | 44.45 | 68.250 | 24.539 |
| No. 50 MMT | 7:24 | 16°35.68′ | 69.85 | 101.600 | 40.208 |
| No. 60 MMT | 7:24 | 16°35.68′ | 107.95 | 161.925 | 60.721 |

### 1.10 Taper systems (inch units)

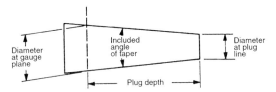

## 1.10.1 Self-holding tapers

| Designation of taper | Taper per foot on diameter | Taper per inch on diameter | Included angle of taper | Diameter at gauge plane | Plug depth | Diameter at plug line |
|---|---|---|---|---|---|---|
| No. 2 metric | 0.600 0 | 0.050 0 | 2°51.85′ | 0.078 7 (2 mm) | 0.472 4 (12 mm) | 0.055 1 |
| No. 3 metric | 0.600 0 | 0.050 0 | 2°51.85′ | 0.118 1 (3 mm) | 0.669 3 (17 mm) | 0.084 6 |
| No. 4 metric | 0.600 0 | 0.050 0 | 2°51.85′ | 0.157 5 (4 mm) | 0.905 5 (23 mm) | 0.112 2 |
| No. 1 B & S | 0.502 0 | 0.041 8 | 2°23.79′ | 0.239 2 | 15/16 | 0.200 0 |
| No. 2 B & S | 0.502 0 | 0.041 8 | 2°23.79′ | 0.299 7 | 13/16 | 0.250 0 |
| No. 3 B & S | 0.502 0 | 0.041 8 | 2°23.79′ | 0.375 2 | 11/2 | 0.312 5 |
| No. 1 Morse | 0.598 6 | 0.049 9 | 2°51.45′ | 0.475 0 | 21/8 | 0.369 0 |
| No. 2 Morse | 0.599 4 | 0.049 9 | 2°51.68′ | 0.700 0 | 29/16 | 0.572 0 |
| No. 3 Morse | 0.602 3 | 0.050 2 | 2°52.52′ | 0.938 0 | 33/16 | 0.778 0 |
| No. 4 Morse | 0.623 3 | 0.051 9 | 2°58.51′ | 1.231 0 | 41/16 | 1.020 0 |
| No. 5 Morse | 0.631 5 | 0.052 6 | 3°00.87′ | 1.748 0 | 53/16 | 1.475 0 |
| No. 6 Morse | 0.625 6 | 0.052 1 | 2°59.19′ | 2.494 0 | 71/4 | 2.116 0 |

*Note:* B & S = Brown and Sharpe taper system.

23

## 1.10.2 Quick-release tapers (milling machines)

| Desig-nation of taper | Taper per foot on diameter | Taper per inch on diameter | Included angle of taper | Diameter at gauge plane | Plug depth | Diameter at plug line |
|---|---|---|---|---|---|---|
| No. 30 MMT | 3.500 | 0.291 7 | 16°35.56′ | 1.250 | 1.875 | 0.703 1 |
| No. 40 MMT | 3.500 | 0.291 7 | 16°35.56′ | 1.750 | 2.687 | 0.966 1 |
| No. 50 MMT | 3.500 | 0.291 7 | 16°35.56′ | 2.750 | 4.000 | 1.583 3 |
| No. 60 MMT | 3.500 | 0.291 7 | 16°35.56′ | 4.250 | 6.375 | 2.390 6 |

## 1.11 Chordal distances on pitch circles

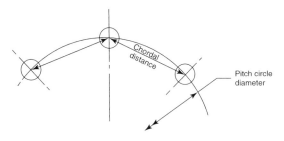

To calculate the chordal distance, for any given number of chords, multiply the pitch circle diameter by the factor given in the following table.

### Example 1.11.1

Calculate the chordal distance for the equal spacing of 8 holes (8 chords) on a pitch circle of 100 mm diameter.

From the table the factor for 8 chords is 0.382 7, therefore the chordal distance = 100 mm × 0.382 7 = 38.27 mm

| No. of chords | Multiply dia. by | No. of chords | Multiply dia. by | No. of chords | Multiply dia. by | No. of chords | Multiply dia. by | No. of chords | Multiply dia. by |
|---|---|---|---|---|---|---|---|---|---|
| 3 | 0.8660 | 23 | 0.1362 | 43 | 0.0730 | 63 | 0.0499 | 83 | 0.0378 |
| 4 | 0.7071 | 24 | 0.1305 | 44 | 0.0713 | 64 | 0.0491 | 84 | 0.0374 |
| 5 | 0.5878 | 25 | 0.1253 | 45 | 0.0698 | 65 | 0.0483 | 85 | 0.0370 |
| 6 | 0.5000 | 26 | 0.1205 | 46 | 0.0682 | 66 | 0.0476 | 86 | 0.0365 |
| 7 | 0.4339 | 27 | 0.1161 | 47 | 0.0668 | 67 | 0.0469 | 87 | 0.0361 |
| 8 | 0.3827 | 28 | 0.1120 | 48 | 0.0654 | 68 | 0.0462 | 88 | 0.0357 |
| 9 | 0.3420 | 29 | 0.1081 | 49 | 0.0641 | 69 | 0.0455 | 89 | 0.0353 |
| 10 | 0.3090 | 30 | 0.1045 | 50 | 0.0628 | 70 | 0.0449 | 90 | 0.0349 |
| 11 | 0.2817 | 31 | 0.1012 | 51 | 0.0616 | 71 | 0.0442 | 91 | 0.0345 |
| 12 | 0.2588 | 32 | 0.0980 | 52 | 0.0604 | 72 | 0.0436 | 92 | 0.0341 |
| 13 | 0.2393 | 33 | 0.0951 | 53 | 0.0592 | 73 | 0.0430 | 93 | 0.0338 |
| 14 | 0.2225 | 34 | 0.0923 | 54 | 0.0581 | 74 | 0.0424 | 94 | 0.0334 |
| 15 | 0.2079 | 35 | 0.0896 | 55 | 0.0571 | 75 | 0.0419 | 95 | 0.0331 |
| 16 | 0.1951 | 36 | 0.0872 | 56 | 0.0561 | 76 | 0.0413 | 96 | 0.0327 |
| 17 | 0.1838 | 37 | 0.0848 | 57 | 0.0551 | 77 | 0.0408 | 97 | 0.0324 |
| 18 | 0.1736 | 38 | 0.0826 | 58 | 0.0541 | 78 | 0.0403 | 98 | 0.0321 |
| 19 | 0.1646 | 39 | 0.0805 | 59 | 0.0532 | 79 | 0.0398 | 99 | 0.0317 |
| 20 | 0.1564 | 40 | 0.0785 | 60 | 0.0523 | 80 | 0.0393 | 100 | 0.0314 |
| 21 | 0.1490 | 41 | 0.0765 | 61 | 0.0515 | 81 | 0.0388 | | |
| 22 | 0.1423 | 42 | 0.0747 | 62 | 0.0507 | 82 | 0.0383 | | |

## 1.12 Useful workshop formulae

### 1.12.1 Heights above keyways

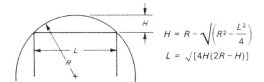

$$H = R - \sqrt{\left(R^2 - \frac{L^2}{4}\right)}$$

$$L = \sqrt{[4H(2R - H)]}$$

### 1.12.2 Radii on bolt ends

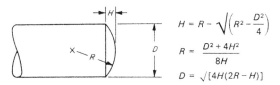

$$H = R - \sqrt{\left(R^2 - \frac{D^2}{4}\right)}$$

$$R = \frac{D^2 + 4H^2}{8H}$$

$$D = \sqrt{[4H(2R - H)]}$$

### 1.12.3 Hexagon: distance across corners

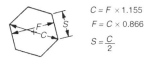

$$C = F \times 1.155$$

$$F = C \times 0.866$$

$$S = \frac{C}{2}$$

where: $F$ = distance across flats (A/F)

$C$ = distance across corners

### 1.12.4 Square: distance across corners

$$F = C \times 0.707$$

$$C = F \times 1.414$$

### 1.12.5 Helix angles

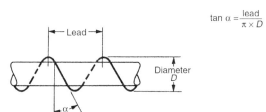

$$\tan \alpha = \frac{\text{lead}}{\pi \times D}$$

### 1.12.6 Cutting speeds (inch units)

$$N = \frac{12S}{\pi d}$$

where: $N$ = spindle speed in rev/min
$S$ = cutting speed in ft/min
$d$ = cutter or work diameter (in)
$\pi$ = 3.142

### 1.12.7 Cutting speeds (metric units)

$$N = \frac{1000S}{\pi d}$$

where: $N$ = spindle speed in rev/min
$S$ = cutting speed in m/min
$d$ = cutter or work diameter (mm)
$\pi$ = 3.142

### 1.12.8 Typical cutting speeds for HSS tools

| Material | ft/min | m/min |
|---|---|---|
| Aluminum | 230–325 | 70–100 |
| Brass | 115–165 | 35–50 |
| Bronze (phosphor) | 65–115 | 20–35 |
| Cast iron (grey) | 80–130 | 25–40 |
| Copper | 115–145 | 35–45 |
| Steel (mild) | 95–130 | 30–40 |
| Steel (medium carbon) | 65–95 | 20–30 |
| Steel (alloy: high-tensile) | 15–25 | 5–8 |
| Thermosetting plastics | 65–95 | 20–30 |
| (Low speed due to abrasive properties of the filter material.) | | |

- For carbide tipped tools, see manufacturers' literature.
- If the calculated spindle speed is not available on the machine gearbox, always use the next lower speed, never use a higher speed.
- The above are average values when using a coolant. Experience may show that under some circumstances higher or lower speeds may be desirable.

## Example 1.12.1

Calculate the spindle speed when turning a grey iron casting 12 inches diameter using a high-speed steel cutting tool.

$$N = \frac{12S}{\pi d}$$
$$= \frac{12 \times 90}{3.142 \times 12}$$
$$\simeq \underline{29 \text{ rev/min}}$$

where: $S = 90$ ft/min (previous table)
$\quad\quad\quad d = 12$ inches

Since the nearest gearbox speed lower than this would be selected $\pi = 3$ would be a suitable approximation and the answer would then be 30 rev/min.

## Example 1.12.2

Calculate the spindle speed in rev/min for a high-speed steel drill 12 mm diameter, cutting mild steel.

$$N = \frac{1000S}{\pi d}$$

where: $N =$ spindle speed in rev/min
$\quad\quad\quad S =$ cutting speed in m/min
$\quad\quad\quad d =$ drill diameter (mm)
$\quad\quad\quad \pi = 3.14$

From the above table, a suitable cutting speed ($S$) for mild steel is 30 m/min, thus:

$$N = \frac{1000 \times 30}{3.14 \times 12}$$
$$\underline{796.2 \text{ rev/min}}$$

A spindle speed between 750 and 800 rev/min would be satisfactory.

To avoid having to make calculations under workshop conditions the following tables may be found helpful.

**CUTTING SPEEDS**
Approximate

| Dia. Ins. | 30<br>9 | 40<br>12 | 50<br>15 | 60<br>18 | 70<br>21 | 80<br>24 | 90<br>27 | 100<br>30 |
|---|---|---|---|---|---|---|---|---|
| Ft/min / M/min | | | | *Revolutions per minute* | | | | |
| 1/64 | 7328 | 9760 | 12224 | 14656 | 17088 | 19520 | 22000 | 24448 |
| 1/32 | 3664 | 4880 | 6112 | 7328 | 8544 | 9760 | 10998 | 12224 |
| 3/64 | 2448 | 3264 | 4064 | 4896 | 5696 | 6528 | 7328 | 8130 |
| 1/16 | 1832 | 2440 | 3056 | 3664 | 4272 | 4880 | 5496 | 6112 |
| 5/64 | 1464 | 1952 | 2448 | 2928 | 3424 | 3904 | 4400 | 4896 |
| 3/32 | 1224 | 1632 | 2032 | 2448 | 2848 | 3264 | 3664 | 4078 |
| 1/8 | 916 | 1220 | 1528 | 1832 | 2136 | 2440 | 2750 | 3056 |
| 5/32 | 732 | 976 | 1224 | 1464 | 1712 | 1952 | 2200 | 2448 |
| 3/16 | 612 | 816 | 1016 | 1224 | 1424 | 1632 | 1832 | 2040 |
| 7/32 | 524 | 700 | 872 | 1048 | 1224 | 1400 | 1570 | 1744 |
| 1/4 | 458 | 610 | 764 | 916 | 1068 | 1220 | 1376 | 1528 |
| 5/16 | 366 | 488 | 612 | 732 | 856 | 976 | 1100 | 1224 |
| 3/8 | 306 | 408 | 508 | 612 | 712 | 816 | 916 | 1020 |
| 7/16 | 262 | 350 | 436 | 524 | 612 | 700 | 784 | 872 |

*(continued)*

(*continued*)

| Inch Series | | | | | | | | | Inch Series |
|---|---|---|---|---|---|---|---|---|---|
| | | | **CUTTING SPEEDS** Approximate | | | | | | |
| Ft/min | 30 | 40 | 50 | 60 | 70 | 80 | 90 | 100 | |
| M/min | 9 | 12 | 15 | 18 | 21 | 24 | 27 | 30 | |
| Dia. Ins. | | | | *Revolutions per minute* | | | | | |
| 1/2 | 229 | 305 | 382 | 458 | 534 | 610 | 688 | 764 | |
| 9/16 | 204 | 272 | 340 | 408 | 476 | 544 | 612 | 680 | |
| 5/8 | 183 | 244 | 306 | 366 | 428 | 488 | 550 | 612 | |
| 11/16 | 167 | 222 | 278 | 334 | 388 | 444 | 500 | 556 | |
| 3/4 | 153 | 204 | 254 | 306 | 356 | 408 | 458 | 510 | |
| 13/16 | 141 | 188 | 234 | 282 | 330 | 376 | 424 | 470 | |
| 7/8 | 131 | 175 | 218 | 262 | 306 | 350 | 392 | 436 | |
| 15/16 | 122 | 163 | 204 | 244 | 286 | 326 | 366 | 408 | |
| 1 | 114 | 152 | 191 | 229 | 267 | 305 | 344 | 382 | |
| 1 1/8 | 102 | 136 | 170 | 204 | 238 | 272 | 306 | 340 | |
| 1 1/4 | 91.5 | 122 | 153 | 183 | 214 | 244 | 275 | 306 | |
| 1 3/8 | 83.5 | 111 | 139 | 167 | 194 | 222 | 250 | 278 | |
| 1 1/2 | 76.5 | 102 | 127 | 153 | 178 | 204 | 229 | 255 | |
| 1 5/8 | 70.5 | 94 | 117 | 141 | 165 | 188 | 212 | 235 | |

| | | | | | | | | |
|---|---|---|---|---|---|---|---|---|
| 1 3/4 | 65.5 | 87.5 | 109 | 131 | 153 | 175 | 196 | 218 |
| 1 7/8 | 61 | 81.5 | 102 | 122 | 143 | 163 | 183 | 204 |
| 2 | 57.5 | 76.5 | 95.5 | 114 | 133 | 152 | 172 | 191 |
| 2 1/8 | 54 | 72 | 90 | 108 | 126 | 144 | 162 | 180 |
| 2 1/4 | 51 | 68 | 85.5 | 102 | 119 | 136 | 153 | 170 |
| 2 3/8 | 48.5 | 64.5 | 80.5 | 96.5 | 113 | 129 | 145 | 161 |
| 2 1/2 | 46 | 61 | 76.5 | 91.5 | 107 | 122 | 138 | 153 |
| 2 5/8 | 43.5 | 58 | 72.5 | 87 | 102 | 116 | 131 | 145 |
| 2 3/4 | 41.5 | 55.5 | 69.5 | 83.5 | 97 | 111 | 125 | 139 |
| 2 7/8 | 39.5 | 53 | 66 | 79 | 92.5 | 106 | 119 | 132 |
| 3 | 38 | 51 | 63.5 | 76.5 | 89 | 102 | 114 | 127 |
| 3 1/4 | 35 | 47 | 58.5 | 70 | 82 | 93.5 | 105 | 117 |
| 3 1/2 | 32.5 | 43.5 | 54.5 | 65.5 | 76.5 | 87.5 | 98 | 109 |
| 3 3/4 | 30.5 | 41 | 51 | 61 | 71.5 | 81.5 | 92 | 102 |
| 4 | 28.5 | 38 | 48 | 57.5 | 67 | 76.5 | 86 | 95.5 |
| 5 | 23 | 30.5 | 38 | 46 | 53.5 | 61 | 69 | 76.5 |
| 6 | 19 | 25.5 | 32 | 38 | 44.5 | 51 | 57 | 63.5 |
| 7 | 16.5 | 22 | 27.5 | 32.5 | 38 | 43.5 | 49 | 54.5 |
| 8 | 14.5 | 19 | 24 | 28.5 | 33.5 | 38 | 43 | 48 |
| 9 | 12.5 | 17 | 21 | 25.5 | 29.5 | 34 | 38 | 42.5 |

| Metric Series | | | | | | | Metric Series |
| --- | --- | --- | --- | --- | --- | --- | --- |
| | | | | | | | 100 30 |

**CUTTING SPEEDS**
Approximate

| Metric Series | | | | | | | | | |
| --- | --- | --- | --- | --- | --- | --- | --- | --- | --- |
| Ft/min<br>M/min | 30<br>9 | 40<br>12 | 50<br>15 | 60<br>18 | 70<br>21 | 80<br>24 | 90<br>27 | 100<br>30 | |
| Dia. mm | | | | | Revolutions per minute | | | | |
| 0.5 | 5817 | 7756 | 9695 | 11634 | 13573 | 15512 | 17451 | 19390 | |
| 1.0 | 2909 | 3878 | 4847 | 5817 | 6786 | 7756 | 8725 | 9695 | |
| 1.5 | 1942 | 2589 | 3237 | 3884 | 4532 | 5179 | 5826 | 6474 | |
| 2.0 | 1456 | 1942 | 2427 | 2912 | 3397 | 3883 | 4369 | 4854 | |
| 3.0 | 970 | 1294 | 1617 | 1940 | 2264 | 2587 | 2911 | 3234 | |
| 4.0 | 728 | 970 | 1213 | 1455 | 1698 | 1940 | 2183 | 2425 | |
| 5.0 | 582 | 777 | 970 | 1164 | 1359 | 1553 | 1747 | 1941 | |
| 6.0 | 485 | 647 | 808 | 970 | 1132 | 1294 | 1455 | 1617 | |
| 7.0 | 416 | 555 | 693 | 832 | 970 | 1109 | 1248 | 1386 | |
| 8.0 | 364 | 485 | 606 | 728 | 849 | 970 | 1091 | 1213 | |
| 9.0 | 324 | 431 | 539 | 647 | 755 | 862 | 970 | 1078 | |
| 10.0 | 291 | 388 | 485 | 582 | 679 | 776 | 873 | 970 | |
| 11.0 | 265 | 353 | 441 | 529 | 617 | 706 | 794 | 882 | |

| | | | | | | | | |
|---|---|---|---|---|---|---|---|---|
| 12.0 | 243 | 324 | 404 | 485 | 566 | 647 | 728 | 808 |
| 13.0 | 224 | 299 | 373 | 448 | 522 | 597 | 672 | 746 |
| 14.0 | 208 | 277 | 346 | 416 | 485 | 554 | 623 | 693 |
| 15.0 | 194 | 259 | 323 | 388 | 453 | 517 | 582 | 647 |
| 16.0 | 182 | 243 | 303 | 364 | 424 | 485 | 546 | 606 |
| 17.0 | 171 | 228 | 285 | 342 | 399 | 456 | 513 | 571 |
| 18.0 | 162 | 216 | 269 | 323 | 377 | 431 | 485 | 539 |
| 19.0 | 153 | 204 | 255 | 306 | 357 | 408 | 459 | 511 |
| 20.0 | 146 | 194 | 242 | 291 | 340 | 388 | 436 | 485 |
| 21.0 | 139 | 185 | 231 | 277 | 323 | 370 | 416 | 462 |
| 22.0 | 133 | 177 | 220 | 265 | 309 | 353 | 397 | 441 |
| 23.0 | 127 | 169 | 211 | 253 | 295 | 337 | 380 | 422 |
| 24.0 | 121 | 162 | 202 | 242 | 283 | 323 | 364 | 404 |
| 25.0 | 117 | 155 | 194 | 233 | 272 | 310 | 349 | 388 |

## 1.13 Solution of triangles

### 1.13.1 Pythagoras

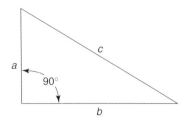

$$c^2 = a^2 + b^2 \text{ or } c = \sqrt{(a^2 + b^2)}$$
$$a^2 = c^2 - b^2 \text{ or } a = \sqrt{(c^2 - b^2)}$$
$$b^2 = c^2 - a^2 \text{ or } b = \sqrt{(c^2 - a^2)}$$

e.g. Let $a = 3$, $b = 4$ and $c = 5$
 Then $3^2 + 4^2 = 5^2$ or $9 + 16 = 25$

*Note:* The 3 : 4 : 5 ratio is useful for setting out square carrier.

### 1.13.2 Trigonometry (right-angled triangles)

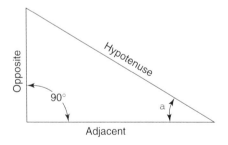

$$\text{Sine } \alpha = \frac{\text{opposite}}{\text{hypotenuse}}$$
$$\text{Cosine } \alpha = \frac{\text{adjacent}}{\text{hypotenuse}}$$
$$\text{Tangent } \alpha = \frac{\text{opposite}}{\text{adjacent}}$$

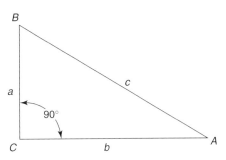

$$\text{Sin } A = a/c \quad | \quad a = b \times \tan A \quad | \quad c = \dfrac{a}{\sin A}$$
$$\text{Sin } B = b/c \quad | \quad a = c \times \cos B$$
$$\text{Cos } A = b/c \quad | \quad a = c \times \sin A$$
$$\text{Cos } B = a/c \quad | \quad b = c \times \tan B \quad | \quad C = \dfrac{b}{\cos A}$$
$$\text{Tan } A = a/b \quad | \quad b = c \times \cos A$$
$$\text{Tan } B = b/a \quad | \quad b = c \times \sin B$$

### 1.13.3 Trigonometry (any triangle)

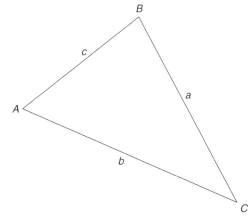

(i) cosine rule

$$a^2 = b^2 + c^2 - 2bc \cos A$$
$$b^2 = a^2 + c^2 - 2ac \cos B$$
$$c^2 = b^2 + a^2 - 2ab \cos C$$

(ii) sine rule

$$\frac{a}{\sin A} = \frac{b}{\sin B} = \frac{c}{\sin C}, \ \text{ e.g. } \sin A = \frac{a \sin B}{b}$$

(iii) Area

$$\text{Area} = \tfrac{1}{2}ab \sin C = \tfrac{1}{2}bc \sin A$$
$$= \tfrac{1}{2}ca \sin B$$

Also: $\quad \text{Area} = \sqrt{[s(s-a)(s-b)(s-c)]}$

where $s = \tfrac{1}{2}(a+b+c)$

## 1.14 Sine-bar (principle)

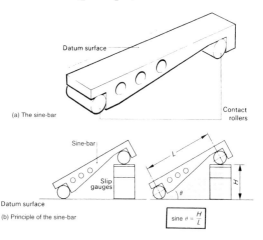

Datum surface

(a) The sine-bar

Contact rollers

Sine-bar

Slip gauges

Datum surface

(b) Principle of the sine-bar

$$\text{sine } \theta = \frac{H}{L}$$

The *sine-bar* provides a simple means of measuring angles to a high degree of accuracy. Figure (a), above, shows a typical sine-bar, and for accurate results it is essential that:

(a) the contact rollers must be of equal diameter and true geometric cylinders;
(b) the distance between the roller axes must be precise and known, and these axes must be mutually parallel;
(c) the upper surface of the bar must be flat and parallel with the roller axes, and equidistant from each.

The principle of the sine-bar is shown in Fig. (b) above. The sine-bar, slip gauges and datum surface on which they stand form a right-angled triangle. The sine-bar itself forms the hypotenuse of that triangle and the slip gauges form the side *opposite* the required angle.

$$\text{Since: sine } \theta = \frac{\text{opposite side}}{\text{hypotenuse}}$$

$$\text{Then: sine } \theta = \frac{\text{height of slip gauges}}{\text{length of sine-bar}}$$

$$= \frac{H}{L}$$

---

**Example 1.14.1**

Calculate the slip gauges required to give an angle of 25° when using a 250 mm sine-bar.

$$\text{sine } \theta = \frac{H}{L}$$
$$H = L \text{ sine } \theta$$
$$= 250 \times 0.422\,6$$
$$= \underline{105.65 \text{ mm}}$$

where: $\theta = 25°$

$L = 250$ mm

*Note:* The four-figure mathematical tables used by students are only of limited accuracy. Except when working examples for practice, always use an electronic calculator or the ready-worked *sine-bar constants* found in Section 1.16.

---

(Courtesy Addison Wesley Longman.)

## 1.15 Sine-bar (use of)

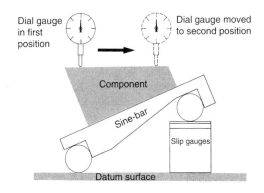

Dial gauge in first position

Dial gauge moved to second position

Component

Sine-bar

Slip gauges

Datum surface

The above figure shows how the sine-bar is used to check small components that may be mounted upon it. The dial test indicator (DTI) is mounted upon a suitable stand such as a universal surface gauge (scribing block) or a vernier height gauge (the latter is more rigid and gives more consistant readings). It is moved over the component into the first position as shown above and zeroed. The stand and DTI is then slid along the datum surface to the second position as shown and the DTI reading is noted.

## Method 1
The height of the slip gauges is adjusted until the DTI reads zero at both ends of the component. The actual angle is then calculated as explained in Example 1.14.1 and any deviation from the specified angle is the error.

## Method 2
The sine-bar is set to the specified angle. The DTI will then indicate any error as a 'run' of so many hundredths of a millimetre along the length of the component. Providing the DTI was set to zero in the first position, the error will be shown as a plus or minus reading at the second position.

Examination of natural sine tables will show that as the angle increases, the accuracy of the tables decreases. Therefore, when measuring angles over $45°$ the component is turned over – if possible – so that the *complementary angle* can be used as shown below. In Fig. (a) the angle $\theta°$ is considerably over $45°$ and it will not be possible to obtain sufficient accuracy from natural sine tables. In Fig. (b) the component is re-positioned and the sine-bar is set to the complementary angle of $90° - \theta°$. The sine of this smaller angle can be obtained more accurately from the tables and the angle $\theta°$ can be calculated.

(Courtesy of Addison Wesley Longman.)

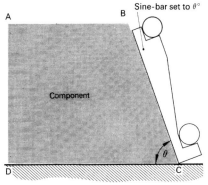

(a) Incorrect use of sine-bar

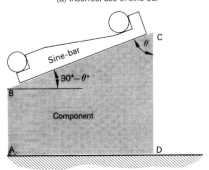

(b) Correct use of sine-bar

39

## 1.16 Sine-bar constants (250 mm)

*Note:* For use with a 125 mm sine-bar the following constants are halved.

(Dimensions in millimetres)

| Min. | 0° | 1° | 2° | 3° | 4° | 5° | 6° | 7° | 8° |
|---|---|---|---|---|---|---|---|---|---|
| 0 | 0.00000 | 4.36310 | 8.72487 | 13.08399 | 17.43912 | 21.78894 | 26.13212 | 30.46734 | 34.79328 |
| 1 | 0.07272 | 4.38125 | 8.79756 | 13.15661 | 17.51664 | 21.86138 | 26.20444 | 30.53952 | 34.86529 |
| 2 | 0.14543 | 4.50851 | 8.87021 | 13.22922 | 17.58419 | 21.93381 | 26.27674 | 30.61676 | 34.93730 |
| 3 | 0.21870 | 4.58123 | 8.94290 | 13.30185 | 17.65675 | 22.00626 | 26.34908 | 30.68386 | 35.00931 |
| 4 | 0.29090 | 4.65396 | 9.01560 | 13.37448 | 17.72930 | 22.07872 | 26.42141 | 30.75605 | 35.08132 |
| 5 | 0.36359 | 4.72664 | 9.08824 | 13.44710 | 17.80181 | 22.15114 | 26.49371 | 30.82820 | 35.15330 |
| 6 | 0.43633 | 4.79936 | 9.16093 | 13.51970 | 17.87436 | 22.22357 | 26.56602 | 30.90037 | 35.22531 |
| 7 | 0.50906 | 4.87221 | 9.233614 | 13.59232 | 17.94689 | 22.29601 | 26.63832 | 30.97253 | 35.29730 |
| 8 | 0.58176 | 4.94476 | 9.30626 | 13.66492 | 18.01942 | 22.36842 | 26.71062 | 31.04468 | 35.36928 |
| 9 | 0.65450 | 5.01748 | 9.37894 | 13.75449 | 18.09196 | 22.44087 | 26.78294 | 31.11685 | 35.44128 |
| 10 | 0.72723 | 5.09021 | 9.45163 | 13.81016 | 18.16451 | 22.51331 | 26.85525 | 31.18902 | 35.51329 |
| 11 | 0.79993 | 5.16288 | 9.52428 | 13.88277 | 18.23702 | 22.58571 | 26.92753 | 31.26116 | 35.58525 |
| 12 | 0.87266 | 5.23561 | 9.59695 | 13.95538 | 18.30955 | 22.65815 | 26.99984 | 31.33331 | 35.65723 |
| 13 | 0.94540 | 5.30831 | 9.66962 | 14.02799 | 18.38208 | 22.73057 | 27.07214 | 31.40546 | 35.72921 |
| 14 | 1.01809 | 5.38100 | 9.74227 | 14.10058 | 18.45459 | 22.80297 | 27.14441 | 31.47759 | 35.80117 |

| | | | | | | | | | |
|---|---|---|---|---|---|---|---|---|---|
| 15 | 1.090 83 | 5.453 72 | 9.814 95 | 14.173 20 | 18.527 12 | 22.875 40 | 27.216 72 | 31.549 74 | 35.873 16 |
| 16 | 1.163 56 | 5.526 44 | 9.887 63 | 14.245 82 | 18.599 66 | 22.947 84 | 27.289 02 | 31.621 89 | 35.945 14 |
| 17 | 1.236 27 | 5.599 12 | 9.960 27 | 14.318 39 | 18.672 15 | 23.020 22 | 27.361 29 | 31.694 00 | 36.017 09 |
| 18 | 1.308 99 | 5.671 83 | 10.032 95 | 14.391 01 | 18.744 68 | 23.092 65 | 27.433 58 | 31.766 15 | 36.089 05 |
| 19 | 1.381 73 | 5.744 54 | 10.105 61 | 14.463 61 | 18.817 21 | 23.165 06 | 27.505 86 | 31.838 28 | 36.161 01 |
| 20 | 1.454 42 | 5.817 09 | 10.178 27 | 14.536 06 | 18.889 70 | 23.237 45 | 27.578 12 | 31.910 41 | 36.232 95 |
| 21 | 1.527 15 | 5.889 41 | 10.250 93 | 14.608 81 | 18.962 23 | 23.309 87 | 27.650 42 | 31.982 54 | 36.304 92 |
| 22 | 1.599 89 | 5.962 66 | 10.323 61 | 14.681 42 | 19.034 75 | 23.382 29 | 27.722 71 | 32.054 68 | 36.376 88 |
| 23 | 1.672 58 | 6.035 33 | 10.396 24 | 14.753 98 | 19.107 23 | 23.454 66 | 27.794 95 | 32.126 77 | 36.448 80 |
| 24 | 1.745 32 | 6.108 04 | 10.468 91 | 14.826 59 | 19.179 76 | 23.527 08 | 27.867 23 | 32.198 89 | 36.520 76 |
| 25 | 1.818 05 | 6.180 76 | 10.541 57 | 14.899 19 | 19.252 27 | 23.599 48 | 27.939 50 | 32.271 02 | 36.592 69 |
| 26 | 1.890 61 | 6.253 43 | 10.614 21 | 14.971 76 | 19.324 75 | 23.671 86 | 28.011 75 | 32.343 11 | 36.664 62 |
| 27 | 1.963 48 | 6.325 14 | 10.686 88 | 15.044 37 | 19.397 27 | 23.744 27 | 28.084 03 | 32.415 24 | 36.736 57 |
| 28 | 2.036 21 | 6.398 86 | 10.759 55 | 15.116 93 | 19.469 79 | 23.816 67 | 28.156 30 | 32.487 36 | 36.808 51 |
| 29 | 2.108 90 | 6.471 53 | 10.832 19 | 15.189 55 | 19.542 26 | 23.889 05 | 28.228 55 | 32.559 43 | 36.880 43 |
| 30 | 2.181 63 | 6.544 23 | 10.904 85 | 15.262 13 | 19.614 77 | 23.961 44 | 28.300 80 | 32.631 55 | 36.952 35 |
| 31 | 2.254 37 | 6.616 64 | 10.977 50 | 15.334 72 | 19.687 27 | 24.033 83 | 28.373 06 | 32.703 65 | 37.024 28 |
| 32 | 2.327 06 | 6.689 62 | 11.050 15 | 15.407 29 | 19.759 75 | 24.106 19 | 28.445 29 | 32.775 73 | 37.096 05 |
| 33 | 2.399 80 | 6.762 33 | 11.122 80 | 15.479 89 | 19.832 26 | 24.178 59 | 28.517 56 | 32.847 84 | 37.168 11 |
| 34 | 2.472 52 | 6.834 73 | 11.195 47 | 15.552 48 | 19.904 77 | 24.250 99 | 28.589 82 | 32.919 94 | 37.240 04 |

*(continued)*

**1.16** (continued)

| Min. | 0° | 1° | 2° | 3° | 4° | 5° | 6° | 7° | 8° |
|---|---|---|---|---|---|---|---|---|---|
| 35 | 2.545 21 | 6.907 70 | 11.268 10 | 15.625 03 | 19.977 23 | 24.323 33 | 28.662 04 | 32.991 99 | 37.311 92 |
| 36 | 2.617 95 | 6.980 41 | 11.340 74 | 15.697 63 | 20.049 73 | 24.395 72 | 28.734 29 | 33.064 09 | 37.838 36 |
| 37 | 2.690 67 | 7.053 12 | 11.413 41 | 15.770 21 | 20.122 22 | 24.468 10 | 28.806 53 | 33.136 18 | 37.455 74 |
| 38 | 2.763 63 | 7.125 78 | 11.486 03 | 15.842 77 | 20.194 69 | 24.540 46 | 28.878 75 | 33.208 24 | 37.527 49 |
| 39 | 2.836 10 | 7.198 49 | 11.558 68 | 15.915 36 | 20.267 19 | 24.612 84 | 28.951 00 | 33.280 33 | 37.599 53 |
| 40 | 2.908 83 | 7.271 18 | 11.631 34 | 15.988 08 | 20.339 81 | 24.685 22 | 29.023 24 | 33.352 42 | 37.671 44 |
| 41 | 2.981 52 | 7.343 86 | 11.703 96 | 16.060 50 | 20.412 13 | 24.757 52 | 29.095 44 | 33.424 46 | 37.743 32 |
| 42 | 3.054 25 | 7.416 56 | 11.776 61 | 16.133 08 | 20.484 63 | 24.829 94 | 29.167 68 | 33.496 55 | 37.815 21 |
| 43 | 3.126 98 | 7.489 25 | 11.849 25 | 16.205 65 | 20.557 11 | 24.902 30 | 29.239 88 | 33.568 61 | 37.887 10 |
| 44 | 3.199 68 | 7.561 93 | 11.921 89 | 16.278 20 | 20.629 56 | 24.974 64 | 29.312 12 | 33.640 66 | 37.958 95 |
| 45 | 3.272 40 | 7.634 63 | 11.994 53 | 16.350 78 | 20.702 05 | 25.047 02 | 29.384 35 | 33.712 73 | 38.030 85 |
| 46 | 2.908 96 | 7.707 34 | 12.067 18 | 16.423 36 | 20.774 52 | 25.119 38 | 29.456 58 | 33.784 80 | 38.102 73 |
| 47 | 3.417 82 | 7.780 00 | 12.139 81 | 16.495 90 | 20.846 98 | 25.191 72 | 29.528 78 | 33.856 83 | 38.174 59 |
| 48 | 3.490 55 | 7.852 70 | 12.212 44 | 16.568 48 | 20.919 46 | 25.264 07 | 29.600 99 | 33.928 89 | 38.246 46 |
| 49 | 3.563 27 | 7.925 38 | 12.285 08 | 16.641 00 | 20.991 92 | 25.336 42 | 29.673 20 | 34.000 94 | 38.318 32 |
| 50 | 3.635 60 | 7.998 05 | 12.357 69 | 16.713 58 | 21.064 25 | 25.408 76 | 29.745 39 | 34.072 97 | 38.390 17 |
| 51 | 3.708 69 | 8.070 75 | 12.430 34 | 16.786 16 | 21.136 85 | 25.481 11 | 29.817 61 | 34.145 03 | 38.462 04 |
| 52 | 3.781 42 | 8.143 44 | 12.502 98 | 16.858 73 | 21.209 33 | 25.553 47 | 29.889 83 | 34.217 08 | 38.533 91 |
| 53 | 3.854 10 | 8.216 10 | 12.575 60 | 16.931 25 | 21.281 76 | 25.625 79 | 29.962 01 | 34.289 10 | 38.605 75 |

| Min. | 9° | 10° | 11° | 12° | 13° | 14° | 15° | 16° | 17° |
|---|---|---|---|---|---|---|---|---|---|
| 54 | 3.926 83 | 8.288 80 | 12.648 23 | 17.003 82 | 21.354 23 | 25.698 13 | 30.034 21 | 34.361 14 | 38.677 60 |
| 55 | 3.999 54 | 8.361 49 | 12.720 87 | 17.076 38 | 21.426 69 | 25.770 48 | 30.106 41 | 34.433 17 | 38.749 44 |
| 56 | 4.072 24 | 8.434 14 | 12.793 49 | 17.148 91 | 21.499 125 | 25.842 79 | 30.178 58 | 34.505 18 | 38.821 27 |
| 57 | 4.165 80 | 8.506 84 | 12.866 12 | 17.221 48 | 21.571 59 | 25.915 135 | 30.250 78 | 34.577 22 | 38.893 12 |
| 58 | 4.217 69 | 8.579 53 | 12.938 74 | 17.294 04 | 21.644 06 | 25.987 48 | 30.322 99 | 34.649 26 | 38.964 97 |
| 59 | 4.290 39 | 8.652 19 | 13.011 36 | 17.366 57 | 21.716 47 | 26.059 79 | 30.395 15 | 34.721 25 | 39.036 77 |
| Min. | 9° | 10° | 11° | 12° | 13° | 14° | 15° | 16° | 17° |
| 0 | 39.108 62 | 43.412 04 | 47.702 25 | 51.977 92 | 56.237 76 | 60.480 47 | 64.704 76 | 68.909 34 | 73.092 93 |
| 1 | 39.180 44 | 43.483 66 | 47.773 63 | 52.049 06 | 56.308 62 | 60.551 03 | 64.775 00 | 68.979 24 | 73.162 47 |
| 2 | 39.252 25 | 43.555 26 | 47.844 99 | 52.120 18 | 56.379 46 | 60.621 57 | 64.845 23 | 69.049 12 | 73.231 99 |
| 3 | 39.324 08 | 43.626 88 | 47.987 76 | 52.191 30 | 56.450 32 | 60.692 14 | 64.915 47 | 69.119 03 | 73.301 32 |
| 4 | 39.395 91 | 43.698 49 | 48.023 44 | 52.262 42 | 56.521 16 | 60.762 68 | 64.985 69 | 69.188 92 | 73.371 07 |
| 5 | 39.467 71 | 43.770 08 | 48.059 13 | 52.333 53 | 56.591 99 | 60.833 22 | 65.055 91 | 69.258 80 | 73.440 57 |
| 6 | 39.539 52 | 43.841 68 | 48.130 49 | 52.404 64 | 56.662 83 | 60.903 75 | 65.126 13 | 69.328 66 | 73.510 08 |
| 7 | 39.611 32 | 43.913 28 | 48.201 85 | 52.475 75 | 56.733 66 | 60.974 28 | 65.196 34 | 69.398 53 | 73.579 59 |
| 8 | 39.683 11 | 43.984 85 | 48.273 21 | 52.546 84 | 56.804 46 | 61.044 79 | 65.266 52 | 69.468 38 | 73.649 07 |
| 9 | 39.754 92 | 44.056 45 | 48.344 56 | 52.617 94 | 56.875 29 | 61.111 53 | 65.533 67 | 69.538 25 | 73.718 58 |

*(continued)*

**1.16** (*continued*)

| Min. | 9° | 10° | 11° | 12° | 13° | 14° | 15° | 16° | 17° |
|---|---|---|---|---|---|---|---|---|---|
| 10 | 39.826 73 | 44.128 03 | 48.415 91 | 52.689 03 | 56.946 11 | 61.185 85 | 65.409 29 | 69.608 10 | 73.788 06 |
| 11 | 39.898 51 | 44.199 61 | 48.487 24 | 52.760 12 | 57.016 91 | 61.256 34 | 65.477 11 | 69.677 94 | 73.857 54 |
| 12 | 39.970 30 | 44.271 18 | 48.558 59 | 52.831 20 | 57.087 72 | 61.326 85 | 65.547 29 | 69.747 78 | 73.927 01 |
| 13 | 40.042 08 | 44.342 76 | 48.629 92 | 52.902 28 | 57.158 52 | 61.397 32 | 65.617 47 | 69.817 61 | 73.996 48 |
| 14 | 40.113 85 | 44.414 31 | 48.701 25 | 52.973 34 | 57.229 29 | 61.467 82 | 65.687 63 | 69.887 42 | 74.065 90 |
| 15 | 40.185 64 | 44.485 89 | 48.772 58 | 53.044 42 | 57.300 10 | 61.538 32 | 65.757 80 | 69.957 25 | 74.135 39 |
| 16 | 40.257 43 | 44.557 45 | 48.843 90 | 53.115 48 | 57.370 90 | 61.608 81 | 65.827 98 | 70.027 07 | 74.204 86 |
| 17 | 40.329 19 | 44.629 00 | 48.915 22 | 53.186 54 | 57.441 66 | 61.679 28 | 65.898 11 | 70.096 87 | 74.274 28 |
| 18 | 40.400 95 | 44.700 55 | 48.986 54 | 53.257 60 | 57.512 43 | 61.749 75 | 65.968 26 | 70.166 68 | 74.343 72 |
| 19 | 40.472 72 | 44.772 10 | 49.057 85 | 53.328 65 | 57.583 20 | 61.820 22 | 66.038 41 | 70.236 45 | 74.413 15 |
| 20 | 40.544 47 | 44.843 63 | 49.129 14 | 53.399 68 | 57.653 95 | 61.890 67 | 66.108 53 | 70.306 12 | 74.482 43 |
| 21 | 40.616 24 | 44.915 19 | 49.200 45 | 53.470 74 | 57.724 73 | 61.961 14 | 66.178 67 | 70.376 05 | 74.551 99 |
| 22 | 40.688 01 | 44.986 74 | 49.271 75 | 53.541 79 | 57.795 48 | 62.031 59 | 66.248 80 | 70.445 83 | 74.621 40 |
| 23 | 40.759 73 | 45.058 24 | 49.343 04 | 53.612 80 | 57.866 23 | 62.102 02 | 66.318 91 | 70.515 60 | 74.690 80 |
| 24 | 40.831 49 | 45.129 79 | 49.413 35 | 53.683 83 | 57.936 98 | 62.172 47 | 66.389 03 | 70.585 36 | 74.760 20 |
| 25 | 40.903 24 | 45.201 31 | 49.485 62 | 53.754 86 | 58.007 72 | 62.242 91 | 66.459 14 | 70.655 13 | 74.829 59 |
| 26 | 40.974 96 | 45.272 82 | 49.556 89 | 53.825 86 | 58.078 44 | 62.313 32 | 66.529 22 | 70.724 87 | 74.898 97 |
| 27 | 41.046 71 | 45.344 35 | 49.628 18 | 53.896 89 | 58.149 18 | 62.383 76 | 66.599 40 | 70.794 63 | 74.968 40 |
| 28 | 41.118 45 | 45.415 88 | 49.699 45 | 53.967 90 | 58.219 92 | 62.454 18 | 66.669 43 | 70.864 37 | 75.037 73 |
| 29 | 41.190 17 | 45.487 37 | 49.770 72 | 54.038 90 | 58.290 62 | 62.524 59 | 66.739 51 | 70.934 11 | 75.107 09 |

| | | | | | | | | | |
|---|---|---|---|---|---|---|---|---|---|
| 30 | 41.26190 | 45.55888 | 49.84198 | 54.10990 | 58.36134 | 62.59500 | 66.80960 | 71.00384 | 75.17645 |
| 31 | 41.33363 | 45.63039 | 49.91324 | 54.18090 | 58.43210 | 62.66541 | 66.87967 | 71.07356 | 75.24580 |
| 32 | 41.40533 | 45.70187 | 49.98449 | 54.25188 | 58.50274 | 62.73579 | 66.94972 | 71.14327 | 75.31514 |
| 33 | 41.47706 | 45.77238 | 50.05575 | 54.32288 | 58.57346 | 62.80619 | 67.01980 | 71.21299 | 75.38450 |
| 34 | 41.54879 | 45.84288 | 50.12699 | 54.39386 | 58.64417 | 62.87658 | 67.08986 | 71.28271 | 75.45383 |
| 35 | 41.62048 | 45.91635 | 50.19824 | 54.46484 | 58.71484 | 62.94696 | 67.15991 | 71.35240 | 75.52315 |
| 36 | 41.69219 | 45.98784 | 50.26948 | 54.53581 | 58.78553 | 63.01734 | 67.22995 | 71.42209 | 75.59247 |
| 37 | 41.76389 | 46.05932 | 50.34072 | 54.60678 | 58.85621 | 63.08771 | 67.29997 | 71.49178 | 75.66179 |
| 38 | 41.83557 | 46.13078 | 50.41194 | 54.67774 | 58.92687 | 63.15806 | 67.37002 | 71.56145 | 75.73108 |
| 39 | 41.90728 | 46.20226 | 50.48317 | 54.74870 | 58.99755 | 63.22844 | 67.44006 | 71.63114 | 75.80040 |
| 40 | 41.97899 | 46.27375 | 50.55439 | 54.81966 | 59.06822 | 63.29879 | 67.51008 | 71.70082 | 75.86971 |
| 41 | 42.05066 | 46.34519 | 50.62561 | 54.89060 | 59.13888 | 63.36914 | 67.58010 | 71.77047 | 75.93898 |
| 42 | 42.12234 | 46.41665 | 50.69682 | 54.96155 | 59.20953 | 63.43949 | 67.65013 | 71.84013 | 76.00827 |
| 43 | 42.19403 | 46.48811 | 50.76803 | 55.03249 | 59.28019 | 63.50983 | 67.72012 | 71.90978 | 76.07754 |
| 44 | 42.26569 | 46.55955 | 50.83922 | 55.10343 | 59.35082 | 63.58014 | 67.79010 | 71.97943 | 76.14681 |
| 45 | 42.33738 | 46.63101 | 50.91044 | 55.17436 | 59.42147 | 63.65049 | 67.86011 | 72.04906 | 76.21608 |
| 46 | 42.40906 | 46.70247 | 50.98164 | 55.24529 | 59.49211 | 63.72081 | 67.93010 | 72.11870 | 76.28533 |
| 47 | 42.48071 | 46.77389 | 51.05282 | 55.31621 | 59.56274 | 63.79111 | 68.00008 | 72.18833 | 76.35458 |
| 48 | 42.55237 | 46.84533 | 51.12401 | 55.38712 | 59.63336 | 63.86144 | 68.07006 | 72.25795 | 76.42383 |
| 49 | 42.62404 | 46.91676 | 51.19519 | 55.45804 | 59.70399 | 63.93175 | 68.14003 | 72.32757 | 76.49306 |

*(continued)*

**1.16** (*continued*)

| Min. | 9° | 10° | 11° | 12° | 13° | 14° | 15° | 16° | 17° |
|---|---|---|---|---|---|---|---|---|---|
| 50 | 42.695 68 | 46.988 17 | 51.266 36 | 55.528 93 | 59.774 59 | 64.002 03 | 68.210 00 | 72.397 16 | 76.562 28 |
| 51 | 42.767 34 | 47.059 61 | 51.337 55 | 55.598 48 | 59.845 21 | 64.072 34 | 68.280 00 | 72.466 78 | 76.631 52 |
| 52 | 42.839 00 | 47.131 03 | 51.408 72 | 55.670 75 | 59.915 82 | 64.142 65 | 68.349 91 | 72.536 37 | 76.700 74 |
| 53 | 42.910 63 | 47.202 45 | 51.479 88 | 55.741 63 | 59.986 41 | 64.219 10 | 68.419 86 | 72.605 96 | 76.769 95 |
| 54 | 42.982 28 | 47.273 86 | 51.551 05 | 55.812 53 | 60.057 01 | 64.283 20 | 68.489 81 | 72.675 55 | 76.839 15 |
| 55 | 43.053 91 | 47.345 27 | 51.622 21 | 55.883 41 | 60.127 60 | 64.353 47 | 68.559 74 | 72.745 13 | 76.908 35 |
| 56 | 43.125 53 | 47.416 66 | 51.693 34 | 55.954 29 | 60.198 17 | 64.423 74 | 68.629 67 | 72.814 70 | 76.977 54 |
| 57 | 43.197 18 | 47.488 07 | 51.764 50 | 56.025 17 | 60.268 77 | 64.494 00 | 68.699 60 | 72.884 27 | 77.046 73 |
| 58 | 43.268 80 | 47.559 48 | 51.835 65 | 56.096 04 | 60.339 34 | 64.564 26 | 68.769 52 | 72.953 83 | 77.115 91 |
| 59 | 43.340 42 | 47.630 86 | 51.906 79 | 56.166 90 | 60.409 91 | 64.345 13 | 68.839 43 | 73.023 38 | 77.185 08 |

| Min. | 18° | 19° | 20° | 21° | 22° | 23° | 24° | 25° | 26° |
|---|---|---|---|---|---|---|---|---|---|
| 0 | 77.254 25 | 81.392 04 | 85.505 04 | 89.591 99 | 93.651 65 | 97.682 78 | 101.684 16 | 105.654 57 | 109.592 79 |
| 1 | 77.323 41 | 81.460 80 | 85.573 37 | 89.659 87 | 93.719 07 | 97.749 72 | 101.750 59 | 105.720 47 | 109.658 14 |
| 2 | 77.392 56 | 81.529 53 | 85.641 69 | 89.727 74 | 93.786 49 | 97.816 65 | 101.817 01 | 105.786 36 | 109.723 49 |
| 3 | 77.461 71 | 81.598 29 | 85.710 01 | 89.795 63 | 93.853 89 | 97.883 57 | 101.883 42 | 105.852 25 | 109.788 83 |
| 4 | 77.530 86 | 81.667 03 | 85.778 34 | 89.863 50 | 93.921 29 | 97.950 48 | 101.949 83 | 105.918 13 | 109.854 16 |
| 5 | 77.599 98 | 81.735 75 | 85.846 63 | 89.931 35 | 93.988 68 | 98.017 38 | 102.016 23 | 105.983 99 | 109.919 48 |

| | | | | | | | | | |
|---|---|---|---|---|---|---|---|---|---|
| 6 | 77.669 11 | 81.180 45 | 85.914 92 | 89.999 20 | 94.056 07 | 98.084 28 | 102.082 62 | 106.049 86 | 109.984 79 |
| 7 | 77.382 29 | 81.187 32 | 85.983 21 | 90.067 05 | 94.123 44 | 98.151 17 | 102.148 99 | 106.115 71 | 110.050 09 |
| 8 | 77.807 33 | 81.941 88 | 86.051 48 | 90.134 88 | 94.190 79 | 98.218 03 | 102.215 36 | 106.181 55 | 110.115 39 |
| 9 | 77.876 45 | 82.010 60 | 86.119 77 | 90.202 71 | 94.258 17 | 98.284 92 | 102.281 73 | 106.247 38 | 110.180 67 |
| 10 | 77.945 55 | 82.079 31 | 86.188 04 | 90.270 53 | 94.325 52 | 98.351 78 | 102.348 08 | 106.313 20 | 110.245 95 |
| 11 | 78.014 64 | 82.147 98 | 86.256 30 | 90.338 34 | 94.392 86 | 98.418 63 | 102.414 42 | 106.379 02 | 110.311 21 |
| 12 | 78.083 73 | 82.216 66 | 86.324 55 | 90.406 14 | 94.460 20 | 98.485 48 | 102.480 76 | 106.444 82 | 110.376 46 |
| 13 | 78.152 81 | 82.285 34 | 86.392 80 | 90.473 94 | 94.527 52 | 98.552 32 | 102.547 09 | 106.510 62 | 110.441 71 |
| 14 | 78.221 87 | 82.353 99 | 86.461 02 | 90.541 71 | 94.594 84 | 98.619 14 | 102.613 40 | 106.576 41 | 110.506 94 |
| 15 | 78.290 95 | 82.422 66 | 86.529 26 | 90.609 51 | 94.662 15 | 98.685 96 | 102.679 71 | 106.642 19 | 110.572 17 |
| 16 | 78.360 01 | 82.491 31 | 86.597 49 | 90.677 29 | 94.729 46 | 98.752 78 | 101.746 02 | 106.707 96 | 110.637 39 |
| 17 | 78.429 07 | 82.559 96 | 86.665 70 | 90.745 05 | 94.796 75 | 98.819 58 | 102.812 31 | 106.773 71 | 110.702 60 |
| 18 | 78.498 11 | 82.628 60 | 86.733 91 | 90.812 81 | 94.864 04 | 98.886 38 | 102.878 59 | 106.839 47 | 110.767 80 |
| 19 | 78.567 16 | 82.697 23 | 86.802 12 | 90.880 56 | 94.931 32 | 98.953 16 | 102.944 86 | 106.905 21 | 110.832 99 |
| 20 | 78.636 05 | 82.765 84 | 86.870 29 | 90.948 29 | 94.998 59 | 99.019 94 | 103.011 12 | 106.970 93 | 110.898 16 |
| 21 | 78.705 22 | 82.834 47 | 86.938 50 | 91.016 04 | 95.065 85 | 99.086 71 | 103.077 39 | 107.036 67 | 110.963 34 |
| 22 | 78.774 24 | 82.903 08 | 87.006 68 | 91.083 77 | 95.133 11 | 99.153 48 | 103.143 64 | 107.102 38 | 111.028 50 |
| 23 | 78.843 23 | 82.971 69 | 87.074 85 | 91.151 48 | 95.200 35 | 99.220 23 | 103.209 88 | 107.168 09 | 111.093 65 |
| 24 | 78.912 26 | 83.040 28 | 87.143 01 | 91.219 61 | 95.267 59 | 99.286 97 | 103.276 11 | 107.233 78 | 111.158 79 |

(continued)

**1.16** (continued)

| Min. | 18° | 19° | 20° | 21° | 22° | 23° | 24° | 25° | 26° |
|---|---|---|---|---|---|---|---|---|---|
| 25 | 78.981 26 | 83.108 87 | 87.211 17 | 91.286 90 | 95.334 83 | 99.353 71 | 103.342 33 | 107.299 47 | 111.223 93 |
| 26 | 79.050 24 | 83.177 44 | 87.279 32 | 91.354 58 | 95.402 05 | 99.420 44 | 103.408 54 | 107.365 14 | 111.289 05 |
| 27 | 79.119 24 | 83.246 03 | 87.347 46 | 91.422 29 | 95.469 26 | 99.487 16 | 103.474 75 | 107.430 82 | 111.354 17 |
| 28 | 79.188 22 | 83.314 61 | 87.415 60 | 91.489 97 | 95.536 47 | 99.553 87 | 103.540 95 | 107.496 48 | 111.419 27 |
| 29 | 79.257 18 | 83.383 16 | 87.483 72 | 91.557 64 | 95.603 67 | 99.620 57 | 103.607 13 | 107.562 13 | 111.484 37 |
| 30 | 79.326 16 | 83.451 72 | 87.551 85 | 91.625 31 | 95.670 86 | 99.687 27 | 103.673 31 | 107.627 77 | 111.549 45 |
| 31 | 79.395 13 | 83.520 26 | 87.619 95 | 91.692 97 | 95.738 04 | 99.753 95 | 103.739 48 | 107.693 41 | 111.614 53 |
| 32 | 79.464 07 | 83.588 79 | 87.688 05 | 91.760 61 | 95.805 21 | 99.820 63 | 103.805 63 | 107.775 90 | 111.679 58 |
| 33 | 79.533 03 | 83.657 34 | 87.756 16 | 91.828 26 | 95.872 38 | 99.887 30 | 103.871 79 | 107.824 65 | 111.744 66 |
| 34 | 79.601 98 | 83.725 86 | 87.824 27 | 91.895 89 | 95.939 54 | 99.953 97 | 103.937 94 | 107.890 25 | 111.809 71 |
| 35 | 79.670 89 | 83.794 38 | 87.892 34 | 91.963 52 | 96.006 69 | 100.020 61 | 104.004 07 | 107.955 85 | 111.874 74 |
| 36 | 79.739 80 | 83.862 89 | 87.960 41 | 92.031 14 | 96.073 83 | 100.087 26 | 104.070 20 | 108.021 44 | 111.939 77 |
| 37 | 79.808 75 | 83.931 40 | 88.028 48 | 92.098 75 | 96.140 97 | 100.153 89 | 104.136 32 | 108.087 02 | 112.004 79 |
| 38 | 79.877 66 | 83.999 88 | 88.096 53 | 92.166 35 | 96.207 96 | 100.220 51 | 104.202 41 | 108.152 58 | 112.069 80 |
| 39 | 79.946 57 | 84.068 39 | 88.164 60 | 92.233 95 | 96.275 21 | 100.287 14 | 104.268 52 | 108.218 15 | 112.134 80 |
| 40 | 80.015 48 | 84.136 88 | 88.232 66 | 92.301 54 | 96.342 33 | 100.353 76 | 104.334 62 | 108.283 70 | 112.199 81 |
| 41 | 80.084 40 | 84.205 34 | 88.300 68 | 92.369 12 | 96.409 42 | 100.420 35 | 104.400 69 | 108.349 24 | 112.264 78 |
| 42 | 80.152 35 | 84.273 82 | 88.368 71 | 92.436 69 | 96.476 51 | 100.486 94 | 104.466 77 | 108.414 77 | 112.329 75 |
| 43 | 80.222 13 | 84.342 28 | 88.436 73 | 92.504 25 | 96.543 60 | 100.555 53 | 104.532 83 | 108.480 30 | 112.394 71 |

| Min. | 27° | 28° | 29° | 30° | 31° | 32° | 33° | 34° | 35° |
|---|---|---|---|---|---|---|---|---|---|
| 44 | 80.29099 | 84.41072 | 88.50474 | 92.57180 | 96.61067 | 100.62010 | 104.59888 | 108.54580 | 112.45967 |
| 45 | 80.35987 | 84.47918 | 88.57276 | 92.63936 | 96.67774 | 100.68667 | 104.66493 | 108.61314 | 112.52461 |
| 46 | 80.42870 | 84.54762 | 88.64077 | 92.70733 | 96.74480 | 100.75323 | 104.73097 | 108.67681 | 112.58955 |
| 47 | 80.49758 | 84.61605 | 88.70874 | 92.77443 | 96.81185 | 100.81978 | 104.79700 | 108.74230 | 112.65447 |
| 48 | 80.56642 | 84.68448 | 88.77674 | 92.84196 | 96.87890 | 100.88632 | 104.86322 | 108.80778 | 112.71939 |
| 49 | 80.63526 | 84.75290 | 88.84472 | 92.90948 | 96.94593 | 100.95286 | 104.92903 | 108.87324 | 112.78429 |
| 50 | 80.704081 | 84.82130 | 88.91268 | 92.97699 | 97.01295 | 101.01938 | 104.99502 | 108.93870 | 112.84918 |
| 51 | 80.77292 | 84.88972 | 88.98065 | 93.04449 | 97.07998 | 101.08590 | 105.06103 | 109.00415 | 112.91407 |
| 52 | 80.84174 | 84.95812 | 89.04861 | 93.11192 | 97.14699 | 101.15241 | 105.12703 | 109.06959 | 112.97895 |
| 53 | 80.91055 | 85.02650 | 89.11656 | 93.17947 | 97.21399 | 101.21891 | 105.19299 | 109.13503 | 113.04382 |
| 54 | 80.97935 | 85.09489 | 89.18450 | 93.24695 | 97.28099 | 101.28540 | 105.25895 | 109.20045 | 113.10868 |
| 55 | 81.04815 | 85.16327 | 89.25243 | 93.31442 | 97.34797 | 101.35188 | 105.32491 | 109.26586 | 113.17352 |
| 56 | 81.11693 | 85.23163 | 89.32036 | 93.38188 | 97.41494 | 101.41835 | 105.39086 | 109.33126 | 113.23836 |
| 57 | 81.18573 | 85.29999 | 89.38828 | 93.44933 | 97.48192 | 101.48482 | 105.45680 | 109.39666 | 113.30319 |
| 58 | 81.25451 | 85.36835 | 89.45619 | 93.51678 | 97.54888 | 101.55128 | 105.52273 | 109.46204 | 113.36802 |
| 59 | 81.32327 | 85.43669 | 89.52409 | 93.58422 | 97.61584 | 101.61772 | 105.58865 | 109.52742 | 113.43282 |
| Min. | 27° | 28° | 29° | 30° | 31° | 32° | 33° | 34° | 35° |
| 0 | 113.49762 | 117.36789 | 121.20241 | 125.00000 | 128.75952 | 132.47982 | 136.15976 | 139.79823 | 143.39411 |
| 1 | 113.56242 | 117.43210 | 121.26600 | 125.06297 | 128.82185 | 132.54148 | 136.22074 | 139.85851 | 143.45367 |

(continued)

**1.16** (continued)

| Min. | 27° | 28° | 29° | 30° | 31° | 32° | 33° | 34° | 35° |
|---|---|---|---|---|---|---|---|---|---|
| 2 | 113.627 18 | 117.496 29 | 121.329 59 | 125.125 94 | 128.884 17 | 132.603 14 | 136.281 70 | 139.918 77 | 143.513 22 |
| 3 | 113.691 97 | 117.560 48 | 121.391 17 | 125.188 89 | 128.946 48 | 132.642 68 | 136.342 68 | 139.979 04 | 143.572 77 |
| 4 | 113.756 73 | 117.624 65 | 121.456 74 | 125.251 83 | 129.008 77 | 132.726 42 | 136.403 63 | 140.038 93 | 143.632 29 |
| 5 | 113.821 48 | 117.688 82 | 121.520 30 | 125.314 76 | 129.071 06 | 132.788 03 | 136.464 56 | 140.099 52 | 143.691 81 |
| 6 | 113.886 23 | 117.752 97 | 121.583 85 | 125.377 68 | 129.133 33 | 132.849 64 | 136.525 49 | 140.159 75 | 143.751 31 |
| 7 | 113.950 96 | 117.817 12 | 121.647 38 | 125.440 59 | 129.195 59 | 132.911 24 | 136.586 41 | 140.219 96 | 143.810 81 |
| 8 | 114.015 68 | 117.881 25 | 121.710 91 | 125.503 49 | 129.257 84 | 132.972 83 | 136.647 31 | 140.280 15 | 143.870 28 |
| 9 | 114.080 40 | 117.945 38 | 121.774 43 | 125.566 38 | 129.320 09 | 133.034 41 | 136.708 20 | 140.340 35 | 143.929 75 |
| 10 | 114.145 12 | 118.009 49 | 121.837 93 | 125.629 26 | 129.382 32 | 133.095 97 | 136.769 08 | 140.400 53 | 143.989 21 |
| 11 | 114.209 80 | 118.073 60 | 121.901 43 | 125.692 13 | 129.444 54 | 133.157 52 | 136.829 95 | 140.460 69 | 144.048 65 |
| 12 | 114.274 48 | 118.137 69 | 121.964 90 | 125.754 99 | 129.506 75 | 133.219 07 | 136.890 81 | 140.520 84 | 144.108 08 |
| 13 | 114.339 16 | 118.201 78 | 122.028 39 | 125.817 83 | 129.568 95 | 133.280 60 | 136.951 66 | 140.580 99 | 144.167 50 |
| 14 | 114.403 82 | 118.265 85 | 122.091 84 | 125.880 67 | 129.631 14 | 133.342 11 | 137.012 47 | 140.641 14 | 144.226 90 |
| 15 | 114.468 48 | 118.329 92 | 122.155 31 | 125.943 49 | 129.693 32 | 133.403 63 | 137.073 31 | 140.701 23 | 144.286 30 |
| 16 | 114.533 13 | 118.393 97 | 122.218 76 | 126.006 31 | 129.755 48 | 133.465 13 | 137.134 12 | 140.761 30 | 144.345 66 |
| 17 | 114.597 76 | 118.458 02 | 122.282 19 | 126.069 11 | 129.817 63 | 133.526 61 | 137.194 92 | 140.821 43 | 144.405 05 |
| 18 | 114.662 39 | 118.522 05 | 122.345 61 | 126.131 91 | 129.879 78 | 133.588 09 | 137.255 71 | 140.881 51 | 144.464 41 |
| 19 | 114.720 06 | 118.586 08 | 122.409 03 | 126.194 69 | 129.941 91 | 133.649 55 | 137.316 48 | 140.941 58 | 144.523 75 |
| 20 | 114.791 60 | 118.650 08 | 122.472 43 | 126.257 45 | 130.004 03 | 133.711 00 | 137.377 24 | 141.001 64 | 144.583 08 |
| 21 | 114.856 21 | 118.714 10 | 122.535 82 | 126.320 22 | 130.066 14 | 133.772 44 | 137.438 00 | 141.061 69 | 144.642 41 |
| | | | | 126.382 97 | 130.128 24 | 133.833 88 | 137.498 74 | 141.121 73 | 144.701 71 |

| | | | | | | | | |
|---|---|---|---|---|---|---|---|---|
| 23 | 114.985 38 | 118.842 06 | 122.662 58 | 126.445 71 | 130.190 33 | 133.895 29 | 137.559 47 | 141.181 74 |
| 24 | 115.049 95 | 118.906 05 | 122.725 94 | 126.508 44 | 130.252 41 | 133.956 70 | 137.620 19 | 141.241 75 |
| 25 | 115.114 51 | 118.970 02 | 122.789 29 | 126.571 16 | 130.314 47 | 134.018 09 | 137.680 89 | 141.301 75 |
| 26 | 115.179 04 | 119.033 96 | 122.852 63 | 126.633 87 | 130.376 53 | 134.079 48 | 137.741 58 | 141.361 73 |
| 27 | 115.243 59 | 119.097 92 | 122.915 96 | 126.696 56 | 130.438 57 | 134.140 85 | 137.802 27 | 141.421 71 |
| 28 | 115.308 12 | 119.161 85 | 122.979 28 | 126.759 25 | 130.500 61 | 134.202 21 | 137.862 94 | 141.481 67 |
| 29 | 115.372 64 | 119.225 77 | 123.042 59 | 126.821 93 | 130.562 63 | 134.263 56 | 137.923 60 | 141.541 62 |
| 30 | 115.437 15 | 119.289 69 | 123.105 89 | 126.884 59 | 130.624 64 | 134.324 90 | 137.984 25 | 141.601 56 |
| 31 | 115.501 65 | 119.353 59 | 123.169 18 | 126.947 25 | 130.686 64 | 134.386 23 | 138.044 88 | 141.661 49 |
| 32 | 115.566 14 | 119.417 49 | 123.232 46 | 127.009 89 | 130.748 63 | 134.447 54 | 138.105 51 | 141.721 40 |
| 33 | 115.630 63 | 119.481 37 | 123.295 72 | 127.072 52 | 130.810 61 | 134.508 85 | 138.166 12 | 141.781 30 |
| 34 | 115.695 10 | 119.545 25 | 123.358 98 | 127.135 14 | 130.872 58 | 134.570 14 | 138.226 72 | 141.841 19 |
| 35 | 115.759 56 | 119.609 11 | 123.422 23 | 127.197 75 | 130.934 53 | 134.631 40 | 138.287 31 | 141.901 07 |
| 36 | 115.824 01 | 119.672 97 | 123.485 47 | 127.260 35 | 130.996 48 | 134.692 70 | 138.347 89 | 141.960 94 |
| 37 | 115.888 50 | 119.736 81 | 123.548 69 | 127.322 94 | 131.058 41 | 134.753 96 | 138.408 45 | 142.020 79 |
| 38 | 115.952 88 | 119.800 63 | 123.611 91 | 127.385 52 | 131.120 33 | 134.815 19 | 138.469 01 | 142.080 63 |
| 39 | 116.017 30 | 119.884 47 | 123.675 11 | 127.448 09 | 131.182 24 | 134.876 44 | 138.529 55 | 142.140 46 |
| 40 | 116.081 72 | 119.928 28 | 123.738 32 | 127.510 66 | 131.244 15 | 134.937 68 | 138.590 08 | 142.200 28 |
| 41 | 116.146 12 | 119.992 08 | 123.801 49 | 127.573 19 | 131.306 03 | 134.998 88 | 138.650 60 | 142.260 09 |
| 42 | 116.210 51 | 120.055 87 | 123.864 67 | 127.635 73 | 131.367 91 | 135.060 80 | 138.711 07 | 142.319 88 |
| 43 | 116.274 90 | 120.119 66 | 123.927 83 | 127.698 25 | 131.429 78 | 135.121 27 | 138.771 60 | 142.379 66 |
| 44 | 116.339 25 | 120.183 42 | 123.990 97 | 127.760 76 | 131.491 64 | 135.182 45 | 138.832 09 | 142.439 43 |
| 45 | 116.403 60 | 120.247 19 | 124.054 13 | 127.823 27 | 131.553 48 | 135.243 62 | 138.892 56 | 142.499 19 |
| 46 | 116.467 99 | 120.310 95 | 124.117 25 | 127.885 76 | 131.615 32 | 135.304 78 | 138.953 02 | 142.558 94 |

| |
|---|
| 144.761 01 |
| 144.820 29 |
| 144.879 57 |
| 144.938 82 |
| 144.998 07 |
| 145.057 31 |
| 145.116 53 |
| 145.175 74 |
| 145.234 94 |
| 145.294 12 |
| 145.353 30 |
| 145.412 46 |
| 145.471 61 |
| 145.530 74 |
| 145.589 87 |
| 145.648 98 |
| 145.708 08 |
| 145.767 17 |
| 145.826 24 |
| 145.885 30 |
| 145.944 35 |
| 146.003 39 |
| 146.062 42 |
| 146.121 43 |

(continued)

**1.16** (continued)

| Min. | 27° | 28° | 29° | 30° | 31° | 32° | 33° | 34° | 35° |
|---|---|---|---|---|---|---|---|---|---|
| 47 | 116.53233 | 120.37469 | 124.18038 | 127.94824 | 131.67714 | 135.36590 | 139.01347 | 142.61867 | 146.18043 |
| 48 | 116.59666 | 120.43840 | 124.24349 | 128.01072 | 131.73895 | 135.42705 | 139.07390 | 142.67839 | 146.23942 |
| 49 | 116.66098 | 120.50214 | 124.30659 | 128.07318 | 131.80075 | 135.48818 | 139.13433 | 142.73810 | 146.29839 |
| 50 | 116.72529 | 120.56585 | 124.36968 | 128.13562 | 131.86253 | 135.54928 | 139.19474 | 142.79780 | 146.35736 |
| 51 | 116.78960 | 120.62955 | 124.43276 | 128.19806 | 131.92432 | 135.61038 | 139.25514 | 142.85748 | 146.41631 |
| 52 | 116.85390 | 120.69325 | 124.49583 | 128.26049 | 131.98609 | 135.67147 | 139.31551 | 142.91716 | 146.47525 |
| 53 | 116.91818 | 120.75693 | 124.55889 | 128.32291 | 132.04784 | 135.73255 | 139.37588 | 142.97682 | 146.53418 |
| 54 | 116.98245 | 120.82060 | 124.62193 | 128.38531 | 132.10959 | 135.79361 | 139.43628 | 143.03647 | 146.59309 |
| 55 | 117.04671 | 120.88426 | 124.68497 | 128.44771 | 132.17132 | 135.85467 | 139.49663 | 143.09611 | 146.65199 |
| 56 | 117.11096 | 120.94791 | 124.74799 | 128.51009 | 132.23303 | 135.91571 | 139.55697 | 143.15573 | 146.71088 |
| 57 | 117.17522 | 121.01155 | 124.81102 | 128.57247 | 132.29470 | 135.97674 | 139.61730 | 143.21534 | 146.76976 |
| 58 | 117.23945 | 121.07518 | 124.87402 | 128.63483 | 132.35650 | 136.03776 | 139.67763 | 143.27495 | 146.82862 |
| 59 | 117.30368 | 121.13880 | 124.93702 | 128.69718 | 132.41810 | 136.09876 | 139.73793 | 143.33453 | 146.88747 |

| Min. | 36° | 37° | 38° | 39° | 40° | 41° | 42° | 43° | 44° | 45° |
|---|---|---|---|---|---|---|---|---|---|---|
| 0 | 146.94631 | 150.45376 | 153.91537 | 157.33010 | 160.69902 | 164.01476 | 167.28265 | 170.49959 | 173.66460 | 176.77670 |
| 1 | 147.00514 | 150.51183 | 153.97267 | 157.38661 | 160.75260 | 164.06963 | 167.33669 | 170.55277 | 173.71690 | 176.82811 |
| 2 | 147.06390 | 150.56989 | 154.02995 | 157.44310 | 160.80829 | 164.12450 | 167.39071 | 170.60593 | 173.76919 | 176.87951 |
| 3 | 147.12276 | 150.62793 | 154.08723 | 157.49959 | 160.86397 | 164.17935 | 167.44470 | 170.65908 | 173.82146 | 176.93089 |

| | | | | | | | | | | |
|---|---|---|---|---|---|---|---|---|---|---|
| 4 | 147.181 55 | 150.685 97 | 154.144 49 | 157.556 06 | 160.919 63 | 164.234 18 | 167.498 71 | 170.712 22 | 173.873 72 | 176.982 27 |
| 5 | 147.240 32 | 150.743 99 | 154.201 74 | 157.612 51 | 160.975 27 | 164.289 00 | 167.552 69 | 170.765 34 | 173.925 97 | 177.033 62 |
| 6 | 147.299 09 | 150.802 00 | 154.258 97 | 157.668 95 | 161.030 91 | 164.343 81 | 167.606 66 | 170.818 44 | 173.978 20 | 177.084 96 |
| 7 | 147.357 84 | 150.859 99 | 154.316 19 | 157.725 38 | 161.086 53 | 164.398 61 | 167.660 61 | 170.871 54 | 174.030 42 | 177.136 29 |
| 8 | 147.416 58 | 150.917 97 | 154.373 40 | 157.781 80 | 161.142 13 | 164.453 38 | 167.714 54 | 170.924 61 | 174.082 62 | 177.187 59 |
| 9 | 147.475 31 | 150.975 95 | 154.430 59 | 157.838 20 | 161.197 73 | 164.508 15 | 167.768 47 | 170.977 68 | 174.134 80 | 177.238 90 |
| 10 | 147.534 03 | 151.033 90 | 154.487 78 | 157.894 59 | 161.253 31 | 164.562 90 | 167.822 37 | 171.030 72 | 174.186 98 | 177.290 17 |
| 11 | 147.592 73 | 151.091 85 | 154.544 94 | 157.950 96 | 161.308 87 | 164.617 41 | 167.876 25 | 171.083 76 | 174.239 13 | 177.341 43 |
| 12 | 147.651 42 | 151.149 78 | 154.602 10 | 158.007 33 | 161.364 42 | 164.672 37 | 167.930 15 | 171.136 78 | 174.291 28 | 177.392 68 |
| 13 | 147.710 10 | 151.207 70 | 154.659 24 | 158.063 68 | 161.420 00 | 164.727 08 | 167.984 01 | 171.189 78 | 174.343 40 | 177.443 92 |
| 14 | 147.768 76 | 151.265 60 | 154.716 37 | 158.120 09 | 161.475 48 | 164.781 77 | 168.037 86 | 171.242 77 | 174.395 52 | 177.495 14 |
| 15 | 147.827 41 | 151.323 50 | 154.773 49 | 158.176 33 | 161.530 99 | 164.836 45 | 168.091 70 | 171.295 75 | 174.447 60 | 177.546 34 |
| 16 | 147.886 05 | 151.381 38 | 154.830 60 | 158.232 64 | 161.586 49 | 164.891 12 | 168.145 53 | 171.348 71 | 174.499 70 | 177.597 54 |
| 17 | 147.944 68 | 151.439 25 | 154.887 68 | 158.288 94 | 161.641 96 | 164.945 78 | 168.199 32 | 171.401 66 | 174.551 77 | 177.648 71 |
| 18 | 148.003 30 | 151.497 10 | 154.944 76 | 158.345 22 | 161.697 45 | 165.000 42 | 168.253 13 | 171.454 59 | 174.603 82 | 177.699 86 |
| 19 | 148.061 90 | 151.554 94 | 155.001 82 | 158.401 49 | 161.752 90 | 165.055 04 | 168.306 91 | 171.507 51 | 174.655 86 | 177.751 01 |
| 20 | 148.120 49 | 151.612 77 | 155.058 87 | 158.457 41 | 161.808 34 | 165.109 66 | 168.360 68 | 171.560 41 | 174.707 87 | 177.802 14 |
| 21 | 148.179 07 | 151.670 59 | 155.115 91 | 158.513 98 | 161.863 77 | 165.164 26 | 168.414 43 | 171.613 30 | 174.759 89 | 177.853 26 |
| 22 | 148.237 63 | 151.728 39 | 155.172 94 | 158.570 21 | 161.919 19 | 165.218 84 | 168.468 17 | 171.666 17 | 174.811 89 | 177.904 36 |
| 23 | 148.296 18 | 151.786 18 | 155.229 95 | 158.626 43 | 161.974 59 | 165.273 41 | 168.521 89 | 171.719 03 | 174.863 87 | 177.955 44 |
| 24 | 148.354 72 | 151.843 96 | 155.286 95 | 158.682 63 | 162.029 98 | 165.327 97 | 168.575 60 | 171.771 88 | 174.915 84 | 178.006 51 |
| 25 | 148.413 25 | 151.901 73 | 155.343 93 | 158.738 82 | 162.085 35 | 165.382 51 | 168.629 29 | 171.824 71 | 174.967 79 | 178.057 57 |

*(continued)*

**1.16** (*continued*)

| Min. | 36° | 37° | 38° | 39° | 40° | 41° | 42° | 43° | 44° | 45° |
|---|---|---|---|---|---|---|---|---|---|---|
| 26 | 148.471 76 | 151.959 48 | 155.400 90 | 158.794 99 | 162.140 71 | 165.437 04 | 168.682 97 | 171.877 52 | 175.019 72 | 178.108 60 |
| 27 | 148.530 27 | 152.017 22 | 155.457 86 | 158.851 16 | 162.196 06 | 165.491 55 | 168.736 64 | 171.903 26 | 175.071 64 | 178.159 63 |
| 28 | 148.588 76 | 152.074 94 | 155.514 81 | 158.907 30 | 162.251 39 | 165.546 05 | 168.790 29 | 171.983 11 | 175.123 55 | 178.210 64 |
| 29 | 148.647 23 | 152.132 66 | 155.571 74 | 158.963 43 | 162.306 71 | 165.600 54 | 168.843 93 | 172.035 89 | 175.175 44 | 178.261 63 |
| 30 | 148.705 69 | 152.190 36 | 155.628 66 | 159.019 56 | 162.362 01 | 165.655 01 | 168.897 55 | 172.088 64 | 175.227 32 | 178.312 61 |
| 31 | 148.764 15 | 152.248 05 | 155.685 57 | 159.075 66 | 162.417 30 | 165.709 47 | 168.951 16 | 172.141 39 | 175.279 18 | 178.363 58 |
| 32 | 148.822 59 | 152.305 72 | 155.742 46 | 159.131 76 | 162.472 58 | 165.763 91 | 169.004 76 | 172.194 11 | 175.331 01 | 178.414 52 |
| 33 | 148.881 01 | 152.363 38 | 155.799 34 | 159.187 84 | 162.527 84 | 165.818 35 | 169.058 34 | 172.246 83 | 175.382 86 | 178.465 46 |
| 34 | 148.939 43 | 152.421 03 | 155.856 20 | 159.243 90 | 162.583 10 | 165.872 76 | 169.119 04 | 172.299 53 | 175.434 68 | 178.516 38 |
| 35 | 148.997 83 | 152.478 67 | 155.913 06 | 159.299 96 | 162.638 33 | 165.927 17 | 169.165 46 | 172.352 22 | 175.486 48 | 187.567 28 |
| 36 | 149.056 22 | 152.536 29 | 155.969 90 | 159.356 00 | 162.693 55 | 165.981 55 | 169.218 99 | 172.404 89 | 175.538 26 | 178.618 17 |
| 37 | 149.114 60 | 152.593 90 | 156.026 73 | 159.412 02 | 162.748 76 | 166.035 93 | 169.272 52 | 172.457 54 | 175.590 04 | 178.669 04 |
| 38 | 149.172 96 | 152.651 50 | 156.083 54 | 159.468 04 | 162.803 96 | 166.090 29 | 169.326 02 | 172.510 18 | 175.641 79 | 178.719 90 |
| 39 | 149.231 31 | 152.709 08 | 156.140 34 | 159.524 04 | 162.859 14 | 166.144 63 | 169.379 52 | 172.562 81 | 175.693 54 | 178.770 74 |
| 40 | 149.289 66 | 152.766 67 | 156.197 13 | 159.580 02 | 162.914 31 | 166.198 98 | 169.433 01 | 172.615 43 | 175.745 27 | 178.821 57 |
| 41 | 149.347 97 | 152.824 21 | 156.253 90 | 159.635 99 | 162.969 46 | 166.253 28 | 169.486 47 | 172.668 02 | 175.796 98 | 178.872 39 |
| 42 | 149.406 29 | 152.881 76 | 156.310 66 | 159.691 95 | 163.024 60 | 166.307 59 | 169.539 92 | 172.720 60 | 175.848 68 | 178.923 18 |
| 43 | 149.464 59 | 152.939 29 | 156.367 41 | 159.747 90 | 163.079 73 | 166.361 88 | 169.593 36 | 172.773 17 | 175.900 36 | 178.973 97 |
| 44 | 149.522 86 | 152.996 81 | 156.424 20 | 159.803 83 | 163.134 84 | 166.416 15 | 169.646 78 | 172.825 73 | 175.952 03 | 179.024 73 |

| | | | | | | | | | |
|---|---|---|---|---|---|---|---|---|---|
| 45 | 149.581 15 | 153.054 32 | 156.480 87 | 159.859 75 | 163.189 94 | 166.470 42 | 169.700 19 | 172.878 26 | 176.003 68 | 179.075 49 |
| 46 | 149.639 41 | 153.111 81 | 156.537 58 | 159.915 66 | 163.245 02 | 166.524 67 | 169.753 58 | 172.930 79 | 176.055 32 | 179.126 22 |
| 47 | 149.697 66 | 153.169 29 | 156.594 27 | 159.971 55 | 163.300 09 | 166.578 90 | 169.806 96 | 172.983 30 | 176.106 94 | 179.176 95 |
| 48 | 149.755 90 | 153.226 76 | 156.650 95 | 160.027 43 | 163.335 52 | 166.633 12 | 169.860 33 | 173.035 79 | 176.158 56 | 179.227 65 |
| 49 | 149.814 12 | 153.284 22 | 156.707 62 | 160.083 29 | 163.410 19 | 166.687 32 | 169.913 68 | 173.088 27 | 176.210 15 | 179.278 34 |
| 50 | 149.987 23 | 153.341 66 | 156.764 28 | 160.139 14 | 163.465 22 | 166.741 51 | 169.967 01 | 173.140 74 | 176.261 73 | 179.329 02 |
| 51 | 149.930 54 | 153.399 09 | 156.820 92 | 160.194 98 | 163.520 24 | 166.795 69 | 170.020 34 | 173.193 19 | 176.313 29 | 179.379 68 |
| 52 | 149.988 72 | 153.456 51 | 156.877 55 | 160.250 80 | 163.575 24 | 166.849 86 | 170.073 65 | 173.245 63 | 176.364 84 | 179.430 33 |
| 53 | 150.046 90 | 153.513 91 | 156.934 16 | 160.306 61 | 163.630 23 | 166.904 00 | 170.126 94 | 173.298 05 | 176.416 37 | 179.480 96 |
| 54 | 150.105 06 | 153.571 30 | 156.990 76 | 160.362 41 | 163.685 20 | 166.958 14 | 170.180 22 | 173.350 46 | 176.467 90 | 179.531 57 |
| 55 | 150.163 21 | 153.628 68 | 157.047 35 | 160.418 19 | 163.740 16 | 167.012 26 | 170.233 48 | 173.402 85 | 176.519 40 | 179.582 18 |
| 56 | 150.221 34 | 153.686 04 | 157.103 93 | 160.473 96 | 163.795 10 | 167.066 37 | 170.286 73 | 173.455 23 | 176.570 89 | 179.632 76 |
| 57 | 150.279 46 | 153.743 39 | 157.160 49 | 160.529 72 | 163.850 04 | 167.120 46 | 170.339 97 | 173.507 59 | 176.622 36 | 179.683 33 |
| 58 | 150.337 58 | 153.800 73 | 157.217 04 | 160.585 46 | 163.904 96 | 167.174 54 | 170.393 19 | 173.559 94 | 176.673 82 | 179.733 89 |
| 59 | 150.395 67 | 153.858 06 | 157.273 58 | 160.641 19 | 163.959 87 | 167.228 60 | 170.446 40 | 173.612 27 | 176.725 27 | 179.784 43 |

## 1.17 Measurement over precision balls and rollers

*Notes:*

(1) Precision balls are used where a *point* contact is required.
(2) Precision rollers are used where a *line* contact is required.

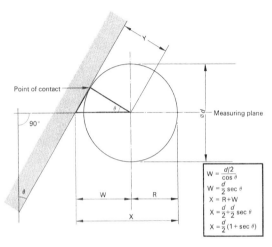

$$W = \frac{d/2}{\cos \theta}$$
$$W = \frac{d}{2} \sec \theta$$
$$X = R + W$$
$$X = \frac{d}{2} + \frac{d}{2} \sec \theta$$
$$X = \frac{d}{2}(1 + \sec \theta)$$

The above figure shows how the distance from the point of measurement to the component can be calculated. It also shows that the point of contact does not always lie in the measuring plane. This figure, together with the following calculations, should be studied carefully as it forms the basis for the subsequent examples.

Referring to the above figure:

$W$ is the hypotenuse of the right-angled triangles
$Y$ is the adjacent side of the angle $\theta°$

Therefore:

$$\frac{Y}{W} = \cos \theta$$

But $Y$ = the radius of the roller

$$\therefore \quad Y = \frac{d}{2}$$

Substituting in equation (1),

$$\frac{d/2}{W} = \cos\theta$$

$$\therefore W = \frac{d/2}{\cos\theta}$$

$$\text{or } W = \frac{d}{2}\sec\theta$$

| Inverse ratios |
| --- |
| $\dfrac{1}{\cos\theta} = \sec\theta$ |
| $\dfrac{1}{\sin\theta} = \cos\theta$ |
| $\dfrac{1}{\tan\theta} = \cot\theta$ |

But $X$ = radius of roller $(R) + W$

where: $R = \dfrac{d}{2}$

$$= \frac{d}{2} + \left(\frac{d}{2}\sec\theta\right)$$

$$= \frac{d}{2}(1 + \sec\theta)$$

(Courtesy of Addison Wesley Longman.)

## 1.18 Measurement of external tapers

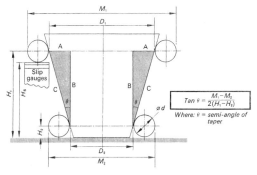

$$Tan\,\theta = \frac{M_1 - M_2}{2(H_1 - H_1)}$$
Where: $\theta$ = semi-angle of taper

### 1.18.1 To find the angle $\theta$ (the semi-angle of taper)

Refer to the above figure. The shaded triangles show that:

$$\tan\theta = \frac{\text{opposite}}{\text{adjacent}} = \frac{A}{B} \qquad (1)$$

and that:

$$A = \frac{D_1 - D_2}{2}$$

$$B = H_1 - H_2$$

57

Thus, substituting in equation (1),

$$\tan \theta = \frac{(D_1 - D_2)/2}{H_1 - H_2}$$
$$= \frac{D_1 - D_2}{2(H_1 - H_2)} \qquad (2)$$

Unfortunately, $D_1$ and $D_2$ cannot be measured directly, so measurements are taken over rollers to give $M_1$ and $M_2$. Conversion from $M_1$ and $M_2$ to $D_1$ and $D_2$ involves the use of the expressions derived previously. (Note the difference between $M$ and $D$ in both instances is equal to $2x$ in Section 1.17)

$$M_1 = D_1 + 2\frac{d}{2}(1 + \text{secant } \theta)$$
$$= D_1 + d(1 + \text{secant } \theta)$$

Similarly:

$$M_2 = D_2 + 2\frac{d}{2}(1 + \text{secant } \theta)$$
$$= D_2 + d(1 + \text{secant } \theta)$$
$$M_1 = D_1 + 2\left[\frac{d}{2}(1 + \sec \theta)\right]$$
$$= D_1 + d(1 + \sec \theta)$$
$$M_2 = D_2 + 2\left[\frac{d}{2}(1 + \sec \theta)\right]$$
$$= D_2 + d(1 + \sec \theta)$$

Since $d(1 + \sec \theta)$ is common to both the above expressions, $M_1 - M_2 = D_1 - D_2$. Substituting in equation (2)

$$\tan \theta = \frac{M_1 - M_2}{2(H_1 - H_2)}$$

---

**Example 1.18.1**

The following data were obtained when checking a taper plug gauge.

Diameter of rollers $= 10$ mm

Micrometer reading over rollers at height $H_1 = 70$ mm

Micrometer reading over rollers at height $H_2 = 65$ mm

Height of slip gauge stack ($H_s$) $= 40$ mm

*Note:* $H_1$, $H_2$ and $H_s$ refer to the above figure.

$$\text{Tan}\,\theta = \frac{M_1 - M_2}{2(H_1 - H_2)}$$

$$= \frac{70 - 65}{2(45 - 5)}$$

$$= \frac{5}{80}$$

$$= 0.0625$$

$\therefore\ \theta = 3°35'$ or angle of taper$(2\theta) = 7°10'$

where: $M_1 = 70\,\text{mm}$
$M_2 = 65\,\text{mm}$
$H_1 = (40 + 5) = 45\,\text{mm}$
$H_2 = (0 + 5)\ \ = \ 5\,\text{mm}$

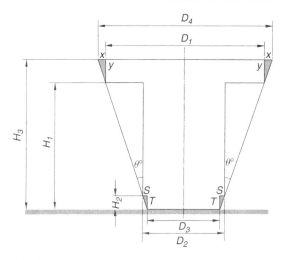

### 1.18.2 To find the major and minor diameters

*Major diameter*
Reference to the above figure shows that:

$D_4$ is obviously bigger than $D_1$ by an amount equal to 2X.

But X can be found knowing height Y and the semi-angle of taper $\theta$.

$$X = Y \tan\theta$$

59

$$\therefore \quad X = (H_3 - H_1)\tan\theta$$

$$\text{and} \quad \underline{2X = 2(H_3 - H_1)\tan\theta}$$

where: $Y = H_3 - H_1$

Thus: $D_4 = D_1 + 2(H_3 - H_1)\tan\theta$

Thus: $\underline{D_4 = M_1 - d(1 + \text{secant }\theta) + 2(H_3 - H_1)\tan\theta}$

where: $D_1 = M_1 - d(1 + \text{secant }\theta)$

*Minor diameter*

Reference to the above figure shows that:

$D_3$ is obviously smaller than $D_2$ by an amount equal to $2S$.

But $S$ can be found knowing the height $T$ and the semi-angle of taper $\theta$.

*Note:* $H_1$ and $H_2$ are not the point of contact of cylinder and taper, but the point at the same level as the roller centres.

$$\therefore \quad S = T\tan\theta$$

$$\therefore \quad S = \frac{d}{2}\tan\theta$$

$$\text{and} \quad \underline{2S = d\tan\theta}$$

where: $T = \dfrac{d}{2}$, when $d$ = roller dismater

Thus: $D_3 = D_2 - d\tan\theta$

Where: $D_2 = M_2 - d(1 + \text{secant }\theta)$ (previously proved)

Thus: $\underline{D_3 = M_2 - d(1 + \text{secant }\theta) - d\tan\theta}$

---

### Example 1.18.2

Calculate the minor diameter ($D_3$) and the major diameter ($D_4$) of a taper plug gauge given the following data:

| | |
|---|---|
| Diameter of rollers | = 10 mm |
| Micrometer reading $M_1$ | = 70 mm |
| Micrometer reading $M_2$ | = 65 mm |
| Height $H_2$ | = 5 mm |
| Height $H_3$ | = 60 mm |
| Height $H_1$ | = 45 mm |

Angle $\theta$ (from Example 1.18.1) $= 3°35'$

Reference should be made to the above figure in the solution of this example.

*To find $D_3$*

$$
\begin{aligned}
D_3 &= M_2 - d(1 + \text{secant } \theta) - d \tan \theta \\
&= 65 - (10 + 10 \text{ secant } 3°35') - 10 \tan 3°35' \\
&= 65 - \left( 10 + \frac{10}{\cos 3°35'} \right) - 10 \tan 3°35' \\
&= 65 - \left( 10 + \frac{10}{0.998\,2} \right) - 10 \times 0.062\,5 \\
&= 65 - 20.02 - 0.625 \\
&= \underline{44.355 \text{ mm diameter}}
\end{aligned}
$$

*To find $D_4$*

$$
\begin{aligned}
D_4 &= M_1 - d(1 + \text{secant } \theta) + 2(H_3 - H_1) \tan \theta \\
&= 70 - 10(1 + \text{secant } 3°35' + 2(60 - 45) \\
&\qquad\qquad\qquad\qquad\qquad\qquad \tan 3°35' \\
&= 70 - \left( 10 + \frac{10}{\cos 3°35'} \right) + 30 \tan 3°35' \\
&= 70 - 20.02 + (30 \times 0.625) \\
&= 70 - 20.02 + 1.875 \\
&= \underline{51.855 \text{ mm diameter}}
\end{aligned}
$$

(Courtesy of Addison Wesley Longman.)

# 1.19 Measurement of internal tapers

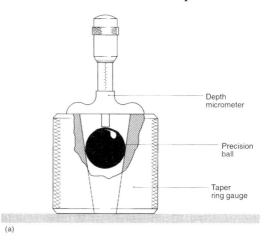

(a)

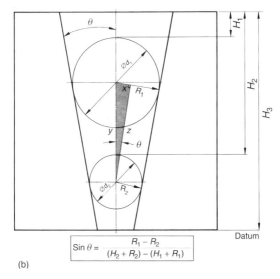

$$\text{Sin } \theta = \frac{R_1 - R_2}{(H_2 + R_2) - (H_1 + R_1)}$$

(b)

## 1.19.1 To find the angle $\theta°$ (the semi-angle of taper)

The above Fig. (a) shows how measurements are taken over two precision balls of different diameters. From a knowledge of the diameter of the balls and the depth to which they sink into the taper, the angle of taper and the major and minor diameters of the bore may be calculated.

It will be seen from Fig. (b) that the semi-angle of taper $\theta$ can be obtained from the triangle $XYZ$.

$$\sin \theta = \frac{X}{Y}$$

But: $\quad X = R_1 - R_2$

where: $\quad R_1 = \frac{d_1}{2}$

$$R_2 = \frac{d_2}{2}$$

and: $\quad Y = (H_2 + R_2) - (H_1 + R_1)$

Thus $\quad \sin \theta = \dfrac{R_1 - R_2}{(H_2 + R_2) - (H_1 + R_1)}$

---

### Example 1.19.1

The following data were obtained when inspecting a taper ring gauge. Calculate the included angle of taper.

$$d_1 = 16 \text{ mm diameter}$$
$$d_2 = 12 \text{ mm diameter}$$
$$H_1 = \phantom{0}6 \text{ mm}$$
$$H_2 = 20 \text{ mm}$$
$$\sin \theta = \frac{R_1 - R_2}{(H_2 + R_1) - (H_1 - R_1)}$$

where: $\quad R_1 = \dfrac{16}{2} = 8 \text{ mm}$

$$R_2 = \frac{12}{2} = 6 \text{ mm}$$

$$= \frac{8 - 6}{(20 + 6) - (6 + 8)}$$

$$= \frac{2}{26 - 14}$$

$$= \frac{2}{12}$$

$$= 0.166\,7$$

$\therefore \quad \underline{\theta = 9°36'}$

Thus the included angle of taper $= (2 \times 9°36') = \underline{19°12'}$.

---

## 1.19.2 To find the major and minor diameters

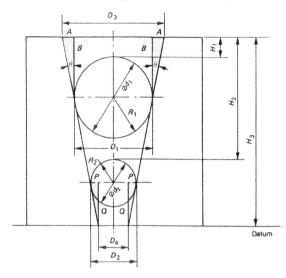

The major diameter $D_3$ and the minor diameter $D_4$ of the taper bore can also be calculated if, in addition to the existing data, the overall height $H_3$ is also known.

The dimensions $D_1$ and $D_2$ are, in fact, twice the dimension $W$ in Section 1.17.

Thus $$D_1 = 2\frac{d_1}{2} \text{ secant } \theta$$
$$= d_1 \text{ secant } \theta$$

Similarly, $D_2 = d_2 \text{ secant } \theta$

*Major diameter*

It will be seen from the above figure that $D_3$ is bigger than $D_1$ by twice the dimension $A$.

$$D_3 = D_1 + 2A$$

But $\dfrac{A}{B} = \tan \theta$

$\therefore \quad A = B \tan \theta \quad$ where: $B = H_1 + R_1$

$\therefore \quad A = (H_1 + R_1) \tan \theta$

Therefore: $\underline{D_3 = D_1 + 2(H_1 + R_1) \tan \theta}$

*Minor diameter*

It will be seen from the above figure that $D_4$ is smaller than $D_2$ by twice the dimension $P$.

$$D_4 = D_2 - 2P$$

But $\dfrac{P}{Q} = \tan \theta$

$\therefore \qquad P = Q \tan \theta \quad$ where: $Q = H_3 - (H_2 + R_2)$

$\therefore \qquad P = [H_3 - (H_2 + R_2)] \tan \theta$

Therefore: $\underline{D_4 = D_2 - 2[H_3 - (H_2 + R_2)] \tan \theta}$

---

**Example 1.19.2**

Reusing the taper ring gauge data from Example 1.19.1 plus the knowledge that $H_3 = 40$ mm and $\theta = 9°36'$, calculate the major diameter ($D_3$) and the minor diameter ($D_4$).

*Major diameter*

$$\begin{aligned}
D_3 &= D_1 + 2(H_1 + R_1) \tan \theta \\
&= d_1 \text{ secant } \theta + 2(H_1 + R_1) \tan \theta \\
&= 16 \text{ secant } 9°36' + 2(6 + 8) \tan 9°36' \\
&= \frac{16}{\cos 9°36'} + 2(6 + 8) \tan 9°36' \\
&= \frac{16}{0.986\,0} + 2 \times 14 \times 0.169\,1 \\
&= 16.23 + 4.735 \\
&= \underline{20.965 \text{ mm diameter}}
\end{aligned}$$

*Minor diameter*

$$\begin{aligned}
D_4 &= D_2 - 2[H_3 - (H_2 + R_2)] \tan \theta \\
&= d_2 \text{ secant } \theta - 2[(H_3 - (H_2 + R_2)] \tan \theta \\
&= 12 \text{ secant } 9°36' \\
&\quad - 2[(40 - (20 + 6)] \tan 9°36' \\
&= \frac{12}{\cos 9°36'} - 2[40 - (20 + 6)] \tan 9°36' \\
&= \frac{12}{0.986\,0} - 2 \times 14 \times 0.169\,1 \\
&= 12.17 - 4.735 \\
&= \underline{7.435 \text{ mm diameter}}
\end{aligned}$$

The examples shown are only an indication of the use of balls and rollers for internal, external and angular measurement. In practice, the applications are limitless, but in every instance the solutions lend themselves to the application of simple trigonometry once the basic triangles have been set up.

(Courtesy of Addison Wesley Longman.)

## 1.20 The dividing head – simple indexing

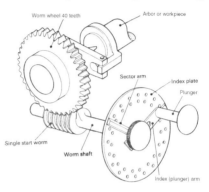

Standard gear ratio = 40 : 1

In the above figure the index arm has to be rotated 40 times to rotate the workpiece once.

(a) The total movement of the index arm for any given number of divisions on the workpiece is given by the expression:

Index arm setting $= \dfrac{40}{N}$

where: $N$ = the number of divisions required

(b) When angular divisions are required, the expression becomes:

Index arm setting $= \dfrac{\text{angle required } (°)}{9}$

since 1/40 of a revolution $= \dfrac{360°}{40} = 9°$

**Example 1.20.1**

Calculate the index arm setting to give 17 equally spaced divisions. The index plate has the following hole circles: 24, 25, 28, 30, 34, 37, 38, 39, 41, 42, 43.

Index arm setting $= \dfrac{40}{N} = \dfrac{40}{17} = 2\frac{6}{17}$

**By inspection, the actual indexing will be two whole turns and 12 holes in the 34-hole circle.**

---

**Example 1.20.2**

Using the same index plate as in the previous example, calculate the index arm setting to give an angular division of the workpiece of 12°18′.

$$\text{Index arm setting} = \frac{\text{angle required}}{9} = \frac{12°18'}{9}$$
$$= \frac{738'}{9 \times 60} = \frac{82}{60} = 1\,{}^{11}\!/_{30}$$

**By inspection, the actual indexing will be one whole turn and 11 holes in the 30-hole circle.**

Note how the angle was converted to minutes so as to have a common system of units in the numerator. Had the angle included seconds, e.g. 12°18′30″, then the angle would be converted into seconds and the denominator would become $9 \times 360$.

---

## 1.20.1 Sector arms

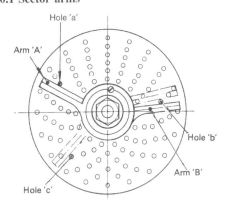

(a) Showing use of dividing head sector arms

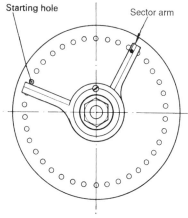

Starting hole

Sector arm

(b) Setting for 152 divisions

To save having to count the holes in the index plate
every time the dividing head is operated, *sector arms*
are provided, as shown in Fig. (a) above. The method
of using the sector arm is as follows:

1. The sector arms are set so that between arm 'A'
   and arm 'B' there is the required number of holes
   *plus the starting hole 'a'*.
2. The plunger and index arm is moved from hole 'a'
   to hole 'b' against sector arm B.
3. The sector arms are rotated so that arm 'A' is now
   against the plunger in hole 'b'.
4. For the next indexing the plunger is moved to hole
   'c' against the newly positioned arm 'B'.
5. The process is repeated for each indexing.

---

**Example 1.20.3**

For 152 divisions around the workpiece the indexing
would be 10 holes in a 38-hole circle.

The sector arms would have 10 holes plus the
starting hole where the indexing plunger is located
between them. That is, a total of 11 holes between
arm A and arm B.

---

(Courtesy of Addison Wesley Longman.)

# 1.21 Differential indexing

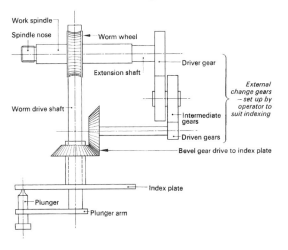

Divisions outside the range of a standard index plate can be obtained by *differential indexing*.

Instead of the index plate being clamped to the body of the dividing head, it is coupled to the work spindle by a gear train as shown in the above figure. Thus, as the index arm is rotated through the required number of turns, the index plate is advanced or retarded through a small amount automatically by the external gears.

The following expression is used to obtain the gear ratio of the drive coupling the work spindle to the index plate:

$$\frac{\text{Driver}}{\text{Driven}} = \frac{N_1 - N_2}{N_2} \times 40$$

where: $N_1$ = required divisions
$N_2$ = actual divisions available on the index plate

---

**Example 1.21.1**

Calculate the gear train to give an indexing of 113 divisions. The index plate available has: 24, 25, 28, 30, 34, 37, 38, 39, 41, 42, 43 holes. The gears available are: 24(2), 28, 32, 40, 48, 56, 64, 72, 86, 100 teeth.

From Example 1.20.1 it will be seen that the required indexing is:

$$\frac{40}{113}$$

but this is not available with the index plate supplied, therefore a near approximation is selected as a basis for calculation. For example:

$$\frac{40}{120} \quad \text{which can be indexed as} \quad \frac{14}{42}$$

That is, 14 holes in a 42-hole circle.

$$\frac{\text{Driver}}{\text{Driven}} = \frac{N_1 - N_2}{N_2} \times 40 \quad \text{where:} \quad N_1 = 113$$

$$N_2 = 120$$

$$= \frac{113 - 120}{120} \times 40$$

$$= \frac{-7}{120} \times 40 \quad (\textit{Note} : \text{The minus sign can be disregarded as it only indicates the direction of rotation.})$$

$$= \frac{7}{3}$$

$$= \frac{56}{24} \quad \text{from the gears available.}$$

Therefore, when indexing 14 holes in a 42-hole circle with a 56-tooth gear driving a 24-tooth gear, the actual number of divisions on the specimen will be 113, and not 120 that would result if the index plate were fixed.

The *negative* sign in the calculation indicates that the plate rotates *with* the index arm. A *positive* sign indicates that the plate rotates *against* the index arm.

(Courtesy of Addison Wesley Longman.)

## 1.22 Helical milling

A *helix* can be defined as the locus (path) of a point travelling around an imaginary cylinder in such a manner that its axial and circumferential velocities maintain a constant ratio. In helical milling, the

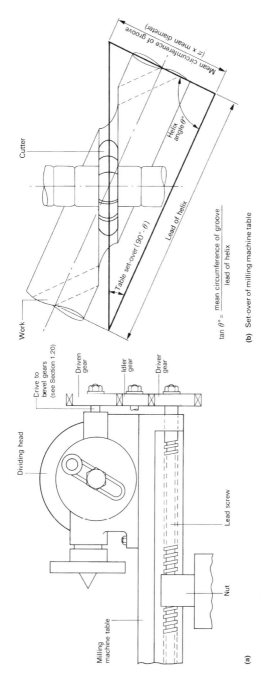

Cutter

Mean circumference of groove
(π × mean diameter)

Helix angle θ°

Table set-over (90° − θ)

Lead of helix

Work

$$\tan \theta° = \frac{\text{mean circumference of groove}}{\text{lead of helix}}$$

(b) Set-over of milling machine table

Drive to
bevel gears
(see Section 1.20)

Driven gear

Idler gear

Driver gear

Dividing head

Lead screw

Nut

Milling
machine table

(a)

table lead provides the axial movement and the dividing head provides the circumferential movement. The method of coupling the table lead screw to the dividing head by a gear train is shown in Fig. (a) above.

If the dividing head was coupled to the milling machine table by gears having a ratio of 1:1 then, because of the worm and worm-wheel in the dividing head, the table lead screw would have to rotate 40 times for the workpiece to rotate once. During those 40 revolutions the table and the work would traverse $40p$ millimetres, where $p$ would be the pitch of the table lead screw. Since a single start lead screw is invariably used, pitch is equal to lead in this instance. This distance of $40p$ millimetres is referred to as the *lead of the machine*. For any given helix the ratio of the gears is

$$\frac{\text{Driver}}{\text{Driven}} = \frac{\text{lead of machine}}{\text{lead of helix to be cut}}$$

---

### Example 1.22.1

Calculate the gear train to cut a helix of 540 mm lead on a milling machine fitted with a table lead screw having a lead of 6 mm.

Lead of machine = $40p = 40 \times 6 = 240$ mm

$$\frac{\text{Driver}}{\text{Driven}} = \frac{\text{lead of machine}}{\text{lead of helix to be cut}} = \frac{240}{540} = \frac{4}{9}$$

From the gears normally available a 32-tooth gear would be used to drive a 72-tooth gear.

---

The number of idler gears introduced between the driver and driven gears will depend upon the 'hand' of the helix being cut. Sometimes the lead being cut cannot be achieved with a *'simple'* gear train and a *'compound'* gear train has to be used – see Sections 1.19 and 1.20.

To prevent the cutter from interfering with the sides of the groove being cut it is necessary to swing the table of the milling machine round until the cutter is lying in the path of the helix as shown above. It is not possible to set over the table of a plain horizontal milling machine, so that helical milling is only possible on *universal* horizontal milling machines which are provided with the requisite table movements.

Even when the table is swung round to the helix angle of the groove it is not possible to mill a groove with straight sides. The only way that straight-sided grooves may be produced is on a vertical milling machine using an end mill or a slot drill. Under these conditions the table does not need to be set over. Unfortunately, the metal removal rate for an end mill is low compared with a side and face milling cutter. For this reason the groove should not be designed with straight sides if quantity production is envisaged.

---

## Example 1.22.2

With reference to Fig. (b) above calculate the angle of set-over when milling a groove with a lead of 540 mm (see Example 1.17.1). The mean diameter of the groove is 40 mm.

$$\tan \theta° = \frac{\text{lead of work}}{\text{mean circumference}}$$

$$= \frac{480}{40 \times \pi}$$

$$= 3.819\,7$$

$$\therefore \quad \theta° = 75.33°$$

$$\text{Set-over angle} = 90° - \theta°$$
$$= 90° - 75.33°$$
$$= \underline{14.67°}$$

*Note*: For practical purposes the table would be set over by 15°.

---

(Courtesy of Addison Wesley Longman.)

## 1.23 Cam milling

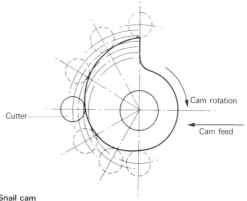

**(a) Snail cam**

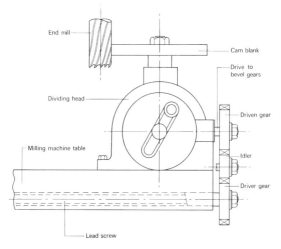

**(b) Set-up for cam milling**

Snail cams of the type shown in Fig. (a) above can be milled using a universal dividing head geared to the table lead screw of a vertical milling machine as shown in Fig. (b) above. As the table feeds the cam blank into the cutter the dividing head rotates the blank. The dividing head has been set with its spindle vertical in Fig. (b) above. The gear ratio to provide a given cam lift can be calculated from the expression:

$$\frac{\text{Driver}}{\text{Driven}} = \frac{\text{lead of machine}}{\text{lift per revolution of cam}}$$

## Example 1.23.1

Calculate the gear ratio to cut a cam that has a lift of 12 mm in 90° rotation if the table lead screw has a pitch (lead) of 6 mm.

Lead of machine = $40p = 40 \times 6$ mm = 240 mm

$$\text{Lift of cam per revolution} = 12 \text{ mm} \times \frac{360°}{90°}$$

$$= 48 \text{ mm}$$

$$\frac{\text{Driver}}{\text{Driven}} = \frac{\text{lead of machine}}{\text{lift per revolution}} = \frac{240}{48} = \frac{5}{1}$$

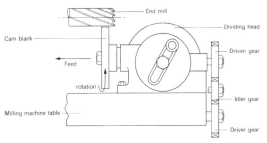

(c) Dividing head horizontal, zero lift generated on cam

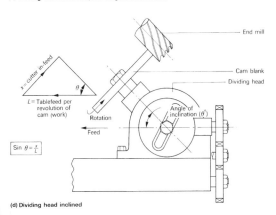

$$\boxed{\sin \theta = \frac{x}{L}}$$

$x =$ cutter in-feed

$L =$ Tablefeed per revolution of cam (work)

Angle of inclination $(\theta°)$

(d) Dividing head inclined

Unfortunately, the ratio of lift to lead rarely works out so conveniently in practice and some means has to be used to obtain intermediate values from the standard gears supplied. With the dividing head spindle set vertically the lift generated on the cam is a maximum for any given gear ratio. However, with the dividing head and the milling machine spindle set horizontally, as shown in Fig. (c) above, only a cylindrical surface will be generated and the lift will be zero. Thus, for some setting intermediate between these extremes will be the required lift.

To cam mill, an inclinable-head vertical milling machine is required. A gear ratio is then selected that gives a lift *larger* than that required, and the machine head and the dividing head are inclined to give the actual lift required, as shown in Fig. (d) above.

$$\text{Sin}\,\theta = \frac{\text{lift per revolution of cam produced } (x)}{\text{table movement per revolution of cam } (L)}$$

and $L = \dfrac{x}{\sin \theta}$

But $L$ is the maximum lift per revolution for any given gear train, and:

$$\frac{\text{Driver}}{\text{Driven}} = \frac{\text{lead of machine}}{L}$$

$$= \frac{\text{lead of machine}}{x/\sin \theta}, \quad \text{when inclined at } \theta°$$

$$= \frac{\text{lead of machine} \times \sin \theta}{x}$$

Thus: $\sin \theta = \dfrac{x}{\text{lead of machine}} \times \dfrac{\text{driver}}{\text{driven}}$

where $x$ = the required lift per revolution of the cam.

---

## Example 1.23.2

Calculate the gears and spindle inclination to cut a cam whose lift is 23.5 mm in 83° on a vertical milling machine whose lead is 240 mm.

$$\text{Lift per revolution of cam} = \frac{23.5 \times 360}{83°}$$

$$= 101.9 \,\text{mm/rev}$$

The nearest convenient gear ratio *greater* than this lift would be:

$$\frac{\text{Driver}}{\text{Driven}} = \frac{\text{lead of machine}}{\text{maximum lift per revolution}}$$

$$= \frac{240}{105}$$

$$= \frac{48}{21}$$

To find the angle of inclination ($\theta$):

$$\text{Sin}\,\theta = \frac{x}{\text{lead of machine}} \times \frac{\text{driver}}{\text{driven}}$$

$$= \frac{101.9}{240} \times \frac{48}{21}$$

$$= \underline{0.970\,5}$$

$$\therefore \quad \theta° = \underline{76°3'}$$

(Courtesy of Addison Wesley Longman.)

## 1.24 Gear trains (simple)

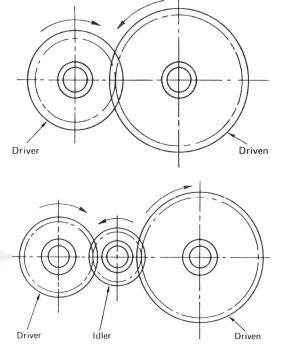

Driver     Driven

Driver     Idler     Driven

77

### 1.24.1 Simple train

(a) Driver and driven gears rotate in opposite directions.
(b) The relative speed of the gears is calculated by the expression

$$\frac{\text{rev/min driver}}{\text{rev/min driven}} = \frac{\text{number of teeth on driven}}{\text{number of teeth on driver}}$$

---

**Example 1.24.1**

Calculate the speed of the driven gear if the driving gear is rotating at 120 rev/min. The driven gear has 150 teeth and the driving gear has 50 teeth.

$$\frac{120}{\text{rev/min driven}} = \frac{150}{50}$$

$$\text{rev/min driven} = \frac{120 \times 50}{150} = 40\,\text{rev/min}$$

---

### 1.24.2 Simple train with idler gear

(a) Driver and driven gears rotate in the same direction if there is an odd number of idler gears, and in the opposite direction if there is an even number of idler gears.
(b) Idler gears are used to change the direction of rotation and/or to increase the centre distance between the driver and driven gears.
(c) The number of idler gears and the number of teeth on the idler gears do not affect the overall relative speed.
(d) The overall relative speed is again calculated using the expression

$$\frac{\text{rev/min driver}}{\text{rev/min driven}} = \frac{\text{number of teeth on driven}}{\text{number of teeth on driver}}$$

## 1.25 Compound gear trains

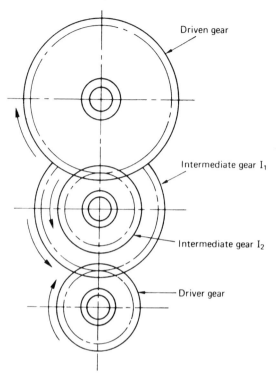

Driven gear

Intermediate gear $I_1$

Intermediate gear $I_2$

Driver gear

a) Unlike the idler gear of a simple train, the intermediate gears of a compound train do influence the overall relative speeds of the driver and driven gears.

b) Both intermediate gears ($I_1$ and $I_2$) are keyed to the same shaft and rotate at the same speed.

c) Driver and driven gears rotate in the same direction. To reverse the direction of rotation an idler gear has to be inserted either between the driver gear and $I_1$ or between $I_2$ and the driven gear.

d) The overall relative speed can be calculated using the expression

$$\frac{\text{rev/min driver}}{\text{rev/min driven}} = \frac{\text{no. of teeth on } I_1}{\text{no. of teeth on driver}}$$
$$\times \frac{\text{no. of teeth on driven}}{\text{no. of teeth on } I_2}$$

**Example 1.25.1**

Calculate the speed of the driven gear given that: the driver rotates at 600 rev/min and has 30 teeth; $I_1$ has 60 teeth; $I_2$ has 40 teeth; and the driven gear has 80 teeth.

$$\frac{\text{rev/min driver}}{\text{rev/min driven}} = \frac{60 \times 80}{30 \times 40} = \frac{4}{1}$$

but speed of driver = 600 rev/min

Therefore:

$$\frac{600 \text{ rev/min}}{\text{rev/min driven}} = \frac{4}{1}$$

$$\text{speed driven} = \frac{600 \times 1}{4}$$

$$= 150 \text{ rev/min}$$

## 1.26 Belt drive (simple)

### 1.26.1 Open belt drive

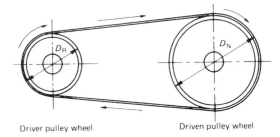

Driver pulley wheel       Driven pulley wheel

(a) Driver and driven pulley wheels rotate in the same direction.

(b) The relative speed of the pulley wheels is calcu lated by the expression

$$\frac{\text{rev/min driver}}{\text{rev/min driven}} = \frac{\text{diameter } D_N \text{ of driven}}{\text{diameter } D_R \text{ of driver}}$$

**Example 1.26.1**

Calculate the speed in rev/min of the driven pulle if the driver rotates at 200 rev/min. Diameter $D_R$

500 mm and diameter $D_N$ is 800 mm.

$$\frac{200\,\text{rev/min}}{\text{rev/min driven}} = \frac{800\,\text{mm}}{500\,\text{mm}}$$

$$\text{rev/min driven} = \frac{200 \times 500}{800} = \underline{125\,\text{rev/min}}$$

### 1.26.2 Crossed belt drive

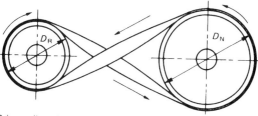

Driver pulley wheel           Driven pulley wheel

(a) Driver and driven pulley wheels rotate in opposite directions.
(b) Crossed belt drives can only be used with flat section belts (long centre distances) or circular section belts (short centre distances).
(c) The relative speed of the pulley wheels is again calculated by the expression

$$\frac{\text{rev/min driver}}{\text{rev/min driven}} = \frac{\text{diameter } D_N \text{ of driven}}{\text{diameter } D_R \text{ of driver}}$$

**Example 1.26.2**

The driver pulley rotates at 500 rev/min and is 600 mm in diameter. Calculate the diameter of the driven pulley if it is to rotate at 250 rev/min.

$$\frac{500\,\text{rev/min}}{250\,\text{rev/min}} = \frac{\text{diameter } D_N}{600\,\text{mm}}$$

$$\text{diameter } D_N = \frac{500 \times 60}{250} = 1200\,\text{mm}$$

## 1.27 Belt drive (compound)

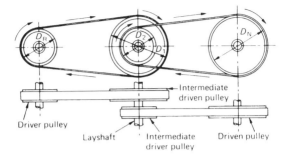

Driver pulley

Layshaft

Intermediate driven pulley

Intermediate driver pulley

Driven pulley

(1) To identify the direction of rotation, the rules for open and crossed belt drives apply (Section 1.26).

(2) The relative speeds of the pulley wheels are calculated by the expression

$$\frac{\text{rev/min driver}}{\text{rev/min driven}} = \frac{\text{diameter } D_1}{\text{diameter } D_R} \times \frac{\text{diameter } D_N}{\text{diameter } D_2}$$

### Example 1.27.1

Calculate the speed in rev/min of the driven pulley if the driver rotates at 600 rev/min. The diameters of the pulley wheels are: $D_R = 250\,\text{mm}$. $D_1 = 750\,\text{mm}$, $D_2 = 500\,\text{mm}$, $D_N = 1000\,\text{mm}$.

$$\frac{600\,\text{rev/min}}{\text{rev/min driven}} = \frac{750\,\text{mm}}{250\,\text{mm}} \times \frac{1000\,\text{mm}}{500\,\text{mm}}$$

$$\text{rev/min driven} = \frac{600 \times 250 \times 500}{750 \times 1000}$$

$$= \underline{100\,\text{rev/min}}$$

## 1.28 Typical belt tensioning devices

Swing bed tensioning device

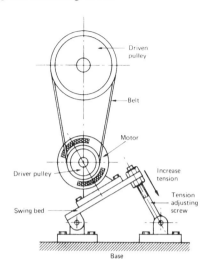

Jockey pulley

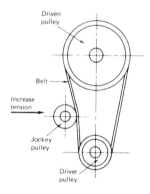

# Slide rail tensioning device

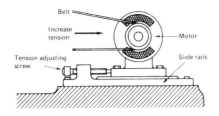

# Part 2
# Threaded Fasteners

## 2.1 Introduction to threaded fasteners

Although dimensioned in 'inch' units the following screw thread tables have been retained for maintenance data and similar applications.

2.1.37  British Standard Whitworth (BSW) screw threads
2.1.38  British Standard Fine (BSF) screw threads
2.1.39  ISO Unified precision internal screw threads: coarse series
2.1.40  ISO Unified precision external screw threads: coarse series
2.1.41  ISO Unified precision internal screw threads: fine series
2.1.42  ISO Unified precision external screw threads: fine series

Although obsolescent, the following 'metric' screw thread table has been retained.

2.1.44  British Association (BA) internal and external screw threads

The tables based upon abstracts from BS 4190 and BS 3692 have become obsolescent and have been replaced by new tables based upon abstracts from the appropriate BSEN standards for screwed fasteners with metric dimension. The fasteners covered by these new standards are as follows.

2.1.12  BSEN 24014: Hexagon head bolts – product grade A and B
2.1.13  BSEN 24016: Hexagon head bolts – product grade C
2.1.14  BSEN 24017: Hexagon head screws – product grade A and B
2.1.15  BSEN 24018: Hexagon head screws – product grade C
2.1.16  BSEN 24032: Hexagon nuts style 1 – product grade A and B
2.1.17  BSEN 24033: Hexagon nuts style 2 – product grade A and B
2.1.18  BSEN 24034: Hexagon nuts style 1 – product grade C

2.1.19 BSEN 24035: Hexagon thin nuts (chamfered) – product grade A and B

2.1.20 BSEN 24036: Hexagon thin nuts (unchamfered) – product grade B

All the above standards refer to screwed fasteners with metric dimensions and have coarse pitch threads. Fine pitch threads will be referred to in due course.

*Notes:*

(i) BSEN 24015 refers to hexagon head bolts with their shanks reduced to effective (pitch) diameter of their threads. These are for specialised applications and have not been included in this pocket book.

(ii) The mechanical property standards contained within BS 4190 and BS 3692 can now be found in BSEN 20898 Part 1 (bolts) and Part 2 (nuts).

Interpretation of the **product grade** is as follows:

• Examine the table in Section 2.3.
• The shank diameter ($d_s$) has a maximum diameter which equals the normal diameter and also a minimum diameter.
• The minimum diameter can have a **product grade A** tolerance or a **product grade B** tolerance. The grade A tolerance is closer (more accurate) than the grade B tolerance.
• Product grade A tolerances apply to fasteners with a size range from M1.6 to M24 inclusive.
• Product grade B tolerances apply to fasteners with a size range from M16 to M64.
• Sizes M16 to M24 inclusive can have product grade A or product grade B tolerances.

*Note* that not only is the product grade defined by the diameters but by the length as well.

---

**Example 2.1.1**

An M5 hexagon head bolt will lie within product grade A tolerances and will have a shank diameter lying between 5.00 mm and 4.82 mm inclusive.

---

**Example 2.1.2**

An M36 hexagon head bolt will lie within product grade B tolerances and will have a shank diameter lying between 36 mm and 35.38 mm inclusive.

**Example 2.1.3**

An M16 hexagon head bolt will have a shank diameter lying between 16 mm diameter and 15.73 mm diameter inclusive if it is to product group A tolerances. If it is to product group B tolerances, it will have a shank diameter lying between 16 mm diameter and 15.57 mm inclusive.

*Note:* The above system of tolerancing applies to all the other dimensions for the fasteners in this table.

- Examine the table in Section 2.4. All the dimensions in this table refer to screwed fasteners with product grade C tolerances.
- Comparing this table with the previous examples shows that the fasteners made to product grade C have much coarser tolerances than those for product grade A and B.
- An M12 bolt to product grade A has a shank diameter ($d_s$) lying between 12 mm and 11.73 mm (a tolerance of 0.27 mm), whereas an M12 bolt to product grade C has a shank diameter ($d_s$) lying between 12.77 mm and 11.3 mm (a tolerance of 1.4 mm). Product grade B does not apply to this size of bolt.

*Note:* The old terminology of 'black' (hot forged) and 'bright/precision' (cold headed or machined from hexagon bar) bolts and nuts no longer applies. However, hot forged (black) bolts and nuts would only be made to product grade C.

All the fasteners listed so far have **coarse pitch** threads. The following tables and standards refer to a corresponding **fine pitch** series of screwed fasteners. These fine pitch series of screwed fasteners are only available in product grades A and B.

2.1.21 BSEN 28765: Hexagon head bolts with metric fine pitch threads – product grade A and B

2.1.22 BSEN 28676: Hexagon head screws with metric fine pitch threads – product grade A and B

2.1.23 BSEN 28673: Hexagon nuts (style 1) with metric fine pitch threads – product grade A and B

2.1.24 BSEN 28674: Hexagon nuts (style 2) with metric fine pitch threads – product grade A and B

2.1.25 BSEN 28675: Hexagon thin nuts with metric fine pitch threads – product grade A and B

## 2.2 Threaded fasteners

### 2.2.1 Drawing proportions

**Bolts and screws**

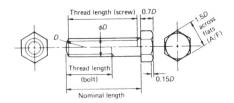

**Studs**

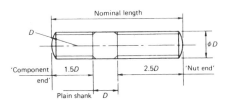

**Standard nut**

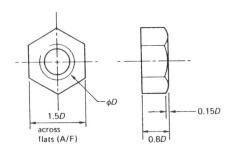

**Thin (lock) nut**

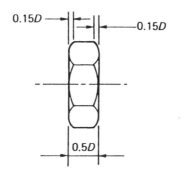

**Plain washer**

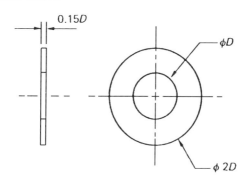

## 2.2.2 Alternative screw heads

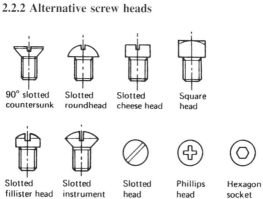

90° slotted countersunk

Slotted roundhead

Slotted cheese head

Square head

Slotted fillister head

Slotted instrument head

Slotted head

Phillips head

Hexagon socket head

### 2.2.3 Alternative scew points

Flat    Dog    Conical    Round    Cup

### 2.2.4 Hexagon socket cap head screw

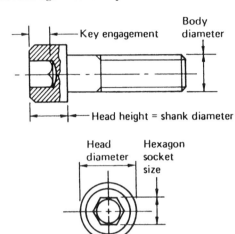

Key engagement

Body diameter

Head height = shank diameter

Head diameter

Hexagon socket size

### 2.2.5 Applications of threaded fasteners

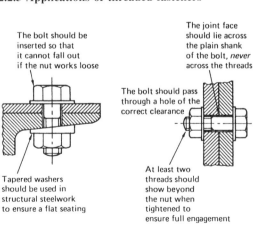

The bolt should be inserted so that it cannot fall out if the nut works loose

Tapered washers should be used in structural steelwork to ensure a flat seating

The joint face should lie across the plain shank of the bolt, *never* across the threads

The bolt should pass through a hole of the correct clearance

At least two threads should show beyond the nut when tightened to ensure full engagement

### 2.2.6 Acme thread form

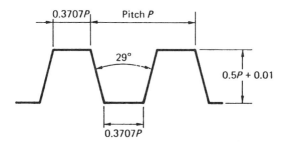

### 2.2.7 Square thread form

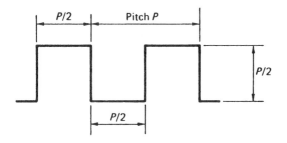

### 2.2.8 Buttress thread form

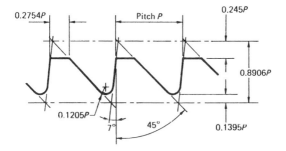

### 2.2.9 V-thread form

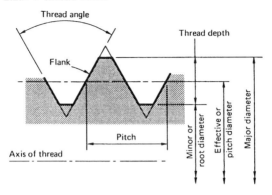

### 2.2.10 Basic Whitworth (55°) thread form: parallel threads

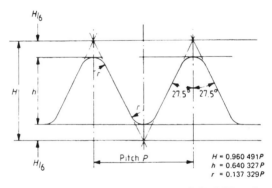

$H = 0.960\ 491P$
$h = 0.640\ 327P$
$r = 0.137\ 329P$

This is the basic thread form for BSW, BSF and BSP screw threads.

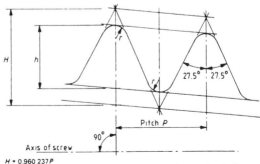

$H = 0.960\ 237P$
$h = 0.640\ 327P$
$r = 0.137\ 278P$

NOTE. The taper is 1 in 16 measured on the diameter (shown exaggerated in the diagram).

## 2.2.11 ISO metric and ISO 60° unified thread forms

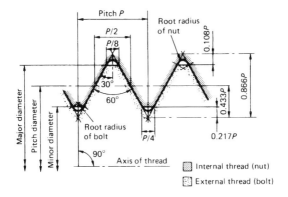

Pitch P

P/2

P/8

Root radius of nut

0.108P

30°

60°

0.866P

0.433P

Root radius of bolt

P/4

0.217P

Major diameter

Pitch diameter

Minor diameter

90°

Axis of thread

▨ Internal thread (nut)

▧ External thread (bolt)

# 2.3 ISO metric hexagon head bolts (coarse thread): preferred sizes: product grade A and B

(Dimensions in millimetres)

| Designated thread size | Product grade | Pitch of thread | Thread | | | Plain Shank | | Hexagon | | | | | |
|---|---|---|---|---|---|---|---|---|---|---|---|---|---|
| | | | Major diameter | Effective diameter | Minor diameter | Maximum diameter (nominal) | Minimum diameter | Across Flats (A/F) | | Across corners (A/C) min. | Thickness | | |
| | | | | | | | | Maximum (nominal) | Minimum | | Nominal | Maximum | Minimum |
| M1.6 | A | 0.35 | 1.60 | 1.37 | 1.17 | 1.6 | 1.46 | 3.2 | 3.02 | 3.41 | 1.1 | 1.225 | 0.975 |
| M2 | A | 0.40 | 2.00 | 1.74 | 1.50 | 2.0 | 1.86 | 4.0 | 3.82 | 4.32 | 1.4 | 1.525 | 1.275 |
| M2.5 | A | 0.45 | 2.50 | 2.21 | 1.15 | 2.50 | 2.36 | 5.0 | 4.82 | 5.45 | 1.7 | 1.825 | 1.575 |
| M3 | A | 0.50 | 3.00 | 2.68 | 2.39 | 3.00 | 2.86 | 5.5 | 5.32 | 6.01 | 2.0 | 2.125 | 1.875 |
| M4 | A | 0.70 | 4.00 | 3.55 | 3.14 | 4.00 | 3.82 | 7.0 | 6.78 | 7.66 | 2.8 | 2.925 | 2.675 |
| M5 | A | 0.80 | 5.00 | 4.48 | 4.02 | 5.00 | 4.82 | 8.0 | 7.78 | 8.79 | 3.5 | 3.650 | 3.350 |
| M6 | A | 1.00 | 6.00 | 5.36 | 4.77 | 6.00 | 5.82 | 10.0 | 9.78 | 11.05 | 4.0 | 4.150 | 3.850 |
| M8 | A | 1.25 | 8.00 | 7.19 | 6.47 | 8.00 | 7.78 | 13.0 | 12.73 | 14.38 | 5.3 | 5.450 | 5.150 |
| M10 | A | 1.50 | 10.00 | 9.03 | 8.16 | 10.00 | 9.78 | 16.0 | 15.73 | 17.77 | 6.4 | 6.580 | 6.220 |

| | | | | | | | | | | | | |
|---|---|---|---|---|---|---|---|---|---|---|---|---|
| M12 | A | 1.75 | 12.00 | 10.86 | 9.85 | 12.00 | 11.73 | 18.0 | 17.73 | 20.03 | 7.5 | 7.680 | 7.320 |
| M16 | A | 2.00 | 16.00 | 14.70 | 13.55 | 16.00 | 15.73 | 24.0 | 23.67 | 26.75 | 10.0 | 10.180 | 9.820 |
| | B | 2.00 | 16.00 | 14.70 | 13.55 | 16.00 | 15.57 | 24.0 | 23.16 | 26.17 | 10.0 | 10.290 | 9.710 |
| M20 | A | 2.50 | 20.00 | 18.38 | 16.93 | 20.00 | 19.67 | 30.00 | 29.67 | 33.53 | 12.5 | 12.715 | 12.285 |
| | B | 2.50 | 20.00 | 18.38 | 16.93 | 20.00 | 19.48 | 30.00 | 29.16 | 32.95 | 12.5 | 12.850 | 12.150 |
| M24 | A | 3.00 | 24.00 | 22.05 | 20.32 | 24.00 | 23.67 | 36.00 | 35.38 | 39.98 | 15.0 | 15.215 | 14.785 |
| | B | 3.00 | 24.00 | 22.05 | 20.32 | 24.00 | 23.48 | 36.00 | 35.00 | 39.55 | 15.0 | 15.350 | 14.650 |
| M30 | B | 3.50 | 30.00 | 27.73 | 25.71 | 30.00 | 29.48 | 46.00 | 45.00 | 50.85 | 18.7 | 19.120 | 18.280 |
| M36 | B | 4.00 | 36.00 | 33.40 | 31.09 | 36.00 | 35.38 | 56.00 | 53.80 | 60.79 | 22.5 | 22.920 | 22.06 |
| M42 | B | 4.50 | 42.00 | 39.08 | 36.48 | 42.00 | 41.38 | 65.00 | 63.10 | 71.30 | 26 | 26.420 | 25.58 |
| M48 | B | 5.00 | 48.00 | 44.75 | 41.87 | 48.00 | 47.38 | 75.00 | 73.10 | 82.60 | 30 | 30.420 | 29.58 |

For further information see BSEN 24014.

(Dimensions in millimetres)

| Designated thread size | Popular length combinations | | | | | | |
|---|---|---|---|---|---|---|---|
| M1.6 | U/head | 12 | 16 | | | | |
| | Thread | 9 | 9 | | | | |
| | Shank | 1.2 | 5.2 | | | | |
| M2 | U/head | 16 | 20 | | | | |
| | Thread | 10 | 10 | | | | |
| | Shank | 4 | 8 | | | | |
| M2.5 | U/head | 16 | 20 | 25 | | | |
| | Thread | 11 | 11 | 11 | | | |
| | Shank | 2.75 | 6.75 | 11.75 | | | |
| M3 | U/head | 20 | 25 | 30 | | | |
| | Thread | 12 | 12 | 12 | | | |
| | Shank | 5.5 | 10.5 | 15.5 | | | |
| M4 | U/head | 25 | 30 | 35 | 40 | | |
| | Thread | 14 | 14 | 14 | 14 | | |
| | Shank | 7.5 | 12.5 | 17.5 | 22.5 | | |
| M5 | U/head | 25 | 30 | 35 | 40 | 45 | 50 |
| | Thread | 16 | 16 | 16 | 16 | 16 | 16 |
| | Shank | 5 | 10 | 15 | 20 | 25 | 30 |

| M6 | U/head | 30 | 35 | 40 | 45 | 50 | 55 | 60 |
|---|---|---|---|---|---|---|---|---|
| | Thread | 18 | 18 | 18 | 18 | 18 | 18 | 18 |
| | Shank | 7 | 12 | 17 | 22 | 27 | 33 | 37 |

| M8 | U/head | 40 | 45 | 50 | 55 | 60 | 65 | 70 | 80 |
|---|---|---|---|---|---|---|---|---|---|
| | Thread | 22 | 22 | 22 | 22 | 22 | 22 | 22 | 22 |
| | Shank | 11.75 | 16.75 | 21.75 | 26.75 | 31.50 | 36.5 | 41.75 | 51.75 |

| M10 | U/head | 45 | 50 | 55 | 60 | 65 | 70 | 80 | 90 | 100 |
|---|---|---|---|---|---|---|---|---|---|---|
| | Thread | 26 | 26 | 26 | 26 | 26 | 26 | 26 | 26 | 26 |
| | Shank | 11.5 | 16.5 | 21.5 | 26.5 | 31.5 | 36.5 | 46.5 | 56.5 | 66.5 |

| M12 | U/head | 50 | 55 | 60 | 65 | 70 | 80 | 90 | 100 | 110 | 120 |
|---|---|---|---|---|---|---|---|---|---|---|---|
| | Thread | 30 | 30 | 30 | 30 | 30 | 30 | 30 | 30 | 30 | 30 |
| | Shank | 11.25 | 16.25 | 21.25 | 26.25 | 31.25 | 41.25 | 51.25 | 61.25 | 71.25 | 81.25 |

| M16 | U/head | 65 | 70 | 80 | 90 | 100 | 110 | 120 | 130 | 140 | 150 | 160 |
|---|---|---|---|---|---|---|---|---|---|---|---|---|
| | Thread | 38 | 38 | 38 | 38 | 38 | 38 | 38 | 44 | 44 | 44 | 44 |
| | Shank | 17 | 22 | 32 | 42 | 52 | 62 | 72 | 76 | 86 | 96 | 106 |

(continued)

**2.3** (continued)

(Dimensions in millimetres)

| Designated thread size | Popular length combinations | | | | | | | | | | |
|---|---|---|---|---|---|---|---|---|---|---|---|
| **M20** | U/head | 80 | 90 | 100 | 110 | 120 | 130 | 140 | 150 | 160 | 180 | 200 |
| | Thread | 46 | 46 | 46 | 46 | 46 | 52 | 52 | 52 | 52 | 52 | 52 |
| | Shank | 21.5 | 31.5 | 41.5 | 51.5 | 61.5 | 65.5 | 75.5 | 85.5 | 95.5 | 115.5 | 135.5 |

| Designated thread size | Popular length combinations | | | | | | | | | | |
|---|---|---|---|---|---|---|---|---|---|---|---|
| **M24** | U/head | 90 | 100 | 110 | 120 | 130 | 140 | 150 | 160 | 180 | 200 |
| | Thread | 54 | 54 | 54 | 54 | 60 | 60 | 60 | 60 | 60 | 60 |
| | Shank | 21 | 31 | 41 | 51 | 56 | 65 | 75 | 85 | 106 | 125 |
| | U/head | 220 | 240 | | | | | | | | |
| | Thread | 73 | 73 | | | | | | | | |
| | Shank | 132 | 152 | | | | | | | | |

| Designated thread size | Popular length combinations | | | | | | | | | | |
|---|---|---|---|---|---|---|---|---|---|---|---|
| **M30** | U/head | 110 | 120 | 130 | 140 | 150 | 160 | 180 | 200 | 220 | 240 |
| | Thread | 66 | 66 | 72 | 72 | 72 | 72 | 72 | 72 | 85 | 85 |
| | Shank | 26.5 | 36.5 | 40.5 | 50.5 | 60.5 | 70.5 | 90.5 | 110.5 | 117.5 | 137.5 |
| | U/head | 260 | 280 | 300 | | | | | | | |
| | Head | 85 | 85 | 85 | | | | | | | |
| | Shank | 157.5 | 177.5 | 197.5 | | | | | | | |

**M36**

| | | | | | | | | | | | | | |
|---|---|---|---|---|---|---|---|---|---|---|---|---|---|
| U/head | 140 | 150 | 160 | 180 | 200 | 220 | 240 | 260 | 280 | 300 | 320 | 340 | 360 |
| Thread | 84 | 84 | 84 | 84 | 84 | 97 | 97 | 97 | 97 | 97 | 97 | 97 | 97 |
| Shank | 36 | 46 | 56 | 76 | 96 | 103 | 123 | 143 | 163 | 183 | 203 | 223 | 243 |

**M42**

| | | | | | | | | | | | | | | | |
|---|---|---|---|---|---|---|---|---|---|---|---|---|---|---|---|
| U/head | 160 | 180 | 200 | 220 | 240 | 260 | 280 | 300 | 320 | 340 | 360 | 380 | 400 | 420 | 440 |
| Thread | 96 | 96 | 96 | 109 | 109 | 109 | 109 | 109 | 109 | 109 | 109 | 109 | 109 | 109 | 109 |
| Shank | 41.5 | 61.5 | 81.5 | 88.5 | 108.5 | 128.5 | 148.5 | 168.5 | 188.5 | 208.5 | 228.5 | 248.5 | 268.5 | 288.5 | 308.5 |

**M48**

| | | | | | | | | | | | | | | | | |
|---|---|---|---|---|---|---|---|---|---|---|---|---|---|---|---|---|
| U/head | 180 | 200 | 220 | 240 | 260 | 280 | 300 | 320 | 340 | 360 | 380 | 400 | 420 | 440 | 460 | 480 |
| Thread | 116 | 116 | 121 | 121 | 121 | 121 | 121 | 121 | 121 | 121 | 121 | 121 | 121 | 121 | 121 | 121 |
| Shank | 47 | 67 | 74 | 94 | 114 | 134 | 154 | 174 | 194 | 214 | 234 | 254 | 274 | 294 | 314 | 334 |

For further information see BSEN 24014.

*Note:* The length under the head of the bolt (U/head) is also the **nominal length**.

## 2.4 ISO metric hexagon head bolts (coarse thread): preferred sizes: product grade C

(Dimensions in millimetres)

| Designated thread size | Pitch of thread | Thread | | | Plain Shank | | Hexagon | | | Thickness | | |
|---|---|---|---|---|---|---|---|---|---|---|---|---|
| | | Major diameter | Effective diameter | Minor diameter | Maximum diameter | Minimum diameter | Across Flats (A/F) | | Across corners (A/C) minimum | Nominal | Maximum | Minimum |
| | | | | | | | Maximum (nominal) | Minimum | | | | |
| M1.6 | | | | | | | | | | | | |
| M2 | | | | | | | | | | | | |
| M2.5 | | | | | Not available in product grade C | | | | | | | |
| M3 | | | | | | | | | | | | |
| M4 | | | | | | | | | | | | |
| M5 | 0.80 | 5.0 | 4.48 | 4.02 | 5.48 | 4.52 | 8.00 | 7.64 | 8.63 | 3.5 | 3.875 | 3.125 |
| M6 | 1.00 | 6.0 | 5.36 | 4.77 | 6.48 | 5.52 | 10.00 | 9.64 | 10.89 | 4.0 | 4.375 | 3.625 |
| M8 | 1.25 | 8.0 | 7.19 | 6.47 | 8.58 | 7.42 | 13.00 | 12.57 | 14.2 | 5.3 | 5.675 | 4.925 |
| M10 | 1.50 | 10.0 | 9.03 | 8.16 | 10.58 | 9.42 | 16.00 | 15.57 | 17.59 | 6.4 | 6.850 | 5.950 |
| M12 | 1.75 | 12.0 | 10.86 | 9.85 | 12.70 | 11.30 | 18.00 | 17.57 | 19.85 | 7.5 | 7.950 | 7.050 |
| M16 | 2.00 | 16.0 | 14.70 | 13.55 | 16.70 | 15.30 | 24.00 | 23.16 | 26.17 | 10.0 | 10.750 | 9.250 |
| M20 | 2.50 | 20.0 | 18.38 | 16.93 | 20.84 | 19.16 | 30.00 | 29.16 | 32.95 | 12.5 | 13.400 | 11.600 |
| M24 | 3.00 | 24.0 | 22.05 | 20.32 | 24.84 | 23.16 | 36.00 | 35 | 39.55 | 15.0 | 15.900 | 14.100 |
| M30 | 3.50 | 30.0 | 27.73 | 25.71 | 30.84 | 29.16 | 46.00 | 45 | 50.85 | 18.7 | 19.750 | 17.650 |
| M36 | 4.00 | 36.0 | 33.40 | 31.09 | 37.00 | 35.0 | 55.00 | 53.8 | 60.79 | 22.5 | 23.550 | 22.500 |
| M42 | 4.50 | 42.0 | 39.08 | 36.48 | 43.00 | 41.00 | 65.00 | 63.1 | 71.30 | 26.0 | 27.050 | 24.950 |
| M48 | 5.00 | 48.0 | 44.75 | 41.87 | 49.00 | 47.00 | 75.00 | 73.1 | 82.60 | 30.0 | 31.050 | 28.950 |

| Designated thread size | Popular length combinations | | | | | | | | | | | | | | | | | | | | | |
|---|---|---|---|---|---|---|---|---|---|---|---|---|---|---|---|---|---|---|---|---|---|---|
| M5 | U/head | 25 | 30 | 35 | 40 | 45 | 50 | | | | | | | | | | | | | | | |
| | Thread | 16 | 16 | 16 | 16 | 16 | 16 | | | | | | | | | | | | | | | |
| | Shank | 5 | 10 | 15 | 20 | 25 | 30 | | | | | | | | | | | | | | | |
| M6 | U/head | | 30 | 35 | 40 | 45 | 50 | 55 | 60 | | | | | | | | | | | | | |
| | Thread | | 18 | 18 | 18 | 18 | 18 | 18 | 18 | | | | | | | | | | | | | |
| | Shank | | 7 | 12 | 17 | 22 | 27 | 32 | 37 | | | | | | | | | | | | | |
| M8 | U/head | | | | 40 | 45 | 50 | 55 | 60 | 65 | 70 | 80 | | | | | | | | | | |
| | Thread | | | | 22 | 22 | 22 | 22 | 22 | 22 | 22 | 22 | | | | | | | | | | |
| | Shank | | | | 11.75 | 16.75 | 21.75 | 26.75 | 31.75 | 36.75 | 41.75 | 51.75 | | | | | | | | | | |
| M10 | U/head | | | | | 45 | 50 | 55 | 60 | 65 | 70 | 80 | 90 | 100 | | | | | | | | |
| | Thread | | | | | 26 | 26 | 26 | 26 | 26 | 26 | 26 | 26 | 26 | | | | | | | | |
| | Shank | | | | | 11.75 | 16.75 | 21.75 | 26.75 | 31.75 | 36.75 | 46.75 | 56.75 | 66.75 | | | | | | | | |
| M12 | U/head | | | | | | | 55 | 60 | 65 | 70 | 80 | 90 | 100 | 110 | 120 | | | | | | |
| | Thread | | | | | | | 30 | 30 | 30 | 30 | 30 | 30 | 30 | 30 | 30 | | | | | | |
| | Shank | | | | | | | 16.25 | 21.25 | 26.25 | 31.25 | 41.25 | 51.25 | 61.25 | 71.25 | 81.25 | | | | | | |
| M16 | U/head | | | | | | | | | 65 | 70 | 80 | 90 | 100 | 110 | 120 | 130 | 140 | | | | |
| | Thread | | | | | | | | | 38 | 38 | 38 | 38 | 38 | 38 | 38 | 44 | 44 | | | | |
| | Shank | | | | | | | | | 17 | 22 | 32 | 42 | 52 | 62 | 72 | 76 | 86 | | | | |
| M20 | U/head | | | | | | | | | | | 80 | 90 | 100 | 110 | 120 | 130 | 140 | 150 | 160 | 180 | 200 |
| | Thread | | | | | | | | | | | 46 | 46 | 46 | 46 | 46 | 52 | 52 | 52 | 52 | 52 | 52 |
| | Shank | | | | | | | | | | | 21.5 | 31.5 | 41.5 | 51.5 | 61.5 | 65.5 | 75.5 | 85.5 | 95.5 | 115.5 | 135.5 |

(continued)

103

**2.4** (continued)

| Designated thread size | Popular length combinations | | | | | | | | | | | | | | |
|---|---|---|---|---|---|---|---|---|---|---|---|---|---|---|---|
| M24 | U/head | 100 | 110 | 120 | 130 | 140 | 150 | 160 | 180 | 200 | 220 | 240 | | | |
| | Thread | 54 | 54 | 54 | 60 | 60 | 60 | 60 | 60 | 60 | 73 | 73 | | | |
| | Shank | 31 | 41 | 51 | 55 | 65 | 75 | 85 | 105 | 125 | 132 | 152 | | | |
| M30 | U/head | 120 | 130 | 140 | 150 | 160 | 180 | 200 | 220 | 240 | 260 | 280 | 300 | | |
| | Thread | 66 | 72 | 72 | 72 | 72 | 72 | 72 | 85 | 85 | 85 | 85 | 85 | | |
| | Shank | 36.5 | 40.5 | 50.5 | 60.5 | 70.5 | 90.5 | 110.5 | 117.5 | 137.5 | 157.5 | 177.5 | 197.5 | | |
| M36 | U/head | 140 | 150 | 160 | 180 | 200 | 220 | 240 | 260 | 280 | 300 | 320 | 340 | 360 | |
| | Thread | 84 | 84 | 84 | 84 | 84 | 97 | 97 | 97 | 97 | 97 | 97 | 97 | 97 | |
| | Shank | 36 | 46 | 56 | 76 | 96 | 103 | 123 | 143 | 163 | 183 | 203 | 223 | 243 | |
| M42 | U/head | 180 | 200 | 220 | 240 | 260 | 280 | 300 | 320 | 340 | 360 | 380 | 400 | 420 | |
| | Thread | 96 | 96 | 109 | 109 | 109 | 109 | 109 | 109 | 109 | 109 | 109 | 109 | 109 | |
| | Shank | 61.5 | 81.5 | 88.5 | 108.5 | 128.5 | 148.5 | 168.5 | 188.5 | 208.5 | 228.5 | 248.5 | 268.5 | 288.5 | |
| M48 | U/head | 200 | 220 | 240 | 260 | 280 | 300 | 320 | 340 | 360 | 380 | 400 | 420 | 440 | |
| | Thread | 108 | 121 | 121 | 121 | 121 | 121 | 121 | 121 | 121 | 121 | 121 | 121 | 121 | |
| | Shank | 67 | 74 | 94 | 114 | 134 | 154 | 174 | 194 | 214 | 234 | 254 | 271 | 294 | |
| | U/head | 460 | 480 | | | | | | | | | | | | |
| | Thread | 121 | 121 | | | | | | | | | | | | |
| | Shank | 314 | 334 | | | | | | | | | | | | |

For further information see: BSEN 24016.

*Note:* This length under the bolt head (U/head) is also the **nominal length**.

## 2.5 ISO metric hexagon head screws (coarse thread): preferred sizes: product grade A and B

*Notes:*

(1) Reference back to Section 2.2.1 shows that hexagon head **bolts** have a plain shank between the head and the thread. It also shows that hexagon head **screws** have the thread running the full length up to the head of the screw. For practical tooling purposes there is a short distance (a + c) immediately under the head to allow the thread to run out and also allow for a small radius. The dimensions (a + c) refer to BSEN 24017.

(2) The thread and hexagon proportions are the same as those shown in Section 2.3. Therefore only the length under the head (**nominal length**) and the dimensions (a + c) are listed here.

| Designated thread size | Grade | Feature | Popular lengths | | | | | | | | |
|---|---|---|---|---|---|---|---|---|---|---|---|
| M1.6 | A | U/head (nom) | 2 | 3 | 4 | 5 | 6 | 8 | 10 | 12 | 16 |
| | | U/head (max) | 2.20 | 3.20 | 4.24 | 5.24 | 6.24 | 8.29 | 10.29 | 12.35 | 16.35 |
| | | U/head (min) | 1.80 | 2.80 | 3.76 | 4.76 | 5.76 | 7.71 | 9.71 | 11.65 | 15.65 |
| | | a + c (max) | 1.05 + 0.25 = 1.30 mm for all screw lengths | | | | | | | | |
| | | a + c (min) | 0.35 + 0.10 = 0.45 mm for all screw lengths | | | | | | | | |
| M2 | A | U/head (nom) | 4 | 5 | 6 | 8 | 10 | 12 | 16 | 20 | |
| | | U/head (max) | 4.24 | 5.24 | 6.24 | 8.29 | 10.29 | 12.35 | 16.35 | 20.42 | |
| | | U/head (min) | 3.76 | 4.76 | 5.76 | 7.71 | 9.71 | 11.65 | 15.65 | 19.58 | |
| | | a + c (max) | 1.20 + 0.25 = 1.45 mm for all screw lengths | | | | | | | | |
| | | a + c (min) | 0.40 + 0.10 = 0.50 mm for all screw lengths | | | | | | | | |
| M2.5 | A | U/head (nom) | 5 | 6 | 8 | 10 | 12 | 16 | 20 | 25 | |
| | | U/head (max) | 5.24 | 6.24 | 8.29 | 10.29 | 12.35 | 16.35 | 20.42 | 25.42 | |
| | | U/head (min) | 4.76 | 5.76 | 7.71 | 9.71 | 11.65 | 15.65 | 19.58 | 24.58 | |
| | | a + c (max) | 1.35 + 0.25 = 1.60 mm for all screw lengths | | | | | | | | |
| | | a + c (min) | 0.45 + 0.10 = 0.55 mm for all screw lengths | | | | | | | | |
| M3 | A | U/head (nom) | 6 | 8 | 10 | 12 | 16 | 20 | 25 | 30 | |
| | | U/head (max) | 6.24 | 8.29 | 10.29 | 12.35 | 16.35 | 20.42 | 25.42 | 30.42 | |
| | | U/head (min) | 5.76 | 7.71 | 9.71 | 11.65 | 15.65 | 19.58 | 24.58 | 29.58 | |
| | | a + c (max) | 1.50 + 0.40 = 1.90 mm for all screw lengths | | | | | | | | |
| | | a + c (min) | 0.50 + 0.15 = 0.65 mm for all screw lengths | | | | | | | | |

**M4  A**

| U/head (nom) | 8 | 10 | 12 | 16 | 20 | 25 | 30 | 35 | 40 |
|---|---|---|---|---|---|---|---|---|---|
| U/head (max) | 8.29 | 10.29 | 12.35 | 16.35 | 20.42 | 25.42 | 30.42 | 35.5 | 40.5 |
| U/head (min) | 7.71 | 9.71 | 11.65 | 15.65 | 19.58 | 24.58 | 29.58 | 34.5 | 39.5 |
| a + c (max) | 2.1 + 0.4 = 2.5 mm for all screw lengths | | | | | | | | |
| a + c (min) | 0.70 + 0.15 = 0.85 mm for all screw lengths | | | | | | | | |

**M5  A**

| U/head (nom) | 10 | 12 | 16 | 20 | 25 | 30 | 35 | 40 | 45 | 50 |
|---|---|---|---|---|---|---|---|---|---|---|
| U/head (max) | 10.29 | 12.35 | 16.35 | 20.42 | 25.42 | 30.42 | 35.5 | 40.5 | 45.5 | 50.5 |
| U/head (min) | 9.71 | 11.65 | 15.65 | 19.58 | 24.58 | 29.58 | 34.5 | 39.5 | 44.5 | 49.5 |
| a + c (max) | 2.4 + 0.5 = 2.9 mm for all screw lengths | | | | | | | | | |
| a + c (min) | 0.80 + 0.15 = 0.95 mm for all screw lengths | | | | | | | | | |

**M6  A**

| U/head (nom) | 12 | 16 | 20 | 25 | 30 | 35 | 40 | 45 | 50 | 55 | 60 |
|---|---|---|---|---|---|---|---|---|---|---|---|
| U/head (max) | 12.35 | 16.35 | 20.42 | 25.42 | 30.42 | 35.5 | 40.5 | 45.5 | 50.5 | 55.6 | 60.6 |
| U/head (min) | 11.65 | 15.65 | 19.58 | 24.58 | 29.58 | 34.5 | 39.5 | 44.5 | 49.5 | 54.4 | 59.4 |
| a + c (max) | 3.0 + 0.5 = 3.5 for all screw lengths | | | | | | | | | | |
| a + c (min) | 1.0 + 0.15 = 1.15 for all screw lengths | | | | | | | | | | |

**M6  B**

| U/head (nom) | 60 |
|---|---|
| U/head (max) | 51.5 |
| U/head (min) | 58.5 |
| a + c | As for product grade A. |

**M8  A**

| U/head (nom) | 16 | 20 | 25 | 30 | 35 | 40 | 45 | 50 | 55 | 60 | 65 | 70 | 80 |
|---|---|---|---|---|---|---|---|---|---|---|---|---|---|
| U/head (max) | 16.35 | 20.42 | 25.42 | 30.42 | 35.5 | 40.5 | 45.5 | 50.5 | 55.6 | 60.6 | 65.6 | 70.6 | 80.6 |
| U/head (min) | 15.65 | 19.58 | 24.58 | 29.58 | 34.5 | 39.5 | 44.5 | 49.5 | 54.4 | 59.4 | 64.4 | 69.4 | 79.4 |
| a + c (max) | 4.0 + 0.6 = 4.6 mm for all screw lengths | | | | | | | | | | | | |
| a + c (min) | 1.25 + 0.15 = 1.40 mm for all screw lengths | | | | | | | | | | | | |

*(continued)*

**2.5** (continued)

| Designated Grade thread size | Grade | Feature | Popular lengths | | | | | | | | | | | | | | | |
|---|---|---|---|---|---|---|---|---|---|---|---|---|---|---|---|---|---|---|
| M8 | B | U/head (nom) | | | | | | | | 60 | 65 | 70 | 80 | | | | |
| | | U/head (max) | | | | | | | | 61.5 | 66.5 | 71.5 | 81.5 | | | | |
| | | U/head (min) | | | | | | | | 58.5 | 63.5 | 68.5 | 78.5 | | | | |
| | | a + c | As for product grade A. | | | | | | | | | | | | | | |
| M10 | A | U/head (nom) | 20 | 25 | 30 | 35 | 40 | 45 | 50 | 55 | 60 | 65 | 70 | 80 | 90 | 100 | |
| | | U/head (max) | 20.42 | 25.42 | 30.42 | 35.5 | 40.5 | 45.5 | 50.5 | 55.6 | 60.6 | 65.6 | 70.6 | 80.6 | 90.7 | 100.7 | |
| | | U/head (min) | 19.58 | 24.58 | 29.58 | 34.5 | 39.5 | 44.5 | 49.5 | 54.4 | 59.4 | 64.4 | 69.4 | 79.4 | 89.3 | 99.3 | |
| | | a + c (max) | 4.5 + 0.6 = 5.1 mm for all screw lengths | | | | | | | | | | | | | | |
| | | a + c (min) | 1.50 + 0.15 = 1.65 mm for all screw lengths | | | | | | | | | | | | | | |
| M10 | B | U/head (nom) | | | | | | | | | 60 | 65 | 70 | 80 | 90 | 100 | |
| | | U/head (max) | | | | | | | | | 61.5 | 66.5 | 71.5 | 81.5 | 91.75 | 101.75 | |
| | | U/head (min) | | | | | | | | | 58.5 | 63.5 | 68.5 | 78.5 | 88.25 | 98.25 | |
| | | a + c | As for product grade A. | | | | | | | | | | | | | | |
| M12 | A | U/head (nom) | | 25 | 30 | 35 | 40 | 45 | 50 | 55 | 60 | 65 | 70 | 80 | 90 | 100 | 110 | 120 |
| | | U/head (max) | | 25.42 | 30.42 | 35.5 | 40.5 | 45.5 | 50.5 | 55.6 | 60.6 | 65.6 | 70.6 | 80.6 | 90.7 | 100.7 | 110.7 | 120.7 |
| | | U/head (min) | | 24.58 | 29.58 | 34.5 | 39.5 | 44.5 | 49.5 | 54.4 | 59.4 | 64.4 | 69.4 | 79.4 | 89.3 | 99.3 | 109.3 | 119.3 |
| | | a + c (max) | 5.3 + 0.6 = 5.9 mm for all screw lengths | | | | | | | | | | | | | | |
| | | a + c (min) | 1.75 + 0.15 = 1.90 mm for all screw lengths | | | | | | | | | | | | | | |

**M12**

**B**

| | | | | | | | | | | | |
|---|---|---|---|---|---|---|---|---|---|---|---|
| U/head (nom) | 60 | 65 | 70 | 80 | 90 | 100 | 110 | 120 | 130 | 140 | 150 |
| U/head (max) | 61.5 | 66.5 | 71.5 | 81.5 | 91.75 | 101.75 | 111.75 | 121.75 | 130.8 | 140.8 | 150.8 |
| U/head (min) | 58.5 | 63.5 | 68.5 | 78.5 | 88.25 | 98.25 | 108.25 | 118.25 | 129.2 | 139.2 | 149.2 |
| a + c | As for product grade A. | | | | | | | | | | |

**M16**

**A**

| | | | | | | | | | | | | | | |
|---|---|---|---|---|---|---|---|---|---|---|---|---|---|---|
| U/head (nom) | 30 | 35 | 40 | 45 | 50 | 55 | 60 | 65 | 70 | 80 | 90 | 100 | 110 | 120 |
| U/head (max) | 30.42 | 35.5 | 40.5 | 45.5 | 50.5 | 55.6 | 60.6 | 65.6 | 70.6 | 80.6 | 90.7 | 100.7 | 110.7 | 120.7 |
| U/head (min) | 29.58 | 34.5 | 39.5 | 44.5 | 49.5 | 54.4 | 59.4 | 64.4 | 69.4 | 79.4 | 89.3 | 99.3 | 109.3 | 119.3 |
| a + c (max) | $6.0 + 0.8 = 6.8$ mm for all screw lengths | | | | | | | | | | | | | |
| a + c (min) | $2.0 + 0.2 = 2.2$ mm for all screw lengths | | | | | | | | | | | | | |

**M16**

**B**

| | | | | | | | | | | | |
|---|---|---|---|---|---|---|---|---|---|---|---|
| U/head (nom) | 60 | 65 | 70 | 80 | 90 | 100 | 110 | 120 | 130 | 140 | 150 |
| U/head (max) | 61.5 | 66.5 | 71.5 | 81.5 | 91.75 | 101.75 | 111.75 | 121.75 | 132 | 142 | 151 |
| U/head (min) | 58.5 | 63.5 | 68.5 | 78.5 | 88.25 | 98.25 | 108.25 | 118.25 | 128 | 138 | 148 |
| a + c | As for product grade A. | | | | | | | | | | |

**M20**

**A**

| | | | | | | | | | | | | | | | |
|---|---|---|---|---|---|---|---|---|---|---|---|---|---|---|---|
| U/head (nom) | 40 | 45 | 50 | 55 | 60 | 65 | 70 | 80 | 90 | 100 | 110 | 120 | 130 | 140 | 150 |
| U/head (max) | 40.5 | 45.5 | 50.5 | 55.6 | 60.6 | 65.6 | 70.6 | 80.6 | 90.7 | 100.7 | 110.7 | 120.7 | 130.8 | 140.8 | 150.8 |
| U/head (min) | 39.5 | 44.5 | 49.5 | 54.5 | 59.5 | 64.4 | 69.4 | 79.4 | 89.3 | 99.3 | 109.3 | 119.3 | 129.2 | 139.2 | 149.2 |
| a + c (max) | $7.5 + 0.8 = 8.3$ mm for all screw lengths | | | | | | | | | | | | | | |
| a + c (min) | $2.5 + 0.2 = 2.7$ mm for all screw lengths | | | | | | | | | | | | | | |

**M20**

**B**

| | | | | | | | | | | | | |
|---|---|---|---|---|---|---|---|---|---|---|---|---|---|
| U/head (nom) | 60 | 80 | 90 | 100 | 110 | 120 | 130 | 140 | 150 | 160 | 180 | 200 |
| U/head (max) | 61.5 | 81.5 | 91.75 | 101.75 | 111.75 | 121.75 | 132 | 142 | 152 | 162 | 182 | 202.3 |
| U/head (min) | 58.5 | 78.5 | 88.25 | 98.25 | 108.25 | 118.25 | 128 | 138 | 148 | 158 | 178 | 197.7 |
| a + c | As for product grade A. | | | | | | | | | | | |

(continued)

**2.5** (continued)

(Dimensions in millimetres)

| Designated thread size | Grade | Feature | 50 | 55 | 60 | 65 | 70 | 80 | 90 | 100 | 110 | 120 | 130 | 140 | 150 | 160 | 180 | 200 |
|---|---|---|---|---|---|---|---|---|---|---|---|---|---|---|---|---|---|---|
| | | | | | | | | Popular lengths | | | | | | | | | | |
| M24 | A | U/head (nom) | 50 | 55 | 60 | 65 | 70 | 80 | 90 | 100 | 110 | 120 | 130 | 140 | 150 | | | |
| | | U/head (max) | 50.5 | 55.6 | 60.6 | 65.6 | 70.6 | 80.6 | 90.7 | 100.7 | 110.7 | 120.7 | 130.8 | 140.8 | 150.8 | | | |
| | | U/head (min) | 49.5 | 54.5 | 59.5 | 64.4 | 69.4 | 79.4 | 89.3 | 99.3 | 109.3 | 119.3 | 129.2 | 139.2 | 149.2 | | | |
| | | a + c (max) | 9.0 + 0.8 = 9.8 mm for all screw lengths | | | | | | | | | | | | | | | |
| | | a + c (min) | 3.0 + 0.2 = 3.2 mm for all screw lengths | | | | | | | | | | | | | | | |
| M24 | B | U/head (nom) | | | 60 | 65 | 70 | 80 | 90 | 100 | 110 | 120 | 130 | 140 | 150 | 160 | 180 | 200 |
| | | U/head (max) | | | 61.5 | 66.5 | 71.5 | 81.5 | 91.75 | 101.75 | 111.75 | 121.75 | 132 | 142 | 152 | 162 | 182 | 202.3 |
| | | U/head (min) | | | 58.5 | 63.5 | 68.5 | 78.5 | 88.25 | 98.25 | 108.25 | 118.25 | 128 | 138 | 148 | 158 | 178 | 197.7 |
| | | a + c | As for product grade A. | | | | | | | | | | | | | | | |
| M30 | A | U/head (nom) | | | 60 | 65 | 70 | 80 | 90 | 100 | 110 | 120 | 130 | 140 | 150 | | | |
| | | U/head (max) | | | 60.6 | 65.6 | 70.6 | 80.6 | 90.7 | 100.7 | 110.7 | 120.7 | 130.8 | 140.8 | 150.8 | | | |
| | | U/head (min) | | | 59.5 | 64.4 | 69.4 | 79.4 | 89.3 | 99.3 | 109.3 | 119.3 | 129.2 | 139.2 | 149.2 | | | |
| | | a + c (max) | 10.5 + 0.8 = 11.3 mm for all screw lengths | | | | | | | | | | | | | | | |
| | | a + c (min) | 3.5 + 0.2 = 3.7 mm for all screw lengths | | | | | | | | | | | | | | | |
| M30 | B | U/head (nom) | | | 60 | 65 | 70 | 80 | 90 | 100 | 110 | 120 | 130 | 140 | 150 | 160 | 180 | 200 |
| | | U/head (max) | | | 61.5 | 66.5 | 71.5 | 81.5 | 91.75 | 101.75 | 111.75 | 121.75 | 132 | 142 | 152 | 162 | 182 | 202.3 |
| | | U/head (min) | | | 58.5 | 63.5 | 68.5 | 78.5 | 88.25 | 98.25 | 108.25 | 118.25 | 128 | 138 | 148 | 158 | 178 | 197.7 |
| | | a + c | As for product grade A. | | | | | | | | | | | | | | | |
| M36 | A | U/head (nom) | | | | | 70 | 80 | 90 | 100 | 110 | 120 | 130 | 140 | 150 | | | |
| | | U/head (max) | | | | | 70.6 | 80.6 | 90.7 | 100.7 | 110.7 | 120.7 | 130.8 | 140.8 | 150.8 | | | |
| | | U/head (min) | | | | | 69.4 | 79.4 | 89.3 | 99.3 | 109.3 | 119.3 | 129.2 | 139.2 | 149.2 | | | |

| Size | Grade | | 60 | 65 | 70 | 80 | 90 | 100 | 110 | 120 | 130 | 140 | 150 | 160 | 180 | 200 |
|---|---|---|---|---|---|---|---|---|---|---|---|---|---|---|---|---|
| M36 | B | U/head (nom) | 60 | 65 | 70 | 80 | 90 | 100 | 110 | 120 | 130 | 140 | 150 | 160 | 180 | 200 |
| | | U/head (max) | 61.5 | 66.5 | 71.5 | 81.5 | 91.75 | 101.75 | 111.75 | 121.75 | 132 | 142 | 152 | 162 | 182 | 202.3 |
| | | U/head (min) | 58.5 | 63.5 | 68.5 | 78.5 | 88.25 | 98.25 | 108.25 | 118.25 | 128 | 138 | 148 | 158 | 178 | 197.7 |
| | | a + c | As for product grade A. | | | | | | | | | | | | | |
| M42 | A | U/head (nom) | | | | 80 | 90 | 100 | 110 | 120 | 130 | 140 | 150 | | | |
| | | U/head (max) | | | | 80.6 | 90.7 | 100.7 | 110.7 | 120.7 | 130.8 | 140.8 | 150.8 | | | |
| | | U/head (min) | | | | 79.4 | 89.3 | 99.3 | 109.3 | 119.3 | 129.2 | 139.2 | 149.2 | | | |
| | | a + c (max) | 13.5 + 1.0 = 14.5 mm for all screw lengths | | | | | | | | | | | | | |
| | | a + c (min) | 4.5 + 0.3 = 4.8 mm for all screw lengths | | | | | | | | | | | | | |
| M42 | B | U/head (nom) | 60 | 65 | 70 | 80 | 90 | 100 | 110 | 120 | 130 | 140 | 150 | 160 | 180 | 200 |
| | | U/head (max) | 61.5 | 66.5 | 71.5 | 81.5 | 91.75 | 101.75 | 111.75 | 121.75 | 132 | 142 | 152 | 162 | 182 | 202.3 |
| | | U/head (min) | 58.5 | 63.5 | 68.5 | 78.5 | 88.25 | 98.25 | 108.25 | 118.25 | 128 | 138 | 148 | 158 | 178 | 197.7 |
| | | a + c | As for product grade A. | | | | | | | | | | | | | |
| M48 | A | U/head (nom) | | | | | 90 | 100 | 110 | 120 | 130 | 140 | 150 | | | |
| | | U/head (max) | | | | | 90.7 | 100.7 | 110.7 | 120.7 | 130.8 | 140.8 | 150.8 | | | |
| | | U/head (min) | | | | | 89.3 | 99.3 | 109.3 | 119.3 | 129.2 | 139.2 | 149.2 | | | |
| | | a + c (max) | 15.0 + 1.0 = 16.0 mm for all screw lengths | | | | | | | | | | | | | |
| | | a + c (min) | 5.0 + 0.3 = 5.3 mm for all screw lengths | | | | | | | | | | | | | |
| M48 | B | U/head (nom) | 60 | 65 | 70 | 80 | 90 | 100 | 110 | 120 | 130 | 140 | 150 | 160 | 180 | 200 |
| | | U/head (max) | 61.5 | 66.5 | 71.5 | 81.5 | 91.75 | 101.75 | 111.75 | 121.75 | 132 | 142 | 152 | 162 | 182 | 202.3 |
| | | U/head (min) | 58.5 | 63.5 | 68.5 | 78.5 | 88.75 | 98.25 | 108.25 | 118.25 | 128 | 138 | 148 | 158 | 178 | 197.7 |
| | | a + c | | | | | | | | | | | | | | |

For further information see BSEN 24017.

## 2.6 ISO metric hexagon head screws (coarse thread): preferred sizes: product grade C

*Notes:*

(1) Reference back to Section 2.2.1 shows that hexagon head **bolts** have a plain shank between the head and the thread. It also shows that hexagon head **screws** have the thread running the full length up to the head of the screws. For practical tooling purposes there is a short distance (a + c) immediately under the head to allow for the thread to run out and also for a small radius. The dimensions (a + c) refer to BSEN 24018.

(2) The thread and hexagon proportions are the same as those shown in Section 2.4. Therefore only the length under the head (**nominal length**) and the dimensions (a + c) are listed here.

(Dimensions in millimetres)

| Designated thread size | Feature | 10 | 12 | 16 | 20 | 25 | 30 | 35 | 40 | 45 | 50 | 55 | 60 | 65 | 70 | 80 | 90 | 100 |
|---|---|---|---|---|---|---|---|---|---|---|---|---|---|---|---|---|---|---|
| M5 | U/head (nom) | 10 | 12 | 16 | 20 | 25 | 30 | 35 | 40 | 45 | 50 | | | | | | | |
| | U/head (max) | 10.75 | 12.9 | 16.9 | 21.05 | 26.05 | 31.05 | 36.25 | 41.25 | 46.25 | 51.25 | | | | | | | |
| | U/head (min) | 9.25 | 11.1 | 15.1 | 18.95 | 23.95 | 28.95 | 33.75 | 38.75 | 43.75 | 48.75 | | | | | | | |
| | a + c (max) | 2.4 + 0.5 = 2.9 mm for all screw lengths | | | | | | | | | | | | | | | | |
| | a + c (min) | 0.8 + 0 = 0.8 mm for all screw lengths | | | | | | | | | | | | | | | | |
| M6 | U/head (nom) | | 12 | 16 | 20 | 25 | 30 | 35 | 40 | 45 | 50 | 55 | 60 | | | | | |
| | U/head (max) | | 12.9 | 16.9 | 21.05 | 26.05 | 31.05 | 36.25 | 41.25 | 46.25 | 51.25 | 56.5 | 61.5 | | | | | |
| | U/head (min) | | 11.1 | 15.1 | 18.95 | 23.95 | 28.95 | 33.75 | 38.75 | 43.75 | 48.75 | 53.5 | 58.5 | | | | | |
| | a + c (max) | 3.0 + 0.5 = 3.5 mm for all screw lengths | | | | | | | | | | | | | | | | |
| | a + c (min) | 1.0 + 0 = 1.0 mm for all screw lengths | | | | | | | | | | | | | | | | |
| M8 | U/head (nom) | | | 16 | 20 | 25 | 30 | 35 | 40 | 45 | 50 | 55 | 60 | 65 | 70 | 80 | | |
| | U/head (max) | | | 16.9 | 21.05 | 26.05 | 31.05 | 36.25 | 41.25 | 46.25 | 51.25 | 56.5 | 61.5 | 66.5 | 71.5 | 81.5 | | |
| | U/head (min) | | | 15.1 | 18.95 | 23.95 | 28.95 | 33.75 | 38.75 | 43.75 | 48.75 | 53.5 | 58.5 | 63.5 | 68.5 | 78.5 | | |
| | a + c (max) | 4.0 + 0.6 = 4.6 mm for all screw lengths | | | | | | | | | | | | | | | | |
| | a + c (min) | 1.25 + 0 = 1.25 mm for all screw lengths | | | | | | | | | | | | | | | | |
| M10 | U/head (nom) | | | | 20 | 25 | 30 | 35 | 40 | 45 | 50 | 55 | 60 | 65 | 70 | 80 | 90 | 100 |
| | U/head (max) | | | | 21.05 | 26.05 | 31.05 | 36.25 | 41.25 | 46.25 | 51.25 | 56.5 | 61.5 | 66.5 | 71.5 | 81.5 | 91.75 | 101.75 |
| | U/head (min) | | | | 18.95 | 23.95 | 28.95 | 33.75 | 38.75 | 43.75 | 48.75 | 53.5 | 58.5 | 63.5 | 68.5 | 78.5 | 88.25 | 98.25 |
| | a + c (max) | 4.5 + 0.6 = 5.1 mm for all screw lengths | | | | | | | | | | | | | | | | |
| | a + c (min) | 1.5 + 0 = 1.5 mm for all screw lengths | | | | | | | | | | | | | | | | |

(continued)

**2.6** (continued)

(Dimensions in millimetres)

| Designated thread size | Feature | | | | | | | | | | | | | | | Popular lengths |
|---|---|---|---|---|---|---|---|---|---|---|---|---|---|---|---|---|

**M12**

| Feature | 25 | 30 | 35 | 40 | 45 | 50 | 55 | 60 | 65 | 70 | 80 | 90 | 100 | 110 | 120 |
|---|---|---|---|---|---|---|---|---|---|---|---|---|---|---|---|
| U/head (nom) | 25 | 30 | 35 | 40 | 45 | 50 | 55 | 60 | 65 | 70 | 80 | 90 | 100 | 110 | 120 |
| U/head (max) | 26.05 | 31.05 | 36.25 | 41.25 | 46.25 | 51.25 | 56.25 | 61.5 | 66.5 | 71.5 | 81.5 | 91.75 | 101.75 | 111.75 | 121.75 |
| U/head (min) | 23.95 | 28.95 | 33.75 | 38.75 | 43.75 | 48.75 | 53.5 | 58.5 | 63.5 | 68.5 | 78.5 | 88.25 | 98.25 | 108.25 | 118.25 |

a + c (max) = 5.3 + 0.6 = 5.9 mm for all screw lengths
a + c (min) 1.75 + 0 = 1.75 mm for all screw lengths

**M16**

| Feature | 30 | 35 | 40 | 45 | 50 | 55 | 60 | 65 | 70 | 80 | 90 | 100 | 110 | 120 | 130 | 140 | 150 | 160 |
|---|---|---|---|---|---|---|---|---|---|---|---|---|---|---|---|---|---|---|
| U/head (nom) | 30 | 35 | 40 | 45 | 50 | 55 | 60 | 65 | 70 | 80 | 90 | 100 | 110 | 120 | 130 | 140 | 150 | 160 |
| U/head (max) | 31.05 | 36.25 | 41.25 | 46.25 | 51.25 | 56.25 | 61.5 | 66.5 | 71.5 | 81.5 | 91.75 | 101.75 | 111.75 | 121.75 | 132 | 142 | 152 | 164 |
| U/head (min) | 28.95 | 33.75 | 38.75 | 43.75 | 48.75 | 53.5 | 58.5 | 63.5 | 68.5 | 78.5 | 88.25 | 98.25 | 108.25 | 118.25 | 128 | 138 | 148 | 156 |

a + c (max) = 6.0 + 0.8 = 6.8 mm for all screw lengths
a + c (min) 2.0 + 0 = 2.0 mm for all screw lengths

**M20**

| Feature | 40 | 45 | 50 | 55 | 60 | 65 | 70 | 80 | 90 | 100 | 110 | 120 | 130 | 140 | 150 | 160 | 180 | 200 |
|---|---|---|---|---|---|---|---|---|---|---|---|---|---|---|---|---|---|---|
| U/head (nom) | 40 | 45 | 50 | 55 | 60 | 65 | 70 | 80 | 90 | 100 | 110 | 120 | 130 | 140 | 150 | 160 | 180 | 200 |
| U/head (max) | 41.25 | 46.25 | 51.25 | 56.25 | 61.5 | 66.5 | 71.5 | 81.5 | 91.75 | 101.75 | 111.75 | 121.75 | 132 | 142 | 152 | 164 | 184 | 204.6 |
| U/head (min) | 38.75 | 43.75 | 48.75 | 53.5 | 58.5 | 63.5 | 68.5 | 78.5 | 88.25 | 98.25 | 108.25 | 118.75 | 128 | 138 | 148 | 156 | 176 | 195.4 |

a + c (max) = 7.5 + 0.8 = 8.3 mm for all screw lengths
a + c (min) 2.5 + 0 = 2.5 mm for all screw lengths

**M24**

| Feature | 50 | 55 | 60 | 65 | 70 | 80 | 90 | 100 | 110 | 120 | 130 | 140 | 150 | 160 | 180 | 200 | 220 | 240 |
|---|---|---|---|---|---|---|---|---|---|---|---|---|---|---|---|---|---|---|
| U/head (nom) | 50 | 55 | 60 | 65 | 70 | 80 | 90 | 100 | 110 | 120 | 130 | 140 | 150 | 160 | 180 | 200 | 220 | 240 |
| U/head (max) | 51.25 | 56.25 | 61.5 | 66.5 | 71.5 | 81.5 | 91.75 | 101.75 | 111.75 | 121.75 | 132 | 142 | 152 | 164 | 184 | 204.6 | 224.6 | 244.6 |
| U/head (min) | 48.75 | 53.5 | 58.5 | 63.5 | 68.5 | 78.5 | 88.25 | 98.25 | 108.25 | 118.75 | 128 | 138 | 148 | 156 | 176 | 195.4 | 215.4 | 235.4 |

a + c (max) = 9.0 + 0.8 = 9.8 mm for all screw lengths

**M30**

| U/head (nom) | 60 | 65 | 70 | 80 | 90 | 100 | 110 | 120 | 130 | 140 | 150 | 160 | 180 | 200 | 220 | 240 | 260 |
|---|---|---|---|---|---|---|---|---|---|---|---|---|---|---|---|---|---|
| U/head (max) | 61.5 | 66.5 | 71.5 | 81.5 | 91.75 | 101.75 | 111.75 | 121.75 | 132 | 142 | 152 | 164 | 184 | 204.6 | 224.6 | 244.6 | 265.2 |
| U/head (min) | 58.5 | 63.5 | 68.5 | 78.5 | 88.25 | 98.25 | 108.25 | 118.75 | 128 | 138 | 148 | 156 | 176 | 195.4 | 215.4 | 235.4 | 254.8 |

| U/head (nom) | 280 | 300 |
|---|---|---|
| U/head (max) | 285.2 | 305.2 |
| U/head (min) | 274.8 | 294.8 |

$a + c$ (max)  $10.5 + 0.8 = 11.3$ mm for all screw lengths
$a + c$ (min)  $3.5 + 0 = 3.5$ mm for all screw lengths

**M36**

| U/head (nom) | 70 | 80 | 90 | 100 | 110 | 120 | 130 | 140 | 150 | 160 | 180 | 200 | 220 | 240 |
|---|---|---|---|---|---|---|---|---|---|---|---|---|---|---|
| U/head (max) | 71.5 | 81.5 | 91.75 | 101.75 | 111.75 | 121.75 | 132 | 142 | 152 | 164 | 184 | 204.6 | 224.6 | 244.6 |
| U/head (min) | 68.5 | 78.5 | 88.25 | 98.25 | 108.25 | 118.25 | 128 | 138 | 148 | 156 | 176 | 195.4 | 215.4 | 235.4 |

| U/head (nom) | 260 | 280 | 300 | 320 | 340 | 360 |
|---|---|---|---|---|---|---|
| U/head (max) | 265.2 | 285.2 | 305.2 | 325.7 | 345.7 | 365.7 |
| U/head (min) | 254.8 | 274.8 | 294.8 | 314.3 | 334.3 | 354.3 |

$a + c$ (max)  $12.0 + 0.8 = 12.8$ mm for all screw lengths
$a + c$ (min)  $4.0 + 0 = 4$ mm for all screw lengths

(continued)

**2.6** (continued)                                                                 (Dimensions in millimetres)

| Designated thread size | Feature | Popular lengths | | | | | | | | | | | | |
|---|---|---|---|---|---|---|---|---|---|---|---|---|---|---|
| M42 | U/head (nom) | 80 | 90 | 100 | 110 | 120 | 130 | 140 | 150 | 160 | 180 | 200 | 220 | 240 |
| | U/head (max) | 81.5 | 91.75 | 101.75 | 111.75 | 121.75 | 132 | 142 | 152 | 164 | 184 | 204.6 | 224.6 | 244.6 |
| | U/head (min) | 78.5 | 88.25 | 98.25 | 108.25 | 118.25 | 128 | 138 | 148 | 156 | 176 | 195.4 | 215.4 | 235.4 |
| | U/head (nom) | 260 | 280 | 300 | 320 | 340 | 360 | 380 | 400 | 420 | | | | |
| | U/head (max) | 265.2 | 285.2 | 305.2 | 325.7 | 345.7 | 365.7 | 385.7 | 405.7 | 426.3 | | | | |
| | U/head (min) | 254.8 | 274.8 | 294.8 | 314.3 | 334.3 | 354.3 | 374.3 | 394.3 | 413.7 | | | | |

$a + c$ (max) $13.5 + 1.0 = 14.5$ for all screw lengths
$a + c$ (min) $4.5 + 0 = 4.5$ for all screw lengths

| Designated thread size | Feature | Popular lengths | | | | | | | | | | | | |
|---|---|---|---|---|---|---|---|---|---|---|---|---|---|---|
| M48 | U/head (nom) | 90 | 100 | 110 | 120 | 130 | 140 | 150 | 160 | 180 | 200 | 220 | 240 | 260 |
| | U/head (max) | 91.75 | 101.75 | 111.75 | 121.75 | 132 | 142 | 152 | 164 | 184 | 204.6 | 224.6 | 244.6 | 265.2 |
| | U/head (min) | 88.25 | 98.25 | 108.25 | 118.25 | 128 | 138 | 148 | 156 | 176 | 195.4 | 215.4 | 235.4 | 254.8 |
| | U/head (nom) | 280 | 300 | 325 | 340 | 360 | 380 | 400 | 420 | 440 | 460 | 480 | | |
| | U/head (max) | 285.2 | 305.2 | 325.7 | 345.7 | 365.7 | 385.7 | 405.7 | 426.3 | 446.3 | 466.3 | 486.3 | | |
| | U/head (min) | 274.8 | 294.8 | 314.3 | 334.3 | 354.3 | 374.3 | 394.3 | 413.7 | 433.7 | 453.7 | 473.7 | | |

$a + c$ (max) $15.0 + 1.0 = 16$ mm for all screw lengths
$a + c$ (min) $5.0 + 0 = 5$ mm for all screw lengths

For further information see BSEN 24018.

## 2.7 ISO metric tapping and clearance drills, coarse thread series

| Nominal size | Tapping drill size (mm) | | Clearance drill size (mm) | | |
|---|---|---|---|---|---|
| | Recommended 80% engagement | Alternative 70% engagement | Close fit | Medium fit | Free fit |
| M1.6 | 1.25 | 1.30 | 1.7 | 1.8 | 2.0 |
| M2 | 1.60 | 1.65 | 2.2 | 2.4 | 2.6 |
| M2.5 | 2.05 | 2.10 | 2.7 | 2.9 | 3.1 |
| M3 | 2.50 | 2.55 | 3.2 | 3.4 | 3.6 |
| M4 | 3.30 | 3.40 | 4.3 | 4.5 | 4.8 |
| M5 | 4.20 | 4.30 | 5.3 | 5.5 | 5.8 |
| M6 | 5.00 | 5.10 | 6.4 | 6.6 | 7.0 |
| M8 | 6.80 | 6.90 | 8.4 | 9.0 | 10.0 |
| M10 | 8.50 | 8.60 | 10.5 | 11.0 | 12.0 |
| M12 | 10.20 | 10.40 | 13.0 | 14.0 | 15.0 |
| M14 | 12.00 | 12.20 | 15.0 | 16.0 | 17.0 |
| M16 | 14.00 | 14.25 | 17.0 | 18.0 | 19.0 |
| M18 | 15.50 | 15.75 | 19.0 | 20.0 | 21.0 |
| M20 | 17.50 | 17.75 | 21.0 | 22.0 | 24.0 |
| M22 | 19.50 | 19.75 | 23.0 | 24.0 | 26.0 |
| M24 | 21.00 | 21.25 | 25.0 | 26.0 | 28.0 |
| M27 | 24.00 | 24.25 | 28.0 | 30.0 | 32.0 |
| M30 | 26.50 | 26.75 | 31.0 | 33.0 | 35.0 |
| M33 | 29.50 | 29.75 | 34.0 | 36.0 | 38.0 |
| M36 | 32.00 | – | 37.0 | 39.0 | 42.0 |
| M39 | 35.00 | – | 40.0 | 42.0 | 45.0 |
| M42 | 37.50 | – | 43.0 | 45.0 | 48.0 |
| M45 | 40.50 | – | 46.0 | 48.0 | 52.0 |
| M48 | 43.00 | – | 50.0 | 52.0 | 56.0 |
| M52 | 47.00 | – | 54.0 | 56.0 | 62.0 |

## 2.8 ISO metric hexagon nuts (coarse thread) style 1: product grade A and B (preferred sizes)

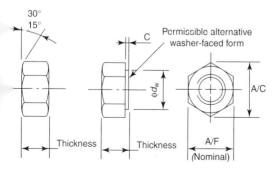

(Dimensions in millimetres)

| Designated thread size | Across corners A/C (min) | Across flats (A/F) | | Thickness | | Washer-faced form (C) thickness | | $d_w$ (min) |
|---|---|---|---|---|---|---|---|---|
| | | Maximum (nom) | Minimum | Maximum | Minimum | Maximum | Minimum | |
| M1.6 | 3.41 | 3.20 | 3.02 | 1.30 | 1.05 | 0.2 | 0.1 | 2.4 |
| M2 | 4.32 | 4 | 3.82 | 1.60 | 1.35 | 0.2 | 0.1 | 3.1 |
| M2.5 | 5.45 | 5 | 4.82 | 2.00 | 1.75 | 0.3 | 0.1 | 4.1 |
| M3 | 6.01 | 5.5 | 5.32 | 2.40 | 2.15 | 0.4 | 0.15 | 4.6 |
| M4 | 7.66 | 7 | 6.78 | 3.2 | 2.9 | 0.4 | 0.15 | 5.9 |
| M5 | 8.79 | 8 | 7.78 | 4.7 | 4.4 | 0.5 | 0.15 | 6.9 |
| M6 | 11.05 | 10 | 9.78 | 5.2 | 4.9 | 0.5 | 0.15 | 8.9 |
| M8 | 14.38 | 13 | 12.73 | 6.8 | 6.44 | 0.6 | 0.15 | 11.6 |
| M10 | 17.77 | 16 | 15.30 | 8.4 | 8.04 | 0.6 | 0.15 | 14.6 |
| M12 | 20.03 | 18 | 17.73 | 10.8 | 10.37 | 0.6 | 0.15 | 16.6 |
| M16 | 26.75 | 24 | 23.67 | 14.8 | 14.1 | 0.8 | 0.2 | 22.5 |
| M20 | 32.95 | 30 | 29.16 | 18.0 | 16.9 | 0.8 | 0.2 | 27.7 |
| M24 | 39.55 | 36 | 35.00 | 21.5 | 20.5 | 0.8 | 0.2 | 33.3 |
| M30 | 50.85 | 46 | 45.00 | 25.6 | 24.3 | 0.8 | 0.2 | 42.8 |
| M36 | 60.79 | 55 | 53.80 | 31.0 | 29.4 | 0.8 | 0.2 | 51.1 |
| M42 | 71.30 | 65 | 63.10 | 34.0 | 32.4 | 1.0 | 0.3 | 60.0 |
| M48 | 82.60 | 75 | 73.10 | 38.0 | 36.4 | 1.0 | 0.3 | 69.5 |

For further information see BSEN 24032.

## 2.9 ISO metric hexagon nuts (coarse thread) style 2: product grade A and B (preferred sizes)

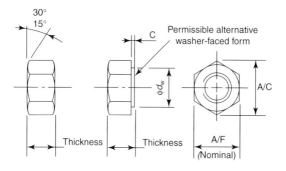

(Dimensions in millimetres)

| Designated thread size | Across corners A/C (min) | Across flats (A/F) | | Thickness | | Washer-faced form (C) thickness | |
|---|---|---|---|---|---|---|---|
| | | Maximum (nom) | Minimum | Maximum | Minimum | Maximum | $d_w$ (min) |
| M5 | 8.79 | 8 | 7.78 | 5.1 | 4.8 | 0.5 | 6.9 |
| M6 | 11.05 | 10 | 9.78 | 5.7 | 5.4 | 0.5 | 8.9 |
| M8 | 14.38 | 13 | 12.73 | 7.5 | 7.14 | 0.6 | 11.6 |
| M10 | 17.77 | 16 | 15.73 | 9.3 | 8.94 | 0.6 | 14.6 |
| M12 | 20.03 | 18 | 17.73 | 12.0 | 11.75 | 0.6 | 16.6 |
| (M14) | 23.35 | 21 | 20.67 | 14.1 | 13.4 | 0.6 | 19.6 |
| M16 | 26.75 | 24 | 23.67 | 16.4 | 15.7 | 0.8 | 22.5 |
| M20 | 32.95 | 30 | 29.16 | 20.3 | 19.0 | 0.8 | 27.7 |
| M24 | 39.55 | 36 | 35.0 | 23.9 | 22.6 | 0.8 | 33.2 |
| M30 | 50.85 | 46 | 45.0 | 28.6 | 27.3 | 0.8 | 42.7 |
| M36 | 60.79 | 55 | 53.8 | 34.7 | 33.1 | 0.8 | 51.1 |

M14 is not a preferred thread size and should be avoided wherever possible.
For further information see BSEN 24033.

## 2.10 ISO metric hexagon nuts (coarse thread) style 1: product grade C (preferred sizes)

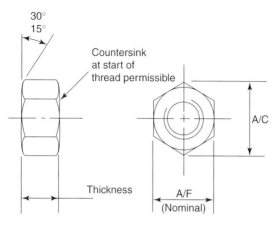

(Dimensions in millimetres)

| Designated thread size | Across corners A/C (min.) | Across Flats (A/F) | | Thickness | |
|---|---|---|---|---|---|
| | | Max. (nom.) | Min. | Max. | Min. |
| M5 | 8.63 | 8 | 7.64 | 5.6 | 4.4 |
| M6 | 10.89 | 10 | 9.64 | 6.1 | 4.6 |
| M8 | 14.20 | 13 | 12.57 | 7.9 | 6.4 |
| M10 | 17.59 | 16 | 15.57 | 9.5 | 8.0 |
| M12 | 19.85 | 18 | 17.57 | 12.2 | 10.4 |
| M16 | 26.17 | 24 | 23.16 | 15.9 | 14.1 |
| M20 | 32.95 | 30 | 29.16 | 19.0 | 16.9 |
| M24 | 39.55 | 36 | 35.00 | 22.3 | 20.2 |
| M30 | 50.85 | 46 | 45.00 | 26.4 | 24.3 |
| M36 | 60.79 | 55 | 53.80 | 31.5 | 28.0 |
| M42 | 72.02 | 65 | 63.10 | 34.9 | 32.4 |
| M48 | 82.60 | 75 | 73.10 | 38.9 | 36.4 |
| M56 | 93.56 | 85 | 82.80 | 45.9 | 43.4 |
| M64 | 104.86 | 95 | 92.80 | 52.4 | 49.4 |

For further information see BSEN 24034.

## 2.11 ISO metric hexagon thin nuts (chamfered) – coarse thread – product grade A and B (also known as lock-nuts)

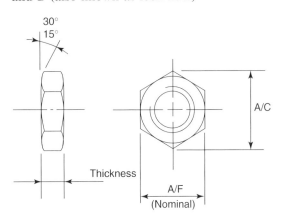

Preferred sizes                                      (Dimensions in millimetres)

| Designated thread size | Across corners A/C (min.) | Across flats (A/F) | | Thickness | |
|---|---|---|---|---|---|
| | | Max. (nom.) | Min. | Max. | Min. |
| M1.6 | 3.41 | 3.2 | 3.02 | 1.0 | 0.75 |
| M2 | 4.32 | 4.0 | 3.82 | 1.2 | 0.95 |
| M2.5 | 5.45 | 5.0 | 4.82 | 1.6 | 1.35 |
| M3 | 6.01 | 5.5 | 5.32 | 1.8 | 1.55 |
| M4 | 7.66 | 7.0 | 6.78 | 2.2 | 1.95 |
| M5 | 8.79 | 8.0 | 7.78 | 2.7 | 2.45 |
| M6 | 11.05 | 10.0 | 9.78 | 3.2 | 2.90 |
| M8 | 14.38 | 13.0 | 12.73 | 4.0 | 3.70 |
| M10 | 17.77 | 16.0 | 15.73 | 5.0 | 4.70 |
| M12 | 20.03 | 18.0 | 17.73 | 6.0 | 5.70 |
| M16 | 26.75 | 24.0 | 23.67 | 8.0 | 7.42 |
| M20 | 32.95 | 30.0 | 29.16 | 10.0 | 9.10 |
| M24 | 39.55 | 36.0 | 35.00 | 12.0 | 10.90 |
| M30 | 50.85 | 46.0 | 45.00 | 15.0 | 13.90 |
| M36 | 60.79 | 55.0 | 53.80 | 18.0 | 16.90 |
| M42 | 71.30 | 65.0 | 63.10 | 21.0 | 19.70 |
| M48 | 82.60 | 75.0 | 73.10 | 24.0 | 22.70 |
| M56 | 93.56 | 85.0 | 82.80 | 28.0 | 26.70 |
| M64 | 104.86 | 95.0 | 92.80 | 32.0 | 30.40 |

For further information see BSEN 24035.

# 2.12 ISO metric hexagon head bolts (fine thread): preferred sizes: product grade A and B

(Dimensions in millimetres)

| Designated thread size | Product Grade | Pitch of Thread | Thread | | | Plain shank | | Hexagon | | | | | |
|---|---|---|---|---|---|---|---|---|---|---|---|---|---|
| | | | Major diameter | Effective diameter | Minor diameter | Maximum diameter (nominal) | Minimum diameter | Across flats (A/F) | | Across corners (A/C) Minimum | Thickness | | |
| | | | | | | | | Maximum (nominal) | Minimum | | Nominal | Maximum | Minimum |
| M8 × 1 | A | 1.0 | 8.0 | 7.35 | 6.77 | 8.0 | 7.78 | 13 | 12.3 | 14.33 | 5.3 | 5.45 | 5.15 |
| M10 × 1 | A | 1.0 | 10.0 | 9.19 | 8.47 | 10.0 | 9.78 | 16 | 15.73 | 17.77 | 6.4 | 6.58 | 6.22 |
| M12 × 1.5 | A | 1.5 | 12.0 | 11.19 | 10.47 | 12.0 | 11.73 | 18 | 17.73 | 20.03 | 7.5 | 7.68 | 7.32 |
| M16 × 1.5 | A | 1.5 | 16.0 | 15.03 | 14.16 | 16.0 | 15.73 | 24 | 23.67 | 26.75 | 10 | 10.18 | 9.82 |
| M16 × 1.5 | B | 1.5 | 16.0 | 15.03 | 14.16 | 16.0 | 15.57 | 24 | 23.16 | 26.17 | 10 | 10.29 | 9.71 |
| M20 × 1.5 | A | 1.5 | 20.0 | 19.03 | 18.16 | 20.0 | 19.67 | 30 | 29.67 | 33.53 | 12.5 | 12.75 | 12.285 |
| M20 × 1.5 | B | 1.5 | 20.0 | 19.03 | 18.16 | 20.0 | 19.48 | 30 | 29.16 | 32.95 | 12.5 | 12.85 | 12.15 |
| M24 × 2 | A | 2.0 | 24.0 | 22.70 | 21.55 | 24.0 | 23.67 | 36 | 35.38 | 39.98 | 15.0 | 15.215 | 14.785 |
| M24 × 2 | B | 2.0 | 24.0 | 22.70 | 21.55 | 24.0 | 23.48 | 36 | 35.0 | 39.55 | 15.0 | 15.35 | 14.65 |
| M30 × 2 | B | 2.0 | 30.0 | 28.70 | 27.55 | 30.0 | 29.48 | 46 | 45.0 | 50.85 | 18.7 | 19.12 | 18.28 |
| M36 × 3 | B | 3.0 | 36.0 | 34.05 | 32.32 | 36.0 | 35.80 | 55 | 53.8 | 60.79 | 22.5 | 22.92 | 22.08 |
| M42 × 3 | B | 3.0 | 42.0 | 40.05 | 38.32 | 42.0 | 41.38 | 65 | 63.1 | 71.3 | 26 | 26.42 | 25.58 |
| M48 × 3 | B | 3.0 | 48.0 | 46.05 | 44.32 | 48.0 | 47.38 | 75 | 73.1 | 82.6 | 30 | 30.42 | 29.58 |
| M56 × 4 | B | 4.0 | 56.0 | 53.40 | 51.09 | 56.0 | 55.26 | 85 | 82.8 | 93.56 | 35 | 35.5 | 34.5 |
| M64 × 4 | B | 4.0 | 64.0 | 61.40 | 59.09 | 64.0 | 63.26 | 95 | 92.8 | 104.86 | 40 | 40.5 | 39.5 |

*Note:* There is no product grade C for the fine thread series.
For further information see BSEN 28765.

123

(Dimensions in millimetres)

Length below and to the right of the broken line – – – refer to and are only available in product grade B.

| Designed at thread size | Feature | Popular length combinations | | | | | | | | | | |
|---|---|---|---|---|---|---|---|---|---|---|---|---|
| M6 × 1 (Grade A) | U/head | 40 | 45 | 50 | 55 | 60 | 65 | 70 | 80 | | | |
| | Thread | 22 | 22 | 22 | 22 | 22 | 22 | 22 | 22 | | | |
| | Shank | 11.75 | 16.75 | 21.75 | 26.75 | 31.75 | 36.75 | 41.75 | 51.75 | | | |
| M10 × 1 (Grade A) | U/head | 45 | 50 | 55 | 60 | 65 | 70 | 80 | 90 | 100 | | |
| | Thread | 26 | 26 | 26 | 26 | 26 | 26 | 26 | 26 | 26 | | |
| | Shank | 11.5 | 16.5 | 21.5 | 26.5 | 31.5 | 36.5 | 46.5 | 56.5 | 66.5 | | |
| M12 × 1.5 (Grade A) | U/head | 50 | 55 | 60 | 65 | 70 | 80 | 90 | 100 | 110 | 120 | |
| | Thread | 30 | 30 | 30 | 30 | 30 | 30 | 30 | 30 | 30 | 30 | |
| | Shank | 11.25 | 16.25 | 21.25 | 26.25 | 31.25 | 41.25 | 51.25 | 61.25 | 71.25 | 81.25 | |
| M16 × 1.5 (Grade A) | U/head | 65 | 70 | 80 | 90 | 100 | 110 | 120 | 130 | 140 | 150 | 160 |
| | Thread | 38 | 38 | 38 | 38 | 38 | 38 | 38 | 44 | 44 | 44 | 44 |
| | Shank | 17 | 22 | 32 | 42 | 52 | 62 | 72 | 76 | 86 | 96 | 106 |
| M20 × 1.5 (Grade A) | U/head | 80 | 90 | 100 | 110 | 120 | 130 | 140 | 150 | 160 | 180 | 200 |
| | Thread | 46 | 46 | 46 | 46 | 46 | 52 | 52 | 52 | 52 | 52 | 52 |
| | Shank | 21.5 | 31.5 | 41.5 | 51.5 | 61.5 | 65.5 | 75.5 | 85.5 | 95.5 | 115.5 | 135.5 |
| M24 × 2 (Grade A) | U/head | 100 | 110 | 120 | 130 | 140 | 150 | 160 | 180 | 200 | 220 | 240 |
| | Thread | 54 | 54 | 54 | 60 | 60 | 60 | 60 | 60 | 60 | 73 | 73 |
| | Shank | 31 | 41 | 51 | 55 | 65 | 75 | 85 | 105 | 125 | 132 | 152 |

**M30 × 2 (Grade B)**

| U/head | 120 | 130 | 140 | 150 | 160 | 180 | 200 | 220 | 240 | 260 | 280 | 300 |
|---|---|---|---|---|---|---|---|---|---|---|---|---|
| Thread | 66 | 72 | 72 | 72 | 72 | 72 | 72 | 85 | 85 | 85 | 85 | 85 |
| Shank | 36.5 | 40.5 | 50.5 | 60.5 | 70.5 | 90.5 | 110.5 | 117.5 | 137.5 | 157.5 | 177.5 | 197.5 |

**M36 × 3 (Grade B)**

| U/head | 140 | 150 | 160 | 180 | 200 | 220 | 240 | 260 | 280 | 300 | 320 | 340 | 360 |
|---|---|---|---|---|---|---|---|---|---|---|---|---|---|
| Thread | 84 | 84 | 84 | 84 | 84 | 97 | 97 | 97 | 97 | 97 | 97 | 97 | 97 |
| Shank | 36 | 46 | 56 | 76 | 96 | 103 | 123 | 143 | 163 | 183 | 203 | 223 | 243 |

**M42 × 3 (Grade B)**

| U/head | 160 | 180 | 200 | 220 | 240 | 260 | 280 | 300 | 320 | 340 | 360 | 380 | 400 | 420 | 440 |
|---|---|---|---|---|---|---|---|---|---|---|---|---|---|---|---|
| Thread | 96 | 96 | 96 | 109 | 109 | 109 | 109 | 109 | 109 | 109 | 109 | 109 | 109 | 109 | 109 |
| Shank | 41.5 | 61.5 | 81.5 | 88.5 | 108.5 | 128.5 | 148.5 | 168.5 | 188.5 | 208.5 | 228.5 | 248.5 | 268.5 | 288.5 | 308.5 |

**M48 × 3 (Grade B)**

| U/head | 200 | 220 | 240 | 260 | 280 | 300 | 320 | 340 | 360 | 380 | 400 | 420 | 440 | 460 | 480 |
|---|---|---|---|---|---|---|---|---|---|---|---|---|---|---|---|
| Thread | 108 | 121 | 121 | 121 | 121 | 121 | 121 | 121 | 121 | 121 | 121 | 121 | 121 | 121 | 121 |
| Shank | 67 | 74 | 94 | 114 | 134 | 154 | 174 | 194 | 214 | 234 | 254 | 274 | 294 | 314 | 334 |

**M56 × 3 (Grade B)**

| U/head | 220 | 240 | 260 | 280 | 300 | 320 | 340 | 360 | 380 | 400 | 420 | 440 | 460 | 480 | 500 |
|---|---|---|---|---|---|---|---|---|---|---|---|---|---|---|---|
| Thread | 137 | 137 | 137 | 137 | 137 | 137 | 137 | 137 | 137 | 137 | 137 | 137 | 137 | 137 | 137 |
| Shank | 55.5 | 75.5 | 95.5 | 115.5 | 135.5 | 155.5 | 175.5 | 195.5 | 215.5 | 235.5 | 255.5 | 275.5 | 295.5 | 315.5 | 335.5 |

**M64 × 4 (Grade B)**

| U/head | 260 | 280 | 300 | 320 | 340 | 360 | 380 | 400 | 420 | 440 | 460 | 480 | 500 |
|---|---|---|---|---|---|---|---|---|---|---|---|---|---|---|
| Thread | 153 | 153 | 153 | 153 | 153 | 153 | 153 | 153 | 153 | 153 | 153 | 153 | 153 |
| Shank | 77 | 97 | 117 | 137 | 157 | 177 | 197 | 217 | 237 | 257 | 277 | 297 | 317 |

*Note:* The length under the head of the bolt (U/head) is also the **nominal length.**
For further information see BSEN 28765.

## 2.13 ISO metric hexagon head screws (fine thread): preferred sizes: product grade A and B

*Notes:*

(1) There is no product grade C for the fine thread series.

(2) Reference back to Section 2.2.1 shows that hexagon head **bolts** have a plain shank between the head and the thread. It also shows that hexagon head **screws** have the thread running the full length up to the head of the screw. For practical tooling purposes there is a short distance $(a + c)$ immediately under the head to allow the thread to run out and also allow for a small radius. The dimensions $(a + c)$ refer to BSEN 28676

$$\begin{bmatrix} a & = & \text{runout of thread} \\ c & = & \text{u/head radius} \end{bmatrix}$$

(3) The thread and hexagon proportions are the same as those shown in the table in Section 2.11. Therefore only the length under the head (**nominal length**) and the dimensions $(a + c)$ are listed here.

| Designated thread size | Grade | Feature | Popular lengths | | | | | | | | | | | | | | | | |
|---|---|---|---|---|---|---|---|---|---|---|---|---|---|---|---|---|---|---|---|
| M8 × 1 | A | U'head (nom.) | 16 | 20 | 25 | 30 | 35 | 40 | 45 | 50 | 55 | 60 | 65 | 70 | 80 | | | |
| | | U/head (max.) | 16.35 | 20.42 | 25.42 | 30.42 | 35.5 | 40.5 | 45.5 | 50.5 | 55.6 | 60.6 | 65.6 | 70.6 | 80.6 | | | |
| | | U/head (min.) | 15.65 | 19.58 | 24.58 | 29.58 | 34.5 | 39.5 | 44.5 | 49.5 | 54.4 | 59.4 | 64.4 | 69.4 | 79.4 | | | |
| | | a + c (max.) | 3.0 + 0.6 = 3.6 mm for all screw lengths | | | | | | | | | | | | | | | |
| | | a + c (min.) | 1.0 + 0.15 = 1.15 mm for all screw lengths | | | | | | | | | | | | | | | |
| M10 × 1 | A | U'head (nom.) | 20 | 25 | 30 | 35 | 40 | 45 | 50 | 55 | 60 | 65 | 70 | 80 | 90 | 100 | | |
| | | U/head (max.) | 20.42 | 25.42 | 30.42 | 35.5 | 40.5 | 45.5 | 50.5 | 55.6 | 60.6 | 65.6 | 70.6 | 80.6 | 90.7 | 100.7 | | |
| | | U/head (min.) | 19.58 | 24.58 | 29.58 | 34.5 | 39.5 | 44.5 | 49.5 | 54.4 | 59.4 | 64.4 | 69.4 | 79.4 | 89.3 | 99.3 | | |
| | | a + c (max.) | 3.0 + 0.6 = 3.6 mm for all screw legths | | | | | | | | | | | | | | | |
| | | a + c (min.) | 1.0 + 0.15 = 1.15 mm for all screw lengths | | | | | | | | | | | | | | | |
| M12 × 1.5 | A | U'head (nom.) | 25 | 30 | 35 | 40 | 45 | 50 | 55 | 60 | 65 | 70 | 80 | 90 | 100 | 110 | 120 | |
| | | U/head (max.) | 25.42 | 30.42 | 35.5 | 40.5 | 45.5 | 50.5 | 55.6 | 60.6 | 65.6 | 70.6 | 80.6 | 90.7 | 100.7 | 110.7 | 120.7 | |
| | | U/head (min.) | 24.58 | 29.58 | 34.5 | 39.5 | 44.5 | 49.5 | 54.4 | 59.4 | 64.4 | 69.4 | 79.4 | 89.3 | 99.3 | 109.3 | 119.3 | |
| | | a + c (max.) | 4.5 + 0.6 = 5.1 mm for all screw lengths | | | | | | | | | | | | | | | |
| | | a + c (min.) | 1.5 + 0.15 = 1.20 mm for all screw lengths | | | | | | | | | | | | | | | |

(continued)

**2.13** (continued)

(Dimensions in millimetres)

Popular lengths (lengths 160, 180 and 200 are Grade B)

| Designated thread size | Grade | Feature | 35 | 40 | 45 | 50 | 55 | 60 | 65 | 70 | 80 | 90 | 100 | 110 | 120 | 130 | 140 | 150 | 160 | 180 | 200 |
|---|---|---|---|---|---|---|---|---|---|---|---|---|---|---|---|---|---|---|---|---|---|
| M16 × 1.5 | A | U/head (nom.) | 35 | 40 | 45 | 50 | 55 | 60 | 65 | 70 | 80 | 90 | 100 | 110 | 120 | 130 | 140 | 150 | 160 | 180 | 200 |
| | | U/head (max.) | 35.5 | 40.5 | 45.5 | 50.5 | 55.6 | 60.6 | 65.6 | 70.6 | 80.6 | 90.7 | 100.7 | 110.7 | 120.7 | 130.8 | 140.8 | 150.8 | 162 | 182 | 202.3 |
| | | U/head (min.) | 34.5 | 39.5 | 44.5 | 49.5 | 54.4 | 59.4 | 64.4 | 69.4 | 79.4 | 89.3 | 99.3 | 109.3 | 119.3 | 129.2 | 139.2 | 149.2 | 158 | 178 | 197.7 |
| | | a + c (max.) | 4.5 + 0.8 = 5.3 mm for all screw lengths | | | | | | | | | | | | | | | | | | |
| | | a + c (min.) | 1.5 + 0.2 = 1.7 mm for all screw lengths | | | | | | | | | | | | | | | | | | |
| M20 × 1.5 | A | U/head (nom.) | | 40 | 45 | 50 | 55 | 60 | 65 | 70 | 80 | 90 | 100 | 110 | 120 | 130 | 140 | 150 | 160 | 180 | 200 |
| | | U/head (max.) | | 40.5 | 45.5 | 50.5 | 55.6 | 60.6 | 65.6 | 70.6 | 80.6 | 90.7 | 100.7 | 110.7 | 120.7 | 130.8 | 140.8 | 150.8 | 162 | 182 | 202.3 |
| | | U/head (min.) | | 39.5 | 44.5 | 49.5 | 54.4 | 59.4 | 64.4 | 69.4 | 79.4 | 89.3 | 99.3 | 109.3 | 119.3 | 129.2 | 139.2 | 149.2 | 158 | 178 | 197.7 |
| | | a + c (max.) | 6.0 + 0.8 = 6.8 mm for all screw lengths | | | | | | | | | | | | | | | | | | |
| | | a + c (min.) | 2.0 + 0.2 = 2.2 mm for all screw lengths | | | | | | | | | | | | | | | | | | |
| M24 × 2.0 | A | U/head (nom.) | | 40 | 45 | 50 | 55 | 60 | 65 | 70 | 80 | 90 | 100 | 110 | 120 | 130 | 140 | 150 | 160 | 180 | 200 |
| | | U/head (max.) | | 40.5 | 45.5 | 50.5 | 55.6 | 60.6 | 65.6 | 70.6 | 80.6 | 90.7 | 100.7 | 110.7 | 120.7 | 130.8 | 140.8 | 150.8 | 162 | 182 | 202.3 |
| | | U/head (min.) | | 39.5 | 44.5 | 49.5 | 54.4 | 59.4 | 64.4 | 69.4 | 79.4 | 89.3 | 99.3 | 109.3 | 119.3 | 129.2 | 139.2 | 149.2 | 158 | 178 | 197.7 |
| | | a + c (max.) | 6.0 + 0.8 = 6.8 mm for all screw lengths | | | | | | | | | | | | | | | | | | |
| | | a + c (min.) | 2.0 + 0.2 = 2.2 mm for all screw lengths | | | | | | | | | | | | | | | | | | |
| M30 × 2 | B | U/head (nom.) | | 40 | 45 | 50 | 55 | 60 | 65 | 70 | 80 | 90 | 100 | 110 | 120 | 130 | 140 | 150 | 160 | 180 | 200 |
| | | U/head (max.) | | 41.25 | 46.25 | 51.25 | 56.5 | 61.5 | 66.5 | 71.5 | 81.5 | 91.75 | 101.75 | 111.75 | 121.75 | 132 | 142 | 152 | 162 | 182 | 202.3 |
| | | U/head (min.) | | 38.75 | 43.75 | 48.75 | 53.5 | 58.5 | 63.5 | 68.5 | 78.5 | 88.25 | 98.25 | 108.25 | 118.25 | 128 | 138 | 148 | 158 | 178 | 197.7 |
| | | a + c (max.) | 6.0 + 0.8 = 6.8 mm for all screw lengths | | | | | | | | | | | | | | | | | | |
| | | a + c (min.) | 2.0 + 0.2 = 2.2 mm for all screw lengths | | | | | | | | | | | | | | | | | | |

(continued)

### M36×3, B

| | 40 | 45 | 50 | 55 | 60 | 65 | 70 | 80 | 90 | 100 | 110 | 120 | 130 | 140 | 150 | 160 | 180 | 200 |
|---|---|---|---|---|---|---|---|---|---|---|---|---|---|---|---|---|---|---|
| U/head (nom.) | 40 | 45 | 50 | 55 | 60 | 65 | 70 | 80 | 90 | 100 | 110 | 120 | 130 | 140 | 150 | 160 | 180 | 200 |
| U/head (max.) | 41.25 | 46.25 | 51.25 | 56.5 | 61.5 | 66.5 | 71.5 | 81.5 | 91.75 | 101.75 | 111.75 | 121.75 | 132 | 142 | 152 | 162 | 182 | 202.3 |
| U/head (min.) | 38.75 | 43.75 | 48.75 | 53.5 | 58.5 | 63.5 | 68.5 | 78.5 | 88.25 | 98.25 | 108.25 | 118.25 | 128 | 138 | 148 | 158 | 178 | 197.7 |
| a + c (max.) | 9.0 + 0.8 = 9.8 mm for all screw lengths | | | | | | | | | | | | | | | | | |
| a + c (min.) | 3.0 + 0.2 = 3.2 mm for all screw lengths | | | | | | | | | | | | | | | | | |

### M42×3, B

| | 90 | 100 | 110 | 120 | 130 | 140 | 150 | 160 | 180 | 200 | 220 | 240 | 260 | 280 | 300 | 320 |
|---|---|---|---|---|---|---|---|---|---|---|---|---|---|---|---|---|
| U/head (nom.) | 90 | 100 | 110 | 120 | 130 | 140 | 150 | 160 | 180 | 200 | 220 | 240 | 260 | 280 | 300 | 320 |
| U/head (max.) | 91.75 | 101.75 | 111.75 | 121.75 | 132 | 142 | 152 | 162 | 182 | 202.3 | 222.3 | 242.3 | 262.3 | 282.6 | 302.6 | 322.85 |
| U/head (min.) | 88.25 | 98.25 | 108.25 | 118.25 | 128 | 138 | 148 | 158 | 178 | 197.7 | 217.7 | 237.7 | 257.7 | 277.4 | 297.4 | 317.15 |
| a + c (max.) | 9.0 + 1.0 = 10 mm for all screw lengths | | | | | | | | | | | | | | | |
| a + c (min.) | 3.0 + 0.3 = 3.3 mm for all screw lengths | | | | | | | | | | | | | | | |

### M42×3 (continued)

| | 340 | 360 | 380 | 400 | 420 |
|---|---|---|---|---|---|
| U/head (nom.) | 340 | 360 | 380 | 400 | 420 |
| U/head (max.) | 342.85 | 362.85 | 382.85 | 402.85 | 423.15 |
| U/head (min.) | 337.15 | 357.15 | 377.15 | 397.15 | 416.85 |
| a + c (max.) | 9.0 + 1.0 = 10 mm for all screw lengths | | | | |
| a + c (min.) | 3.0 + 0.3 = 3.3 mm for all screw lengths | | | | |

### M48×3, B

| | 100 | 110 | 120 | 130 | 140 | 150 | 160 | 180 | 200 | 220 | 240 | 260 | 280 | 300 | 320 |
|---|---|---|---|---|---|---|---|---|---|---|---|---|---|---|---|
| U/head (nom.) | 100 | 110 | 120 | 130 | 140 | 150 | 160 | 180 | 200 | 220 | 240 | 260 | 280 | 300 | 320 |
| U/head (max.) | 101.75 | 111.75 | 121.75 | 132 | 142 | 152 | 162 | 182 | 202.3 | 222.3 | 242.3 | 262.6 | 282.6 | 302.6 | 322.58 |
| U/head (min.) | 98.25 | 108.25 | 118.25 | 128 | 138 | 148 | 158 | 178 | 197.7 | 217.7 | 237.7 | 257.4 | 277.4 | 297.4 | 317.15 |
| a + c (max.) | 9.0 + 1.0 = 10 mm for all screw lengths | | | | | | | | | | | | | | |
| a + c (min.) | 3.0 + 0.3 = 3.3 mm for all screw lengths | | | | | | | | | | | | | | |

### M48×3 (continued)

| | 340 | 360 | 380 | 400 | 420 | 440 | 460 | 480 |
|---|---|---|---|---|---|---|---|---|
| U/head (nom.) | 340 | 360 | 380 | 400 | 420 | 440 | 460 | 480 |
| U/head (max.) | 342.85 | 362.85 | 382.85 | 402.85 | 423.15 | 443.15 | 463.15 | 483.15 |
| U/head (min.) | 337.15 | 357.15 | 377.15 | 397.15 | 416.85 | 436.85 | 456.85 | 476.85 |
| a + c (max.) | 9.0 + 1.0 = 10 mm for all screw lengths | | | | | | | |
| a + c (min.) | 3.0 + 0.3 = 3.3 mm for all screw lengths | | | | | | | |

**2.13** *(continued)*

(Dimensions in millimetres)

**M56 × 4 — Grade B**

| Feature | | | | | | Popular lengths | | | | | | | |
|---|---|---|---|---|---|---|---|---|---|---|---|---|---|
| U/head (nom.) | 120 | 130 | 140 | 150 | 160 | 180 | 200 | 220 | 240 | 260 | 280 | 300 | 320 |
| U/head (max.) | 121.75 | 132 | 142 | 152 | 162 | 182 | 203.3 | 222.3 | 242.3 | 262.6 | 282.6 | 302.6 | 322.58 |
| U/head (min.) | 118.25 | 128 | 138 | 148 | 158 | 178 | 197.7 | 217.7 | 237.7 | 257.4 | 277.4 | 297.4 | 317.15 |

a + c (max.) = 12 + 1 = 13 mm for all screw lengths
a + c (min.) = 4.0 + 0.3 = 4.3 mm for all screw lengths

**M56 × 4 (continued)**

| Feature | | | | | Popular lengths | | | |
|---|---|---|---|---|---|---|---|---|
| U/head (nom.) | 340 | 360 | 380 | 400 | 420 | 440 | 460 | 480 | 500 |
| U/head (max.) | 342.85 | 362.85 | 382.85 | 402.85 | 423.15 | 443.15 | 463.15 | 483.15 | 503.15 |
| U/head (min.) | 337.15 | 357.15 | 377.15 | 397.15 | 416.85 | 436.85 | 456.85 | 476.85 | 496.85 |

a + c (max.) = 12 + 1 = 13 mm for all screw lengths
a + c (min.) = 4.0 + 0.3 = 4.3 mm for all screw lengths

**M64 × 4 — Grade B**

| Feature | | | | | | Popular lengths | | | | | | | |
|---|---|---|---|---|---|---|---|---|---|---|---|---|---|
| U/head (nom.) | 130 | 140 | 150 | 160 | 180 | 200 | 220 | 240 | 260 | 280 | 300 | 320 |
| U/head (max.) | 132 | 142 | 152 | 162 | 182 | 202.3 | 222.3 | 242.3 | 262.6 | 282.6 | 302.6 | 322.58 |
| U/head (min.) | 128 | 138 | 148 | 158 | 178 | 197.7 | 217.7 | 237.7 | 257.4 | 277.4 | 297.4 | 317.15 |

a + c (max.) = 12 + 1 = 13 mm for all screw lengths
a + c (min.) = 4.0 + 0.3 = 4.3 mm for all screw lengths

| U/head (nom.) | 340 | 360 | 380 | 400 | 420 | 440 | 460 | 480 | 500 |
|---|---|---|---|---|---|---|---|---|---|
| U/head (max.) | 342.85 | 362.85 | 382.85 | 402.85 | 423.15 | 443.15 | 463.15 | 483.15 | 503.15 |
| U/head (min.) | 337.15 | 357.15 | 377.15 | 397.15 | 416.85 | 436.85 | 456.85 | 476.85 | 496.85 |

a + c (max.) = 12 + 1 = 13 mm for all screw lengths
a + c (min.) = 4.0 + 0.3 = 4.3 mm for all screw lengths

For further information see BSEN 28676.

## 2.14 ISO metric tapping and clearance drills, fine thread series

| Nominal size | Tapping drill size (mm) | | Clearance drill size (mm) | | |
|---|---|---|---|---|---|
| | Recommended 80% engagement | Alternative 70% engagement | Close fit | Medium fit | Free fit |
| M6 | 5.20 | 5.30 | 6.4 | 6.6 | 7.0 |
| M8 | 7.00 | 7.10 | 8.4 | 9.0 | 10.0 |
| M10 | 8.80 | 8.90 | 10.5 | 11.0 | 12.0 |
| M12 | 10.80 | 10.90 | 13.0 | 14.0 | 15.0 |
| M14 | 12.50 | 12.70 | 15.0 | 16.0 | 17.0 |
| M16 | 14.50 | 14.75 | 17.0 | 18.0 | 19.0 |
| M18 | 16.50 | 16.75 | 19.0 | 20.0 | 21.0 |
| M20 | 18.50 | 18.75 | 21.0 | 22.0 | 24.0 |
| M22 | 20.50 | 20.75 | 23.0 | 24.0 | 26.0 |
| M24 | 22.00 | 22.25 | 25.0 | 26.0 | 28.0 |
| M27 | 25.00 | 25.25 | 28.0 | 30.0 | 32.0 |
| M30 | 28.00 | 28.25 | 31.0 | 33.0 | 35.0 |
| M33 | 31.00 | 31.25 | 34.0 | 36.0 | 38.0 |
| M36 | 33.00 | – | 37.0 | 39.0 | 42.0 |
| M39 | 36.00 | – | 40.0 | 42.0 | 45.0 |
| M42 | 39.00 | – | 43.0 | 45.0 | 48.0 |

## 2.15 ISO metric hexagon nuts (fine thread) style 1: product grade A and B (preferred sizes)

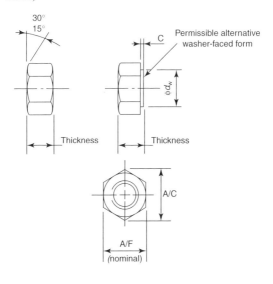

(Dimensions in millimetres)

| Designated thread size | Across corners A/C (min.) | Across flats (A/F) | | Thickness | | Washer-faced form (C) thickness | | |
|---|---|---|---|---|---|---|---|---|
| | | Maximum (nom.) | Minimum | Maximum | Minimum | Maximum | Minimum | $d_w$ (min.) |
| M8 × 1 | 14.38 | 13 | 12.73 | 7.5 | 7.14 | 0.6 | 0.15 | 11.63 |
| M10 × 1 | 17.77 | 16 | 15.73 | 9.3 | 8.94 | 0.6 | 0.15 | 14.63 |
| M12 × 1.5 | 20.03 | 18 | 17.73 | 12.0 | 11.57 | 0.6 | 0.15 | 16.63 |
| M16 × 1.5 | 26.75 | 24 | 23.67 | 16.4 | 15.7 | 0.8 | 0.2 | 22.49 |
| M20 × 1.5 | 32.95 | 30 | 29.16 | 20.3 | 19.0 | 0.8 | 0.2 | 27.70 |
| M24 × 2.0 | 39.55 | 36 | 35 | 23.9 | 22.6 | 0.8 | 0.2 | 33.25 |
| M30 × 2.0 | 50.85 | 46 | 45 | 28.6 | 27.3 | 0.8 | 0.2 | 42.75 |
| M36 × 3.0 | 60.79 | 55 | 53.8 | 34.7 | 33.1 | 0.8 | 0.2 | 51.11 |

For further information see BSEN 28673.

## 2.16 ISO metric hexagon thin nuts (chamfered) – fine thread – product grade A and B (also known as lock-nuts)

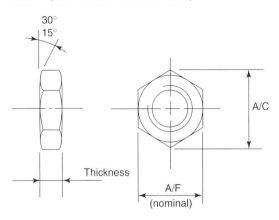

*Preferred sizes*                                              (Dimensions in millimetres)

| Designated thread size | Across corners A/C (min.) | Across flats (A/F) | | Thickness | |
|---|---|---|---|---|---|
| | | max. (nom.) | min. | max. | min. |
| M8 × 1 | 14.38 | 13 | 12.73 | 4.0 | 3.7 |
| M10 × 1 | 17.77 | 16 | 15.73 | 5.0 | 4.7 |
| M12 × 1.5 | 20.03 | 18 | 17.73 | 6.0 | 5.7 |
| M16 × 1.5 | 26.75 | 24 | 23.67 | 8.0 | 7.42 |
| M20 × 1.5 | 32.95 | 30 | 29.16 | 10.0 | 9.1 |
| M24 × 2 | 39.55 | 36 | 35.00 | 12.0 | 10.9 |
| M30 × 2 | 50.85 | 46 | 45.00 | 15.0 | 13.9 |
| M36 × 3 | 60.79 | 55 | 53.80 | 18.0 | 16.9 |
| M42 × 3 | 71.30 | 65 | 63.10 | 21.0 | 19.7 |
| M48 × 3 | 82.60 | 75 | 73.10 | 24.0 | 22.7 |
| M56 × 4 | 93.56 | 85 | 82.80 | 28.0 | 26.7 |
| M64 × 4 | 104.86 | 95 | 92.80 | 32.0 | 30.4 |

For further information see BSEN 28675.

## 2.17 ISO metric hexagon slotted nuts and castle nuts

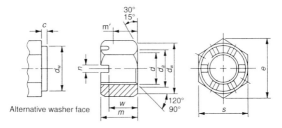

(a) Dimensions of slotted nuts

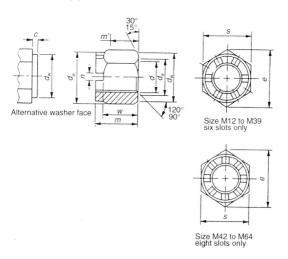

Size M12 to M39
six slots only

Size M42 to M64
eight slots only

(b) Dimensions of castle nuts

Dimensions of hexagon slotted nuts and castle nuts

(Dimensions in millimetres)

| $d$ | M4 | M5 | M6 | M8 | M10 | M12 | (M14) | M16 | (M18) | M20 | (M22) | M24 | (M27) | M30 | (M33) | M36 | (M39) | M42 | (M45) | M48 | (M52) | M56 | (M60) | M64 |
|---|---|---|---|---|---|---|---|---|---|---|---|---|---|---|---|---|---|---|---|---|---|---|---|---|
| $p$ | 0.7 | 0.8 | 1 | 1.25 | 1.5 | 1.75 | 2 | 2 | 2.5 | 2.5 | 2.5 | 3 | 3 | 3.5 | 3.5 | 4 | 4 | 4.5 | 4.5 | 5 | 5 | 5.5 | 5.5 | 6 |
| $c$ | 0.4 | 0.5 | 0.5 | 0.6 | 0.6 | 0.6 | 0.6 | 0.8 | 0.8 | 0.8 | 0.8 | 0.8 | 0.8 | 0.8 | 0.8 | 0.8 | 1 | 1 | 1 | 1 | 1 | 1 | 1 | 1 |
| $d_a$ max. | 4.6 | 5.75 | 6.75 | 8.75 | 10.8 | 13 | 15.1 | 17.3 | 19.5 | 21.6 | 23.7 | 25.9 | 29.1 | 32.4 | 35.6 | 38.9 | 42.1 | 45.4 | 48.6 | 51.8 | 56.2 | 60.5 | 64.8 | 69.1 |
| $d_a$ min. | 4 | 5 | 6 | 8 | 10 | 12 | 14 | 16 | 18 | 20 | 22 | 24 | 27 | 30 | 33 | 36 | 39 | 42 | 45 | 48 | 52 | 56 | 60 | 64 |
| $d_e$ max. | – | – | – | – | – | 16 | 19 | 22 | 25 | 28 | 30 | 34 | 38 | 42 | 46 | 49 | 55 | 58 | 62 | 65 | 70 | 75 | 80 | 85 |
| $d_e$ min. | – | – | – | – | – | 15.57 | 18.48 | 21.48 | 24.8 | 27.16 | 29.16 | 33.2 | 37 | 41 | 45 | 49 | 53.8 | 56.8 | 60.8 | 63.8 | 68.8 | 73.8 | 78.8 | 83.8 |
| $d_w$ min. | 5.9 | 6.9 | 8.9 | 11.6 | 14.6 | 16.6 | 19.6 | 22.5 | 24.8 | 27.7 | 31.4 | 33.2 | 38 | 42.7 | 46.6 | 51.1 | 55.9 | 59.9 | 64.7 | 69.4 | 74.2 | 78.7 | 83.4 | 88.2 |
| $e$ min. | 7.66 | 8.79 | 11.05 | 14.38 | 17.77 | 20.03 | 23.36 | 26.75 | 29.58 | 32.95 | 37.29 | 39.55 | 45.2 | 50.85 | 55.37 | 60.79 | 66.44 | 71.3 | 76.95 | 82.6 | 88.25 | 93.56 | 99.21 | 104.86 |
| $m$ max. | 5 | 6 | 7.5 | 9.5 | 12 | 15 | 16 | 19 | 21 | 22 | 26 | 27 | 30 | 33 | 35 | 38 | 40 | 46 | 48 | 50 | 54 | 57 | 63 | 66 |
| $m$ min. | 4.7 | 5.7 | 7.14 | 9.14 | 11.57 | 14.57 | 15.57 | 18.48 | 20.48 | 21.48 | 25.48 | 25.48 | 29.48 | 32.38 | 34.38 | 37.38 | 39.38 | 45.38 | 47.38 | 49.38 | 53.26 | 56.26 | 62.26 | 65.26 |
| $m'$ min. | 2.32 | 3.52 | 3.92 | 5.15 | 6.43 | 8.3 | 9.68 | 11.28 | 12.08 | 13.52 | 14.48 | 16.16 | 18 | 19.44 | 21.92 | 23.52 | 25.44 | 25.92 | 27.52 | 29.12 | 32.32 | 34.72 | 37.12 | 39.3 |
| $n$ min. | 1.2 | 1.4 | 2 | 2.5 | 2.8 | 3.5 | 3.5 | 4.5 | 4.5 | 5.5 | 5.5 | 5.5 | 5.5 | 7 | 7 | 7 | 7.36 | 9 | 9 | 9 | 9 | 11 | 11 | 11 |
| $n$ max. | 1.45 | 1.65 | 2.25 | 2.75 | 3.05 | 3.8 | 3.8 | 4.8 | 4.8 | 5.8 | 5.8 | 5.8 | 5.8 | 7.36 | 7.36 | 7.36 | 7.36 | 9.36 | 9.36 | 9.36 | 9.36 | 11.43 | 11.43 | 11.43 |
| $s$ max. | 7 | 8 | 10 | 13 | 16 | 18 | 21 | 24 | 27 | 30 | 34 | 36 | 41 | 46 | 50 | 55 | 60 | 65 | 70 | 75 | 80 | 85 | 90 | 95 |
| $s$ min. | 6.78 | 7.78 | 9.78 | 12.73 | 15.73 | 17.73 | 20.67 | 23.67 | 26.16 | 29.16 | 33 | 35 | 40 | 45 | 49 | 53.8 | 58.8 | 63.1 | 68.1 | 73.1 | 78.1 | 82.8 | 87.8 | 92.8 |
| $w$ max. | 2.9 | 4 | 5 | 6.5 | 8 | 10 | 11 | 13 | 15 | 16 | 18 | 19 | 22 | 24 | 26 | 29 | 31 | 34 | 36 | 38 | 42 | 45 | 48 | 51 |
| $w$ min. | 3.2 | 3.7 | 4.7 | 6.14 | 7.64 | 9.64 | 10.57 | 12.59 | 14.57 | 15.57 | 17.57 | 18.48 | 21.48 | 23.48 | 25.48 | 28.48 | 30.38 | 33.38 | 35.38 | 37.38 | 41.38 | 44.38 | 47.38 | 50.26 |

*Note* 1. Non-preferred sizes are shown in brackets.
*Note* 2. Castle nuts shall not be specified below M12.
*Note* 3. Castle nuts above M39 shall have eight slots.

For further information see BS 7764.

## 2.18 Marking threaded fasteners

### 2.18.1 Symbols

Marking symbols are shown in Table (a) below.

### 2.18.2 Identification

#### (a) Hexagon bolts and screws

Hexagon bolts and screws shall be marked with the designation symbol of the property class described in clause 3 of BSEN 20898-1.

The marking is obligatory for all property classes, preferably on the top of the head by indenting or embossing or on the side of the head by indenting (see Fig. (a) below).

Marking is required for hexagon bolts and screws with nominal diameters $d \geq 5$ mm where the shape of the product allows it, preferably on the head.

#### (b) Hexagon socket head cap screws

Hexagon socket head cap screws shall be marked with the designation symbol of the property class described in clause 3 of BSEN 20898-1

The marking is obligatory for property classes equal to or higher than 8.8, preferably on the side of the head by indenting or on the top of the head by indenting or embossing (see Fig. (b) below).

Marking is required for hexagon socket head cap screws with nominal diameters $d \geq 5$ mm where the shape of the product allows it, preferably on the head.

The clock-face marking system as given for nuts in ISO 898-2 may be used as an alternative method on small hexagon socket head cap screws.

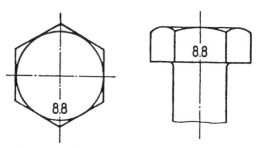

(a) Examples of marking on hexagon bolts and screws

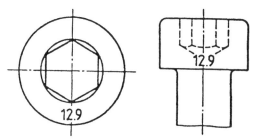

**(b)** Examples of marking on hexagon socket head cap screws

### Table (a) Marking symbols

| Property class | 3.6 | 4.6 | 4.8 | 5.6 | 5.8 | 6.8 | 8.8 | 9.8 | 10.9 | 12.9 |
|---|---|---|---|---|---|---|---|---|---|---|
| Marking symbol | 3.6 | 4.6 | 4.8 | 5.6 | 5.8 | 6.8 | 8.8 | 9.8 | 10.9 | 12.9 |

The full-stop in the marking symbol may be omitted.

### Table (b) Identification marks for studs

| Property class | 8.8 | 9.8 | 10.9 | 12.9 |
|---|---|---|---|---|
| Identification mark | ○ | + | □ | △ |

### (c) Studs

Studs shall be marked with the designation symbol of the property class described in clause 3 of BSEN 20898-1.

The marking is obligatory for property classes equal to or higher than 8.8, preferably on the extreme end of the threaded portion by indenting (see Fig. (c) below). For studs with interference fit, the marking shall be at the nut end.

Marking is required for studs with nominal diameters equal to or greater than 5 mm.

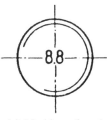

**(c)** Marking of stud

The symbols in Table (b) above are permissible as an alternative identification method.

#### (d) Other types of bolts and screws

The same marking system as described in Sections 2.18.2 (a) and (b) shall be used for other types of bolts and screws of property classes 4.6, 5.6 and all classes equal to or higher than 8.8, as described in the appropriate International Standards or, for special components, as agreed between the interested parties.

### 2.18.3 Marking of left-hand thread

Bolts and screws with left-hand thread shall be marked with the symbol shown in Fig. (d) below either on the top of the head or the point.

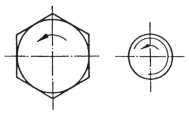

**(d)** Left-hand thread marking

Marking is required for bolts and screws with nominal thread diameters $d \geq 5$ mm.

Alternative marking for left-hand thread may be used for hexagon bolts and screws as shown in Fig. (e) below.

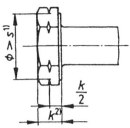

1) *s* is the width across flats.
2) *k* is the height of the head.

(**e**) Alternative left-hand thread marking

## 2.18.4 Alternative marking

Alternative or optional permitted marking as stated in Sections 2.18.1 to 2.18.3 should be left to the choice of the manufacturer.

## 2.18.5 Trade (identification) marking

The trade (identification) marking of the manufacturer is mandatory on all products which are marked with property classes.

For full information on the marking of threaded fasteners see BSEN 20898-1.

# 2.19 ISO metric hexagon socket head screws

## Cap head screws

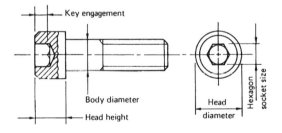

| Nominal size 1st choice | Body diameter and head height | | Head diameter | | Hexagon socket size | Key engagement |
|---|---|---|---|---|---|---|
| | max | min | max | min | | min |
| M3 | 3.00 | 2.86 | 5.50 | 5.20 | 2.50 | 1.30 |
| M4 | 4.00 | 3.82 | 7.00 | 6.64 | 3.00 | 2.00 |
| M5 | 5.00 | 4.82 | 8.50 | 8.14 | 4.00 | 2.70 |
| M6 | 6.00 | 5.82 | 10.00 | 9.64 | 5.00 | 3.30 |
| M8 | 8.00 | 7.78 | 13.00 | 12.57 | 6.00 | 4.30 |
| M10 | 10.00 | 9.78 | 16.00 | 15.57 | 8.00 | 5.50 |
| M12 | 12.00 | 11.73 | 18.00 | 17.57 | 10.00 | 6.60 |
| M16 | 16.00 | 15.73 | 24.00 | 23.48 | 14.00 | 8.80 |
| M20 | 20.00 | 19.67 | 30.00 | 29.48 | 17.00 | 10.70 |
| M24 | 24.00 | 23.67 | 36.00 | 35.38 | 19.00 | 12.90 |

## 90° countersunk head screws

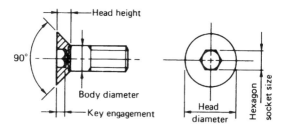

(Dimensions in millimetres)

| Nominal size 1st choice | Body diameter | | Head diameter | | Head height | Hexagon socket size | Key engagement |
|---|---|---|---|---|---|---|---|
| | max | min | max | min | | | min |
| M3 | 3.00 | 2.86 | 6.00 | 5.82 | 1.86 | 2.00 | 1.05 |
| M4 | 4.00 | 3.82 | 8.00 | 7.78 | 2.48 | 2.50 | 1.49 |
| M5 | 5.00 | 4.82 | 10.00 | 9.78 | 3.10 | 3.00 | 1.86 |
| M6 | 6.00 | 5.82 | 12.00 | 11.73 | 3.72 | 4.00 | 2.16 |
| M8 | 8.00 | 7.78 | 16.00 | 15.73 | 4.96 | 5.00 | 2.85 |
| M10 | 10.00 | 9.78 | 20.00 | 19.67 | 6.20 | 6.00 | 3.60 |
| M12 | 12.00 | 11.73 | 24.00 | 23.67 | 7.44 | 8.00 | 4.35 |
| M16 | 16.00 | 15.73 | 32.00 | 29.67 | 8.80 | 10.00 | 4.89 |
| M20 | 20.00 | 19.67 | 40.00 | 35.61 | 10.16 | 12.00 | 5.45 |

For full range and further information see BS 4168 (metric) and BS 2470 (inch).

## 2.20 ISO metric screw threads, miniature series

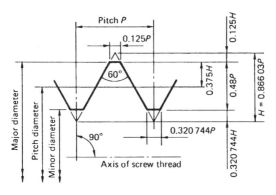

(Dimensions in millimetres)

| Nominal size | | Pitch of thread P | Major diameter | Pitch (effective) diameter | Minor diameter |
|---|---|---|---|---|---|
| 1st choice | 2nd choice | | | | |
| S-0.3 | | 0.080 | 0.300 000 | 0.248 038 | 0.223 200 |
| | S-0.35 | 0.090 | 0.350 000 | 0.291 543 | 0.263 600 |
| S-0.4 | | 0.100 | 0.400 000 | 0.335 048 | 0.304 000 |
| | S-0.45 | 0.100 | 0.450 000 | 0.385 048 | 0.354 000 |
| S-0.5 | | 0.125 | 0.500 000 | 0.418 810 | 0.380 000 |
| | S-0.55 | 0.125 | 0.550 000 | 0.468 810 | 0.430 000 |
| S-0.6 | | 0.150 | 0.600 000 | 0.502 572 | 0.456 000 |
| | S-0.7 | 0.175 | 0.700 000 | 0.586 334 | 0.532 000 |
| S-0.8 | | 0.200 | 0.800 000 | 0.670 096 | 0.608 000 |
| | S-0.9 | 0.225 | 0.900 000 | 0.753 858 | 0.684 000 |
| | S-1 | 0.250 | 1.000 000 | 0.837 620 | 0.760 000 |
| | S-1.1 | 0.250 | 1.100 000 | 0.937 620 | 0.860 000 |
| | S-1.2 | 0.250 | 1.200 000 | 1.037 620 | 0.960 000 |
| | S-1.4 | 0.300 | 1.400 000 | 1.205 144 | 1.112 000 |

For full range and further information see BS 4827.

## 2.21 ISO metric tapping and clearance drills, miniature series

| Nominal size | | Pitch | | Tapping drill size | | Clearance drill size | |
|---|---|---|---|---|---|---|---|
| ISO | ASA B1.10 | | | | Number or | | Number or |
| mm | mm | mm | Threads per inch | mm | fraction | mm | fraction |
| S-0.3 | 0.30 unm | 0.080 | 318 | 0.25 | — | 0.32 | — |
| (S-0.35) | (0.35 unm) | 0.090 | 282 | 0.28 | — | 0.38 | 79 |
| S-0.4 | 0.40 unm | 0.100 | 254 | 0.35 | 80 | 0.45 | 77 |
| (S-0.45) | (0.45 unm) | 0.100 | 254 | 0.38 | 79 | 0.50 | 76 |
| S-0.5 | 0.50 unm | 0.125 | 203 | 0.42 | 78 | 0.55 | 75.74 |
| (S-0.55) | (0.55 unm) | 0.125 | 203 | 0.45 | 77 | 0.60 | 73 |
| S-0.6 | 0.60 unm | 0.150 | 169 | 0.50 | 76 | 0.65 | 72 |
| (S-0.7) | (0.70 unm) | 0.175 | 145 | 0.58 | 74 | 0.78 | $1/32$ in |
| S-0.8 | 0.80 unm | 0.200 | 127 | 0.65 | 72 | 0.88 | 66, 65 |
| (S-0.9) | (0.90 unm) | 0.225 | 113 | 0.72 | 70 | 0.98 | 62 |
| S-1.0 | 1.00 unm | 0.250 | 102 | 0.80 | $1/32$ in | 1.10 | 57 |
| (S-1.1) | (1.10 unm) | 0.250 | 102 | 0.90 | 65 | 1.20 | $3/64$ in |
| S-1.2 | 1.20 unm | 0.250 | 102 | 1.00 | 61 | 1.30 | 55 |
| (S-1.4) | (1.40 unm) | 0.300 | 85 | 1.15 | $3/64$ in | 1.50 | 53 |

## 2.22 ISO metric screw threads: constant pitch series

(Dimensions in millimetres)

| Pitch of thread | Basic major diameter | | | Pitch (effective) diameter | Basic minor diameter | |
|---|---|---|---|---|---|---|
| | 1st choice | 2nd choice | 3rd choice | | External | Internal |
| 0.25 | 2.0 | — | — | 1.84 | 1.69 | 1.73 |
| 0.25 | — | 2.2 | — | 2.04 | 1.89 | 1.93 |
| 0.35 | 2.5 | — | — | 2.27 | 2.07 | 2.12 |
| 0.35 | 3.0 | — | — | 2.77 | 2.57 | 2.62 |
| 0.35 | — | — | 3.5 | 3.27 | 3.07 | 3.12 |
| 0.50 | 4.0 | — | — | 3.68 | 3.39 | 3.46 |
| 0.50 | — | 4.5 | — | 4.18 | 3.86 | 3.96 |
| 0.50 | 5.0 | — | — | 4.68 | 4.39 | 4.46 |
| 0.50 | — | — | 5.5 | 5.18 | 4.86 | 4.96 |
| 0.75 | 6.0 | — | — | 5.51 | 5.08 | 5.19 |
| 0.75 | — | — | 7.0 | 6.51 | 6.08 | 6.19 |
| 0.75 | 8.0 | — | — | 7.51 | 7.08 | 7.19 |
| 0.75 | — | — | 9.0 | 8.51 | 8.08 | 8.19 |
| 0.75 | 10.0 | — | — | 9.51 | 9.08 | 9.19 |
| 0.75 | — | — | 11.0 | 10.51 | 10.08 | 10.19 |
| 1.0 | 8.0 | — | — | 7.35 | 6.77 | 6.92 |
| 1.0 | — | — | 9.0 | 8.35 | 7.77 | 7.92 |
| 1.0 | 10.0 | — | — | 9.35 | 8.77 | 8.92 |
| 1.0 | — | — | 11.0 | 10.35 | 9.77 | 9.92 |
| 1.0 | 12.0 | — | — | 11.35 | 10.77 | 10.92 |
| 1.0 | — | 14.0 | — | 13.35 | 12.77 | 12.92 |

| | | | | | | |
|---|---|---|---|---|---|---|
| 1.0 | — | — | 15.0 | 14.35 | 13.77 | 13.92 |
| 1.0 | 16.0 | — | — | 15.35 | 14.77 | 14.92 |
| 1.0 | — | — | 17.0 | 16.35 | 15.77 | 15.92 |
| 1.0 | — | 18.0 | — | 17.35 | 16.77 | 16.92 |
| 1.0 | 20.0 | — | — | 19.35 | 18.77 | 18.92 |
| 1.0 | — | 22.0 | — | 21.35 | 21.77 | 21.92 |
| 1.0 | 24.0 | — | — | 23.35 | 22.77 | 22.92 |
| 1.0 | — | — | 25.0 | 24.35 | 23.77 | 23.92 |
| 1.0 | — | 27.0 | — | 26.35 | 25.77 | 25.92 |
| 1.0 | — | — | 28.0 | 27.35 | 26.77 | 26.92 |
| 1.0 | 30.0 | — | — | 29.35 | 28.77 | 28.92 |
| | | | | | | |
| 1.25 | 10.0 | — | — | 9.19 | 8.47 | 8.65 |
| 1.25 | 12.0 | — | — | 11.19 | 10.47 | 10.65 |
| 1.25* | — | 14.0* | — | 13.19 | 12.47 | 12.65 |
| | | | | | | |
| 1.5 | 12.0 | — | — | 11.03 | 10.16 | 10.38 |
| 1.5 | — | 14.0 | — | 13.03 | 12.16 | 12.38 |
| 1.5 | — | — | 15.0 | 14.03 | 13.16 | 13.38 |
| 1.5 | 16.0 | — | — | 15.03 | 14.16 | 14.38 |
| 1.5 | — | — | 17.0 | 16.03 | 15.16 | 15.38 |
| 1.5 | — | 18.0 | — | 17.03 | 16.16 | 16.38 |
| 1.5 | 20.0 | — | — | 19.03 | 18.16 | 18.38 |
| 1.5 | — | 22.0 | — | 21.03 | 20.16 | 20.38 |
| 1.5 | 24.0 | — | — | 23.03 | 22.16 | 22.38 |
| 1.5 | — | — | 25.0 | 24.03 | 23.16 | 23.38 |
| 1.5 | — | — | 26.0 | 25.03 | 24.16 | 24.38 |
| 1.5 | — | 27.0 | — | 26.03 | 25.16 | 25.38 |
| 1.5 | — | — | 28.0 | 27.03 | 26.16 | 26.38 |
| 1.5 | 30.0 | — | — | 29.03 | 28.16 | 28.38 |
| 1.5 | — | — | 32.0 | 31.03 | 30.16 | 30.38 |
| 1.5 | — | 33.0 | — | 32.03 | 31.16 | 31.38 |
| 1.5 | — | — | 35.0 | 34.03 | 33.16 | 33.38 |

The 1.5 mm pitch series continues to a maximum diameter of 80 mm.

| | | | | | | |
|---|---|---|---|---|---|---|
| 2.0 | — | 18.0 | — | 16.70 | 15.55 | 15.84 |
| 2.0 | 20.0 | — | — | 18.70 | 17.55 | 17.84 |
| 2.0 | — | 22.0 | — | 20.70 | 19.55 | 19.84 |
| 2.0 | 24.0 | — | — | 22.70 | 21.55 | 21.84 |
| 2.0 | — | — | 25.0 | 23.70 | 22.55 | 22.84 |
| 2.0 | — | — | 26.0 | 24.70 | 23.55 | 23.84 |
| 2.0 | — | 27.0 | — | 25.70 | 24.55 | 24.84 |
| 2.0 | — | — | 28.0 | 26.70 | 25.55 | 25.84 |
| 2.0 | 30.0 | — | — | 28.70 | 27.55 | 27.84 |
| 2.0 | — | — | 32.0 | 30.70 | 29.55 | 29.84 |
| 2.0 | — | 33.0 | — | 31.70 | 30.55 | 30.84 |
| 2.0 | — | — | 35.0 | 33.70 | 32.55 | 32.84 |

The 2.0 mm pitch series continues to a maximum diameter of 150 mm.

| | | | | | | |
|---|---|---|---|---|---|---|
| 3.0 | 30.0 | — | — | 28.05 | 26.32 | 26.75 |
| 3.0 | — | 33.0 | — | 31.05 | 29.32 | 29.75 |
| 3.0 | 36.0 | — | — | 34.05 | 32.32 | 32.75 |
| 3.0 | — | — | 38.0 | 36.05 | 34.32 | 34.75 |
| 3.0 | — | 39.0 | — | 37.05 | 35.32 | 35.75 |
| 3.0 | — | — | 40.0 | 38.05 | 36.32 | 36.75 |
| 3.0 | 42.0 | — | — | 40.05 | 38.32 | 38.75 |
| 3.0 | — | 45.0 | — | 43.05 | 41.32 | 41.75 |
| 3.0 | 48.0 | — | — | 46.05 | 44.32 | 44.75 |
| 3.0 | — | — | 50.0 | 48.05 | 46.32 | 46.75 |
| 3.0 | — | 52.0 | — | 50.05 | 48.32 | 48.75 |
| 3.0 | — | — | 55.0 | 53.05 | 51.32 | 51.75 |

The 3.0 mm pitch series continues to a maximum diameter of 250 mm.

*(continued)*

**2.22** (continued)                    (Dimensions in millimetres)

| Pitch of thread | Basic major diameter | | | Pitch (effective) diameter | Basic minor diameter | |
|---|---|---|---|---|---|---|
| | 1st choice | 2nd choice | 3rd choice | | External | Internal |
| 4.0 | 42.0 | – | – | 39.40 | 37.09 | 37.67 |
| 4.0 | – | 45.0 | – | 42.40 | 40.09 | 40.67 |
| 4.0 | 48.0 | – | – | 45.40 | 43.09 | 43.67 |
| 4.0 | – | – | 50.0 | 47.40 | 45.09 | 45.67 |
| 4.0 | – | 52.0 | – | 49.40 | 47.09 | 47.67 |
| 4.0 | – | – | 55.0 | 52.40 | 50.09 | 50.67 |
| 4.0 | 56.0 | – | – | 53.40 | 51.09 | 51.67 |
| 4.0 | – | – | 58.0 | 55.40 | 53.09 | 53.67 |
| 4.0 | – | 60.0 | – | 57.40 | 55.09 | 55.67 |
| 4.0 | – | – | 62.0 | 59.40 | 57.09 | 57.67 |
| 4.0 | 64.0 | – | – | 61.40 | 59.09 | 59.67 |
| 4.0 | – | – | 65.0 | 62.40 | 60.09 | 60.67 |

The 4.0 mm pitch series continues to a maximum diameter of 300 mm.

| | | | | | | |
|---|---|---|---|---|---|---|
| 6.0 | – | – | 70.0 | 66.10 | 62.64 | 63.50 |
| 6.0 | 72.0 | – | – | 68.10 | 64.64 | 65.50 |
| 6.0 | – | 76.0 | – | 72.10 | 68.64 | 69.50 |
| 6.0 | 80.0 | – | – | 76.10 | 72.64 | 73.50 |
| 6.0 | – | 85.0 | – | 81.10 | 77.64 | 78.50 |
| 6.0 | 90.0 | – | – | 86.10 | 82.64 | 83.50 |
| 6.0 | – | 95.0 | – | 91.10 | 87.64 | 88.50 |
| 6.0 | 100.0 | – | – | 96.10 | 92.64 | 93.50 |
| 6.0 | – | 105.0 | – | 101.10 | 97.64 | 98.50 |
| 6.0 | 110.0 | – | – | 106.10 | 102.64 | 103.50 |
| 6.0 | – | 115.0 | – | 111.10 | 107.64 | 108.50 |
| 6.0 | – | 120.0 | – | 116.10 | 112.64 | 113.50 |
| 6.0 | 125.0 | – | – | 121.10 | 117.64 | 118.50 |

The 6.0 mm pitch series continues to a maximum diameter of 300 mm.

* This size sparking plugs only.
For further information see BS 3643.

## 2.23 ISO pipe thread forms

### 2.23.1 Basic Whitworth thread form: parallel threads

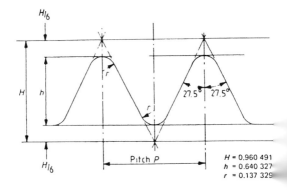

$$H = 0.960\ 491$$
$$h = 0.640\ 327$$
$$r = 0.137\ 329$$

## 2.23.2 Basic Whitworth thread form: taper threads

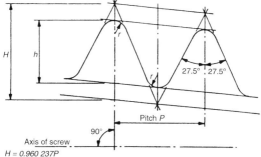

$H = 0.960\ 237P$
$h = 0.640\ 327P$
$r = 0.137\ 278P$

*Note.* The taper is 1 in 16 measured on the diameter (shown exaggerated in the diagram).

## 2.23.3 Terms relating to taper pipe threads

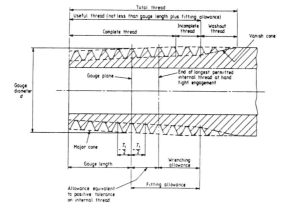

**2.23.4 ISO pipe threads, parallel: basic sizes**

| Nominal (bore) size of pipe* | | Number of threads | Pitch of thread | Depth of thread | | Major diameter | | Pitch (effective) diameter | Minor diameter | Minimum length of thread on pipe end |
|---|---|---|---|---|---|---|---|---|---|---|
| in | mm | per inch | mm | in | mm | in | mm | mm | mm | mm |
| 1/16† | 3 | 28 | 0.907 | 0.0230 | 0.581 | 0.304 | 7.723 | 7.142 | 6.561 | 4.9 |
| 1/8† | 6 | 28 | 0.907 | 0.0230 | 0.581 | 0.383 | 9.728 | 9.147 | 8.566 | 6.5 |
| 1/4 | 8 | 19 | 1.337 | 0.0335 | 0.856 | 0.518 | 13.157 | 12.301 | 11.445 | 9.7 |
| 3/8 | 10 | 19 | 1.337 | 0.0335 | 0.856 | 0.656 | 16.662 | 15.806 | 14.950 | 10.1 |
| 1/2 | 15 | 14 | 1.814 | 0.0455 | 1.162 | 0.805 | 20.455 | 19.793 | 18.631 | 13.2 |
| 5/8 | – | 14 | 1.814 | 0.0455 | 1.162 | 0.902 | 22.911 | 21.749 | 20.587 | 13.9 |
| 3/4 | 20 | 14 | 1.814 | 0.0455 | 1.162 | 1.041 | 26.441 | 25.279 | 24.117 | 14.5 |
| 7/8 | – | 14 | 1.814 | 0.0455 | 1.162 | 1.189 | 30.201 | 29.039 | 27.877 | 15.7 |
| 1 | 25 | 11 | 2.309 | 0.0580 | 1.479 | 1.309 | 33.249 | 31.770 | 30.291 | 16.8 |
| 1 1/8 | – | 11 | 2.309 | 0.0580 | 1.479 | 1.492 | 37.897 | 36.418 | 34.939 | 18.0 |
| 1 1/4 | 32 | 11 | 2.309 | 0.0580 | 1.479 | 1.650 | 41.910 | 40.431 | 38.952 | 19.1 |

| | | | | | | | | | | |
|---|---|---|---|---|---|---|---|---|---|---|
| 1½ | 40 | 11 | 2.309 | 0.0580 | 1.479 | 1.882 | 47.803 | 46.324 | 44.845 | 19.1 |
| 1¾ | – | 11 | 2.309 | 0.0580 | 1.479 | 2.116 | 53.746 | 52.267 | 50.788 | 21.3 |
| 2 | 50 | 11 | 2.309 | 0.0580 | 1.479 | 2.347 | 59.614 | 58.135 | 56.656 | 23.4 |
| 2¼ | – | 11 | 2.309 | 0.0580 | 1.479 | 2.587 | 65.710 | 64.231 | 62.752 | 25.0 |
| 2½ | 65 | 11 | 2.309 | 0.0580 | 1.479 | 2.960 | 75.184 | 73.705 | 72.226 | 26.7 |
| 2¾ | – | 11 | 2.309 | 0.0580 | 1.479 | 3.210 | 81.534 | 80.055 | 78.576 | 28.3 |
| 3 | 80 | 11 | 2.309 | 0.0580 | 1.479 | 3.460 | 87.884 | 84.405 | 84.926 | 29.8 |
| 3½ | 90 | 11 | 2.309 | 0.0580 | 1.479 | 3.950 | 100.330 | 98.851 | 97.372 | 31.4 |
| 4 | 100 | 11 | 2.309 | 0.0580 | 1.479 | 4.450 | 113.030 | 141.551 | 110.072 | 35.8 |
| 4½ | – | 11 | 2.309 | 0.0580 | 1.479 | 4.950 | 125.730 | 124.251 | 122.772 | 35.8 |
| 5 | 125 | 11 | 2.309 | 0.0580 | 1.479 | 5.450 | 138.430 | 136.951 | 135.472 | 40.1 |
| 5½ | – | 11 | 2.309 | 0.0580 | 1.479 | 5.950 | 151.130 | 149.651 | 148.172 | 40.1 |
| 6 | 150 | 11 | 2.309 | 0.0580 | 1.479 | 6.450 | 163.830 | 162.351 | 160.872 | 40.1 |

*These are nominal pipe size equivalents and are *not* inch/metric conversions. For example, for all practical purposes a pipe of 8 mm nominal bore is the same size as ¼ in nominal bore. The actual bore will lie between these nominal sizes and the O/D of this nominal size of pipe will be approximately 14 mm.

†These sizes are no longer recommended.

ISO pipe threads (parallel and tapered) are based upon the previous British Standard pipe (BSP) threads and retain the Whitworth (55°) thread form. For further information see BS 2779.

**2.23.5 ISO pipe threads, tapered: basic sizes**

| Nominal (bore) Size of pipe* | | Number of threads | Pitch of thread | Depth of thread | Basic diameters at gauge plane | | |
|---|---|---|---|---|---|---|---|
| in | mm | per inch | mm | mm | Major (gauge) diameter mm | Pitch (effective) diameter mm | Minor diameter mm |
| 1/8 | 6 | 28 | 0.907 | 0.581 | 9.728 | 9.147 | 8.566 |
| 1/4 | 8 | 19 | 1.337 | 0.856 | 13.157 | 12.301 | 11.445 |
| 3/8 | 10 | 19 | 1.337 | 0.856 | 16.662 | 15.806 | 14.950 |
| 1/2 | 15 | 14 | 1.814 | 1.162 | 20.955 | 19.793 | 18.631 |
| 3/4 | 20 | 14 | 1.814 | 1.162 | 26.441 | 25.279 | 24.117 |
| 1 | 25 | 11 | 2.309 | 1.479 | 33.249 | 31.770 | 30.291 |
| 1 1/4 | 32 | 11 | 2.309 | 1.479 | 41.910 | 40.431 | 38.952 |
| 1 1/2 | 40 | 11 | 2.309 | 1.479 | 47.803 | 46.324 | 44.845 |
| 2 | 50 | 11 | 2.309 | 1.479 | 59.614 | 58.135 | 56.656 |
| 2 1/2 | 65 | 11 | 2.309 | 1.479 | 75.184 | 73.705 | 72.226 |
| 3 | 80 | 11 | 2.309 | 1.479 | 87.884 | 86.405 | 84.926 |
| 4 | 100 | 11 | 2.309 | 1.479 | 113.030 | 111.551 | 110.072 |
| 5 | 125 | 11 | 2.309 | 1.479 | 138.430 | 136.951 | 135.472 |
| | | | 2.309 | 1.479 | 163.830 | 162.351 | 160.872 |

(continued)

| Nominal (bore) size of pipe* | | Gauge length† | | | | Useful thread (min.) | | | Fitting allowance | Wrenching allowance | Position of gauge plane tolerance‡ ± | Diametral tolerance§ ± |
|---|---|---|---|---|---|---|---|---|---|---|---|---|
| in | mm | Basic | Tolerance ± | max. | min. | Basic | max. | min. | | | | |
| 1/8 | 6 | 4 3/8 (4.0) | 1 (0.9) | 5 3/8 (4.9) | 3 3/8 (3.1) | 7 1/8 (6.5) | 8 1/8 (7.4) | 6 1/8 (5.6) | 2 3/4 (2.5) | 1 1/2 (1.4) | 1 1/4 (1.1) | 0.071 |
| 1/4 | 8 | 4 1/2 (5.0) | 1 (1.3) | 5 1/2 (7.3) | 3 1/2 (4.7) | 7 1/4 (9.7) | 8 1/4 (11.0) | 6 1/4 (8.4) | 2 3/4 (3.7) | 1 1/2 (2.0) | 1 1/4 (1.7) | 0.104 |
| 3/8 | 10 | 4 3/4 (6.4) | 1 (1.3) | 5 3/4 (7.7) | 3 3/4 (5.1) | 7 1/2 (10.1) | 8 1/2 (11.4) | 6 1/2 (8.8) | 2 3/4 (3.7) | 1 1/2 (2.0) | 1 1/4 (1.7) | 0.104 |
| 1/2 | 15 | 4 1/2 (8.2) | 1 (1.8) | 5 1/2 (10.0) | 3 1/2 (6.4) | 7 1/4 (13.2) | 8 1/4 (15.0) | 6 1/4 (11.4) | 2 3/4 (5.0) | 1 1/2 (2.7) | 1 1/4 (2.3) | 0.142 |
| 3/4 | 20 | 5 1/4 (9.5) | 1 (1.8) | 6 1/4 (11.3) | 4 1/4 (7.7) | 8 (14.5) | 9 (16.3) | 7 (12.7) | 2 3/4 (5.0) | 1 1/2 (2.7) | 1 1/4 (2.3) | 0.142 |
| 1 | 25 | 4 1/2 (10.4) | 1 (2.3) | 5 1/2 (12.7) | 3 1/2 (8.1) | 7 1/4 (16.8) | 8 3/4 (19.1) | 6 1/4 (14.5) | 2 3/4 (6.4) | 1 1/2 (3.5) | 1 1/4 (2.9) | 0.180 |
| 1 1/4 | 32 | 5 1/2 (12.7) | 1 (2.3) | 6 1/2 (15.0) | 4 1/2 (10.4) | 8 1/4 (19.1) | 9 1/4 (21.4) | 7 1/4 (16.8) | 2 3/4 (6.4) | 1 1/2 (3.5) | 1 1/4 (2.9) | 0.180 |
| 1 1/2 | 40 | 5 1/2 (12.7) | 1 (2.3) | 6 1/2 (15.0) | 4 1/2 (10.4) | 8 1/4 (19.1) | 9 1/4 (21.4) | 7 1/4 (16.8) | 2 3/4 (6.4) | 1 1/2 (3.5) | 1 1/4 (2.9) | 0.180 |

(continued)

**2.23.5** (continued)

| Nominal (bore) size of pipe* | | Gauge length† | | | | Useful thread (min.) | | | Fitting allowance | Wrenching allowance | Position of gauge plane tolerance‡ ± | Diametral tolerance§ ± |
|---|---|---|---|---|---|---|---|---|---|---|---|---|
| in | mm | Basic | Tolerance ± | max. | min. | Basic | max. | min. | | | | |
| 2 | 50 | 6 7/8 (15.9) | 1 (2.3) | 7 7/8 (18.2) | 5 7/8 (15.6) | 10 1/8 (23.4) | 11 1/8 (25.7) | 9 1/8 (21.1) | 3 1/4 (7.5) | 2 (4.6) | 1 1/4 (2.9) | 0.180 |
| 2 1/2 | 65 | 7 9/16 (17.5) | 1 1/2 (3.5) | 9 1/16 (21.0) | 6 1/16 (14.0) | 11 9/16 (26.7) | 13 1/16 (30.2) | 10 1/16 (23.2) | 4 (9.2) | 2 1/2 (5.8) | 1 1/2 (3.5) | 0.216 |
| 3 | 80 | 8 15/16 (20.6) | 1 1/2 (3.5) | 10 7/16 (24.1) | 7 7/16 (17.1) | 12 15/16 (29.8) | 14 7/16 (33.3) | 11 7/16 (26.3) | 4 (9.2) | 2 1/2 (5.8) | 1 1/2 (3.5) | 0.216 |
| 4 | 100 | 11 (25.4) | 1 1/2 (3.5) | 12 1/2 (28.9) | 9 1/2 (21.9) | 15 1/2 (35.8) | 17 (19.3) | 14 (32.3) | 4 1/2 (10.4) | 3 (6.9) | 1 1/2 (3.5) | 0.216 |
| 5 | 125 | 12 3/8 (28.6) | 1 1/2 (3.5) | 13 7/8 (32.1) | 10 7/8 (25.1) | 17 3/8 (40.1) | 18 7/8 (43.6) | 15 7/8 (36.6) | 5 (11.5) | 3 1/2 (8.1) | 1 1/2 (3.5) | 0.216 |
| 6 | 150 | 12 3/8 (28.6) | 1 1/2 (3.5) | 13 7/8 (32.1) | 10 7/8 (25.1) | 17 7/8 (40.1) | 18 7/8 (43.6) | 15 7/8 (36.6) | 5 (11.5) | 3 1/2 (8.1) | 1 1/2 (3.5) | 0.216 |

*Nominal pipe size equivalents, *not* conversions.
† Gauge length in number of turns of thread (( ) = linear equivalent to nearest 0.1 mm).
‡ Tolerance on position of gauge plane relative to face of internally taper threaded parts.
§ Diametral tolerance on parallel internal threads (millimetres).
For further information see BS 2779.

*Note:* The threaded fasteners to be described in the following tables (Sections 2.24 to 2.34) are now obsolete or obsolescent. Therefore they are not recommended for use in new product design or manufacture. However, they are still manufactured and still in widespread use. For this reason they have been included in this book.

## 2.24 British Standard Whitworth (BSW) bolts and nuts

| Nominal size (in) | Threads per inch (TPI) | Diameters | | | | | Hexagon (bolt heads) | | | | | Hexagon (nuts) | | | | |
|---|---|---|---|---|---|---|---|---|---|---|---|---|---|---|---|---|
| | | Pitch (in) | Depth (in) | Major (in) | Effective (in) | Minor (in) | Across flats (A/F) Max. (in) | Min. (in) | Across corners Max. (in) | Head thickness Max. (in) | Min. (in) | Across flats (A/F) Max. (in) | Min. (in) | Across corners Max. (in) | Nut thickness Max. (in) | Min. (in) |
| 1/4 | 20 | 0.05000 | 0.0320 | 0.2500 | 0.2180 | 0.1860 | 0.455 | 0.438 | 0.51 | 0.19 | 0.18 | 0.455 | 0.438 | 0.51 | 0.200 | 0.190 |
| 5/16 | 18 | 0.05536 | 0.0356 | 0.3125 | 0.2769 | 0.2413 | 0.525 | 0.518 | 0.61 | 0.22 | 0.21 | 0.525 | 0.518 | 0.61 | 0.250 | 0.240 |
| 3/8 | 16 | 0.06250 | 0.0400 | 0.3750 | 0.3350 | 0.2950 | 0.600 | 0.592 | 0.69 | 0.27 | 0.26 | 0.600 | 0.592 | 0.69 | 0.312 | 0.302 |
| 7/16 | 14 | 0.07141 | 0.0457 | 0.4375 | 0.3981 | 0.3461 | 0.710 | 0.702 | 0.82 | 0.33 | 0.32 | 0.710 | 0.702 | 0.82 | 0.375 | 0.365 |
| 1/2 | 12 | 0.08333 | 0.0534 | 0.5000 | 0.4466 | 0.3932 | 0.820 | 0.812 | 0.95 | 0.38 | 0.37 | 0.820 | 0.812 | 0.95 | 0.437 | 0.427 |
| 9/16 | 12 | 0.08333 | 0.0534 | 0.5625 | 0.5091 | 0.4557 | 0.920 | 0.912 | 1.06 | 0.44 | 0.43 | 0.920 | 0.912 | 1.06 | 0.500 | 0.490 |
| 5/8 | 11 | 0.09091 | 0.0542 | 0.6250 | 0.5668 | 0.5086 | 1.010 | 1.000 | 1.17 | 0.49 | 0.48 | 1.010 | 1.000 | 1.17 | 0.562 | 0.552 |
| 3/4 | 10 | 0.10000 | 0.0640 | 0.7500 | 0.6860 | 0.3039 | 1.200 | 1.190 | 1.39 | 0.60 | 0.59 | 1.200 | 1.190 | 1.39 | 0.687 | 0.677 |
| 7/8 | 8 | 0.11111 | 0.0711 | 0.8750 | 0.8039 | 0.7328 | 1.300 | 1.288 | 1.50 | 0.66 | 0.65 | 1.300 | 1.288 | 1.50 | 0.750 | 0.740 |

| Size | TPI | 0.12500 | 0.0800 | 1.0000 | 0.8400 | 0.9200 | 1.480 | 1.468 | 1.71 | 0.77 | 0.76 | 1.480 | 1.468 | 1.71 | 0.875 | 0.865 |
|---|---|---|---|---|---|---|---|---|---|---|---|---|---|---|---|---|
| 1 | 8 | 0.12500 | 0.0800 | 1.0000 | 0.8400 | 0.9200 | 1.480 | 1.468 | 1.71 | 0.77 | 0.76 | 1.480 | 1.468 | 1.71 | 0.875 | 0.865 |
| 1 1/8 | 7 | 0.14286 | 0.0915 | 1.-250 | 0.9402 | 1.0335 | 1.670 | 1.658 | 1.93 | 0.88 | 0.87 | 1.670 | 1.658 | 1.93 | 1.000 | 0.990 |
| 1 1/4 | 7 | 0.14286 | 0.0915 | 1.2500 | 1.0670 | 1.1585 | 1.860 | 1.845 | 2.15 | 0.98 | 0.96 | 1.860 | 1.845 | 2.15 | 1.125 | 1.105 |
| 1 1/2 | 6 | 0.16667 | 0.1067 | 1.5000 | 1.2866 | 1.3933 | 2.220 | 2.200 | 2.56 | 1.20 | 1.18 | 2.220 | 2.200 | 2.56 | 1.375 | 1.355 |
| 1 3/4 | 5 | 0.20000 | 0.1281 | 1.7500 | 1.4938 | 1.6219 | 2.580 | 2.555 | 2.98 | 1.42 | 1.40 | 2.580 | 2.555 | 2.98 | 1.625 | 1.605 |
| 2 | 4.5 | 0.20222 | 0.1423 | 2.0000 | 1.7154 | 1.8577 | 2.760 | 2.735 | 3.19 | 1.53 | 1.51 | 2.760 | 2.735 | 3.19 | 1.750 | 1.730 |
| 2 1/4 | 4 | 0.25000 | 0.1601 | 2.2500 | 1.9298 | 2.0899 | – | – | – | – | – | – | – | – | – | – |
| 2 1/2 | 4 | 0.25000 | 0.1601 | 2.5000 | 2.1798 | 2.3399 | – | – | – | – | – | – | – | – | – | – |
| 2 3/4 | 3.5 | 0.28571 | 0.1830 | 2.7500 | 2.3840 | 2.5670 | – | – | – | – | – | – | – | – | – | – |
| 3 | 3.5 | 0.28571 | 0.1830 | 3.0000 | 2.6340 | 2.8170 | – | – | – | – | – | – | – | – | – | – |
| 3 1/2 | 3.25 | 0.30769 | 0.1970 | 3.5000 | 3.1060 | 3.3030 | – | – | – | – | – | – | – | – | – | – |
| 4 | 3 | 0.33333 | 0.2134 | 4.0000 | 3.5732 | 3.7866 | – | – | – | – | – | – | – | – | – | – |
| 4 1/2 | 2.875 | 0.34783 | 0.2227 | 4.5000 | 4.0546 | 4.2773 | – | – | – | – | – | – | – | – | – | – |
| 5 | 2.75 | 0.36364 | 0.2328 | 5.0000 | 4.5344 | 4.7672 | – | – | – | – | – | – | – | – | – | – |

## 2.25 British Standard Whitworth (BSW) tapping and clearance drill sizes

| Size | TPI | Tapping | Clearing |
|------|-----|---------|----------|
| $1/16$ | 60 | 58 | 49 |
| $3/32$ | 48 | 49 | 36 |
| $1/8$ | 40 | 38 | 29 |
| $5/32$ | 32 | 31 | 19 |
| $3/16$ | 24 | 26 | 9 |
| $7/32$ | 24 | 15 | 1 |
| $1/4$ | 20 | 7 | G |
| $5/16$ | 18 | F | P |
| $3/8$ | 16 | O | W |
| $7/16$ | 14 | U | $29/64$ |
| $1/2$ | 12 | $27/64$ | $33/64$ |

Note: if number, letter and fractional size drills are not available, see Table 1.5 for nearest metric alternative.

## 2.26 British Standard Fine (BSF) bolts and nuts

| Nominal size (in) | Threads per inch (TPI) | Pitch (in) | Depth (in) | Major (in) | Effective (in) | Minor (in) | Hexagon (bolt heads) Across flats (AVF) Max. (in) | Min. (in) | Across corners Max. (in) | Head thickness Max. (in) | Min. (in) | Hexagon (nuts) Across flats (A/F) Max. (in) | Min. (in) | Across corners Max. (in) | Nut thickness Max. (in) | Min. (in) |
|---|---|---|---|---|---|---|---|---|---|---|---|---|---|---|---|---|
| 1/4 | 26 | 0.03846 | 0.0246 | 0.2500 | 0.2254 | 0.2008 | 0.455 | 0.438 | 0.51 | 0.19 | 0.18 | 0.455 | 0.438 | 0.51 | 0.200 | 0.190 |
| 5/16 | 22 | 0.04545 | 0.0291 | 0.3125 | 0.2834 | 0.2543 | 0.525 | 0.518 | 0.61 | 0.22 | 0.21 | 0.525 | 0.518 | 0.61 | 0.250 | 0.240 |
| 3/8 | 20 | 0.05000 | 0.0320 | 0.3750 | 0.3430 | 0.3110 | 0.600 | 0.592 | 0.69 | 0.27 | 0.26 | 0.600 | 0.592 | 0.69 | 0.312 | 0.302 |
| 7/16 | 18 | 0.05556 | 0.0356 | 0.4375 | 0.4019 | 0.3663 | 0.710 | 0.708 | 0.82 | 0.33 | 0.32 | 0.710 | 0.706 | 0.82 | 0.375 | 0.365 |
| 1/2 | 16 | 0.06250 | 0.0400 | 0.5000 | 0.4600 | 0.4200 | 0.820 | 0.812 | 0.95 | 0.38 | 0.37 | 0.820 | 0.812 | 0.95 | 0.437 | 0.427 |
| 9/16 | 16 | 0.06250 | 0.0400 | 0.5625 | 0.5225 | 0.4825 | 0.920 | 0.912 | 1.06 | 0.44 | 0.43 | 0.920 | 0.912 | 1.06 | 0.500 | 0.490 |
| 5/8 | 14 | 0.07143 | 0.0457 | 0.6250 | 0.5793 | 0.5335 | 1.010 | 1.000 | 1.17 | 0.49 | 0.48 | 1.010 | 1.000 | 1.17 | 0.562 | 0.552 |
| 3/4 | 12 | 0.08333 | 0.0534 | 0.7500 | 0.6966 | 0.6432 | 1.200 | 1.190 | 1.39 | 0.60 | 0.59 | 1.200 | 1.190 | 1.39 | 0.687 | 0.677 |
| 7/8 | 11 | 0.09091 | 0.0582 | 0.8750 | 0.8168 | 0.7586 | 1.300 | 1.288 | 1.50 | 0.66 | 0.65 | 1.300 | 1.288 | 1.50 | 0.750 | 0.740 |

(continued)

155

**2.26** (continued)

| Nominal size (in) | Threads per inch (TPI) | Diameters | | | | | Hexagon (bolt heads) | | | | | Hexagon (nuts) | | | | |
|---|---|---|---|---|---|---|---|---|---|---|---|---|---|---|---|---|
| | | Pitch (in) | Depth (in) | Major (in) | Effective (in) | Minor (in) | Across flats (A/F) | | Across corners | Head thickness | | Across flats (A/F) | | Across corners | Nut thickness | |
| | | | | | | | Max. (in) | Min. (in) | Max. (in) | Max. (in) | Min. (in) | Max. (in) | Min. (in) | Max. (in) | Max. (in) | Min. (in) |
| 1 | 10 | 0.10000 | 0.0640 | 1.0000 | 0.9360 | 0.8720 | 1.480 | 1.468 | 1.71 | 0.77 | 0.76 | 1.480 | 1.468 | 1.71 | 0.875 | 0.865 |
| 1 1/8 | 9 | 0.11111 | 0.0711 | 1.1250 | 1.0539 | 0.9828 | 1.670 | 1.658 | 1.93 | 0.88 | 0.87 | 1.670 | 1.658 | 1.93 | 1.000 | 0.990 |
| 1 1/4 | 9 | 0.11111 | 0.0711 | 1.2500 | 1.1789 | 1.1078 | 1.860 | 1.845 | 2.15 | 0.98 | 0.96 | 1.860 | 1.845 | 2.15 | 1.125 | 1.105 |
| 1 3/8 | 8 | 0.12500 | 0.0800 | 1.1370 | 1.2950 | 1.2150 | 2.050 | 2.035 | 2.37 | 1.09 | 1.07 | 2.050 | 2.035 | 2.37 | 1.250 | 1.230 |
| 1 1/2 | 8 | 0.12500 | 0.0800 | 1.5000 | 1.4200 | 1.3400 | 2.220 | 2.200 | 2.56 | 1.20 | 1.18 | 2.220 | 2.200 | 2.56 | 1.375 | 1.355 |
| 1 5/8 | 8 | 0.12500 | 0.0800 | 1.6250 | 1.5450 | 1.4650 | — | — | — | — | — | — | — | — | — | — |
| 1 3/4 | 7 | 0.14826 | 0.0915 | 1.7500 | 1.6585 | 1.5670 | 2.580 | 2.555 | 2.98 | 1.42 | 1.40 | 2.580 | 2.555 | 2.98 | 1.625 | 1.605 |
| 2 | 7 | 0.14826 | 0.0915 | 2.0000 | 1.9085 | 1.8170 | 2.760 | 2.735 | 3.19 | 1.53 | 1.51 | 2.760 | 2.735 | 3.19 | 1.750 | 1.730 |
| 2 1/4 | 6 | 0.16667 | 0.1067 | 2.2500 | 2.1433 | 2.0366 | — | — | — | — | — | — | — | — | — | — |

## 2.27 British Standard Fine (BSF) tapping and clearance drill sizes

| Size | TPI | Tapping | Clearing |
|------|-----|---------|----------|
| $7/32$ | 28 | 14 | 1 |
| $1/4$ | 26 | 3 | G |
| $9/32$ | 26 | C | M |
| $5/16$ | 22 | H | P |
| $3/8$ | 20 | $21/64$ | W |
| $7/16$ | 18 | W | $29/64$ |
| $1/2$ | 16 | $7/16$ | $33/64$ |

Note: if number, letter and fractional size drills are not available, see Table 1.5 for nearest metric alternative.

## 2.28 ISO unified precision internal screw threads, coarse series (UNC)

(Dimensions in inches)

| Designation | Major diameter | Pitch (effective) diameter | | Minor diameter | | Hexagon (nut) | | | | |
|---|---|---|---|---|---|---|---|---|---|---|
| | min | max | min | max | min | Max width across flats (A/F) | Max width across corners (A/C) | Nut thickness Thick | Normal | Thin |
| $\frac{1}{4}$-20 UNC-2B | 0.2500 | 0.2223 | 0.2175 | 0.2074 | 0.1959 | 0.4375 | 0.505 | 0.286 | 0.224 | 0.161 |
| $\frac{5}{16}$-18 UNC-2B | 0.3125 | 0.2817 | 0.2764 | 0.2651 | 0.2524 | 0.5000 | 0.577 | 0.333 | 0.271 | 0.192 |
| $\frac{3}{8}$-16 UNC-2B | 0.3750 | 0.3401 | 0.3344 | 0.3214 | 0.3073 | 0.5625 | 0.650 | 0.411 | 0.333 | 0.224 |
| $\frac{7}{16}$-14 UNC-2B | 0.4375 | 0.3972 | 0.3911 | 0.3760 | 0.3602 | 0.6875 | 0.794 | 0.458 | 0.380 | 0.255 |
| $\frac{1}{2}$-13 UNC-2B | 0.5000 | 0.4565 | 0.4500 | 0.4336 | 0.4167 | 0.7500 | 0.866 | 0.567 | 0.442 | 0.317 |
| $\frac{9}{16}$-12 UNC-2B* | 0.5625 | 0.5152 | 0.5084 | 0.4904 | 0.4723 | 0.8750 | 1.010 | 0.614 | 0.489 | 0.349 |
| $\frac{5}{8}$-11 UNC-2B | 0.6250 | 0.5732 | 0.5660 | 0.5460 | 0.5266 | 0.9375 | 1.083 | 0.724 | 0.552 | 0.380 |
| $\frac{3}{4}$-10 UNC-2B | 0.7500 | 0.6927 | 0.6850 | 0.6627 | 0.6417 | 1.1250 | 1.300 | 0.822 | 0.651 | 0.432 |
| $\frac{7}{8}$-9 UNC-2B | 0.8750 | 0.8110 | 0.8028 | 0.7775 | 0.7547 | 1.3125 | 1.515 | 0.916 | 0.760 | 0.494 |
| | 1.000 | 0.9276 | 0.9188 | 0.8897 | 0.8647 | 1.5000 | 1.732 | 1.015 | 0.874 | 0.562 |

| | | | | | | | | | | |
|---|---|---|---|---|---|---|---|---|---|---|
| $1\frac{1}{8}$-7 UNC-2B | 1.1250 | 1.0416 | 1.0322 | 0.9980 | 0.9704 | 1.6875 | 1.948 | 1.176 | 0.989 | 0.629 |
| $1\frac{1}{4}$-7 UNC-2B | 1.2500 | 1.1668 | 1.1572 | 1.1230 | 1.0954 | 1.8750 | 2.165 | 1.275 | 1.087 | 0.744 |
| $1\frac{3}{8}$-6 UNC-2B* | 1.3750 | 1.2771 | 1.2667 | 1.2252 | 1.1946 | 2.0625 | 2.382 | 1.400 | 1.197 | 0.806 |
| $1\frac{1}{2}$-6 UNC-2B | 1.5000 | 1.4022 | 1.3917 | 1.3502 | 1.3196 | 2.2500 | 2.598 | 1.530 | 1.311 | 0.874 |
| $1\frac{3}{4}$-5 UNC-2B | 1.7500 | 1.6317 | 1.6201 | 1.5675 | 1.5335 | 2.6250 | 3.031 | – | 1.530 | 0.999 |
| $2$-$4\frac{1}{2}$ UNC-2B | 2.0000 | 1.8681 | 1.8557 | 1.7952 | 1.7594 | 3.0000 | 3.464 | – | 1.754 | 1.129 |
| $2\frac{1}{4}$-$4\frac{1}{2}$ UNC-2B | 2.2500 | 2.1183 | 2.1057 | 2.0452 | 2.0094 | | | | | |
| $2\frac{1}{2}$-4 UNC-2B | 2.5000 | 2.3511 | 2.3376 | 2.2669 | 2.2294 | | | | | |
| $2\frac{3}{4}$-4 UNC-2B | 2.7500 | 2.6013 | 2.5876 | 2.5169 | 2.4794 | | | | | |
| 3-4 UNC-2B | 3.0000 | 2.8515 | 2.8376 | 2.7669 | 2.7294 | | | | | |
| $3\frac{1}{4}$-4 UNC-2B | 3.2500 | 3.1017 | 3.0876 | 3.0169 | 2.9794 | | | | | |
| $3\frac{1}{2}$-4 UNC-2B | 3.5000 | 3.3519 | 3.3376 | 3.2669 | 3.2294 | | | | | |
| $3\frac{3}{4}$-4 UNC-2B | 3.7500 | 3.6021 | 3.5876 | 3.5169 | 3.4794 | | | | | |
| 4-4 UNC-2B | 4.0000 | 3.8523 | 3.8376 | 3.7669 | 3.7294 | | | | | |

*To be dispensed with wherever possible.
For full range and further information see BS 1768.

**Example**

The interpretation of designation $\frac{1}{2}$-13 UNC-2B is as follows: nominal diameter $\frac{1}{2}$ inch; threads per inch 13; ISO unified thread, coarse series; thread tolerance classification 2B.

# 2.29 ISO unified precision external screw threads, coarse series (UNC)

(Dimensions in inches)

| Designation | Major diameter | | Pitch (effective) diameter | | Minor diameter | | Shank diameter | | Hexagon head (bolt) | | |
|---|---|---|---|---|---|---|---|---|---|---|---|
| | max | min | max | min | max | min | max | min | Max width across flats (A/F) | Max width across corners (A/C) | Max height |
| $\frac{1}{4}$-20 UNC-2A | 0.2489 | 0.2408 | 0.2164 | 0.2127 | 0.1876 | 0.1803 | 0.2500 | 0.2465 | 0.4375 | 0.505 | 0.163 |
| $\frac{5}{16}$-18 UNC-2A | 0.3113 | 0.3026 | 0.2752 | 0.2712 | 0.2431 | 0.2351 | 0.3125 | 0.3090 | 0.5000 | 0.577 | 0.211 |
| $\frac{3}{8}$-16 UNC-2A | 0.3737 | 0.3643 | 0.3331 | 0.3287 | 0.2970 | 0.2881 | 0.3750 | 0.3715 | 0.5625 | 0.650 | 0.243 |
| $\frac{7}{16}$-14 UNC-2A | 0.4361 | 0.4258 | 0.3897 | 0.3850 | 0.3485 | 0.3387 | 0.4375 | 0.4335 | 0.6250 | 0.722 | 0.291 |
| $\frac{1}{2}$-13 UNC-2A | 0.4985 | 0.4876 | 0.4485 | 0.4435 | 0.4041 | 0.3936 | 0.5000 | 0.4960 | 0.7500 | 0.866 | 0.323 |
| $\frac{9}{16}$-12 UNC-2A* | 0.5609 | 0.5495 | 0.5068 | 0.5016 | 0.4587 | 0.4475 | 0.5625 | 0.5585 | 0.8125 | 0.938 | 0.371 |
| $\frac{5}{8}$-11 UNC-2A | 0.6234 | 0.6113 | 0.5644 | 0.5589 | 0.5119 | 0.4999 | 0.6250 | 0.6190 | 0.9375 | 1.083 | 0.403 |
| $\frac{3}{4}$-10 UNC-2A | 0.7482 | 0.7353 | 0.6832 | 0.6773 | 0.6255 | 0.6124 | 0.7500 | 0.7440 | 1.1250 | 1.300 | 0.483 |
| $\frac{7}{8}$-9 UNC-2A | 0.8731 | 0.8592 | 0.8009 | 0.7946 | 0.7368 | 0.7225 | 0.8750 | 0.8670 | 1.3125 | 1.515 | 0.563 |
| 1-8 UNC-2A | 0.9980 | 0.9830 | 0.9168 | 0.9100 | 0.8446 | 0.8288 | 1.0000 | 0.9920 | 1.5000 | 1.732 | 0.627 |

(continued)

**2.29** *(continued)*

(Dimensions in inches)

| Designation | Major diameter | | Pitch (effective) diameter | | Minor diameter | | Shank diameter | | Hexagon head (bolt) | | |
|---|---|---|---|---|---|---|---|---|---|---|---|
| | max | min | max | min | max | min | max | min | Max width across flats (A/F) | Max width across corners (A/C) | Max height |
| $1\frac{1}{8}$-7 UNC-2A | 1.1228 | 1.1064 | 1.0300 | 1.0228 | 0.9475 | 0.9300 | 1.1250 | 1.1170 | 1.6875 | 1.948 | 0.718 |
| $1\frac{1}{4}$-7 UNC-2A | 1.2478 | 1.2314 | 1.1550 | 1.1476 | 1.0725 | 1.0548 | 1.2500 | 1.2420 | 1.8750 | 2.165 | 0.813 |
| $1\frac{3}{8}$-6 UNC-2A* | 1.3726 | 1.3544 | 1.2643 | 1.2563 | 1.1681 | 1.1481 | 1.3750 | 1.3650 | 2.0625 | 2.382 | 0.878 |
| $1\frac{1}{2}$-6 UNC-2A | 1.4976 | 1.4794 | 1.3893 | 1.3812 | 1.2931 | 1.2730 | 1.5000 | 1.4900 | 2.2500 | 2.598 | 0.974 |
| $1\frac{3}{4}$-5 UNC-2A | 1.7473 | 1.7268 | 1.6174 | 1.6085 | 1.5019 | 1.4786 | 1.7500 | 1.7400 | 2.6250 | 3.031 | 1.134 |
| $2$-$4\frac{1}{2}$ UNC-2A | 1.9971 | 1.9751 | 1.8528 | 1.8433 | 1.7245 | 1.6990 | 2.000 | 1.9900 | 3.000 | 3.464 | 1.263 |
| $2\frac{1}{4}$-$4\frac{1}{2}$ UNC-2A | 2.2471 | 2.2251 | 2.1028 | 2.0931 | 1.9745 | 1.9488 | | | | | |
| $2\frac{1}{2}$-4 UNC-2A | 2.4969 | 2.4731 | 2.3345 | 2.3241 | 2.1902 | 2.1618 | | | | | |
| $2\frac{3}{4}$-4 UNC-2A | 2.7468 | 2.7230 | 2.5844 | 2.5739 | 2.4401 | 2.4116 | | | | | |
| 3-4 UNC-2A | 2.9968 | 2.9730 | 2.8344 | 2.8237 | 2.6901 | 2.6614 | | | | | |
| $3\frac{1}{4}$-4 UNC-2A | 3.2467 | 3.2229 | 3.0843 | 3.0734 | 2.9400 | 2.9111 | | | | | |
| $3\frac{1}{2}$-4 UNC-2A | 3.4967 | 3.4729 | 3.3343 | 3.3233 | 3.1900 | 3.1610 | | | | | |
| $3\frac{3}{4}$-4 UNC-2A | 3.7466 | 3.7228 | 3.5842 | 3.5730 | 3.4399 | 3.4107 | | | | | |
| 4-4 UNC-2A | 3.9966 | 3.9728 | 3.8342 | 3.8229 | 3.6899 | 3.6606 | | | | | |

**Example**

The interpretation of designation $\frac{1}{2}$-13 UNC-2A is as follows: nominal diameter $\frac{1}{2}$ inch; threads per inch 13; ISO unified thread, coarse series, thread tolerance classification 2A.

## 2.30 ISO unified tapping and clearance drills, coarse thread series

| Nominal size in | Tapping drill size | | Clearance drill size | |
|---|---|---|---|---|
| | mm | in | mm | Letter or in |
| $\frac{1}{4} \times 20$ | 5.20 | $\frac{13}{64}$ | 6.50 | $\frac{17}{64}$ or F |
| $\frac{5}{16} \times 18$ | 6.60 | $\frac{17}{64}$ | 8.00 | $\frac{21}{64}$ or O |
| $\frac{3}{8} \times 16$ | 8.00 | $\frac{5}{16}$ | 9.80 | $\frac{25}{64}$ or W |
| $\frac{7}{16} \times 14$ | 9.40 | $\frac{3}{8}$ | 11.30 | $\frac{29}{64}$ |
| $\frac{1}{2} \times 13$ | 10.80 | $\frac{27}{64}$ | 13.00 | $\frac{33}{64}$ |
| $\frac{9}{16} \times 12$ | 12.20 | $\frac{31}{64}$ | 14.75 | $\frac{37}{64}$ |
| $\frac{5}{8} \times 11$ | 13.50 | $\frac{17}{32}$ | 16.25 | $\frac{41}{64}$ |
| $\frac{3}{4} \times 10$ | 16.50 | $\frac{21}{32}$ | 19.50 | $\frac{47}{64}$ |
| $\frac{7}{8} \times 9$ | 19.25 | $\frac{49}{64}$ | 20.25 | $\frac{51}{64}$ |
| $1 \times 8$ | 22.25 | $\frac{7}{8}$ | 25.75 | $1\frac{1}{64}$ |
| $1\frac{1}{8} \times 7$ | 25.00 | $\frac{63}{64}$ | 26.00 | $1\frac{9}{64}$ |
| $1\frac{1}{4} \times 7$ | 28.25* | $1\frac{7}{64}$ | 28.25 | $1\frac{17}{64}$ |
| $1\frac{3}{8} \times 6$ | 30.50* | $1\frac{13}{64}$ | 30.75 | $1\frac{25}{64}$ |
| $1\frac{1}{2} \times 6$ | 34.00* | $1\frac{21}{64}$ | 34.00 | $1\frac{33}{64}$ |
| $1\frac{3}{4} \times 5$ | 39.50* | $1\frac{35}{64}$ | 45.00 | $1\frac{49}{64}$ |
| $2 \times 4\frac{1}{2}$ | 45.50* | $1\frac{25}{32}$ | 52.00 | $2\frac{1}{64}$ |

* Nearest standard metric size: approx. 0.25 mm over recommended inch size.

**2.31 ISO unified precision internal screw threads, fine series (UNF)**

(Dimensions in inches)

| Designation | Major diameter min | Pitch (effective) diameter | | Minor diameter | | Hexagon (nut) | | | | |
|---|---|---|---|---|---|---|---|---|---|---|
| | | max | min | max | min | Max width across flats (A/F) | Max width across corners (A/C) | Thick | Normal | Thin |
| | | | | | | | | \| Nut thickness \| | | |
| $\frac{1}{4}$-28 UNF-2B | 0.2500 | 0.2311 | 0.2268 | 0.2197 | 0.2113 | 0.4375 | 0.505 | 0.286 | 0.224 | 0.161 |
| $\frac{5}{16}$-24 UNF-2B | 0.3125 | 0.2902 | 0.2854 | 0.2771 | 0.2674 | 0.5000 | 0.577 | 0.333 | 0.271 | 0.192 |
| $\frac{3}{8}$-24 UNF-2B | 0.3750 | 0.3528 | 0.3479 | 0.3396 | 0.3299 | 0.5625 | 0.650 | 0.411 | 0.333 | 0.224 |
| $\frac{7}{16}$-20 UNF-2B | 0.4375 | 0.4104 | 0.4050 | 0.3949 | 0.3834 | 0.6875 | 0.794 | 0.458 | 0.380 | 0.255 |
| $\frac{1}{2}$-20 UNF-2B | 0.5000 | 0.4731 | 0.4675 | 0.4574 | 0.4459 | 0.7500 | 0.866 | 0.567 | 0.442 | 0.317 |
| $\frac{9}{16}$-18 UNF-2B* | 0.5625 | 0.5323 | 0.5264 | 0.5151 | 0.5024 | 0.8750 | 1.010 | 0.614 | 0.489 | 0.349 |
| $\frac{5}{8}$-18 UNF-2B | 0.6250 | 0.5949 | 0.5889 | 0.5776 | 0.5649 | 0.9375 | 1.083 | 0.724 | 0.552 | 0.380 |

| | | | | | | | | | |
|---|---|---|---|---|---|---|---|---|---|
| $\frac{3}{4}$-14 UNF-2B | 0.7500 | 0.7159 | 0.7094 | 0.6964 | 0.6823 | 1.1250 | 1.300 | 0.822 | 0.651 | 0.432 |
| $\frac{7}{8}$-14 UNF-2B | 0.8750 | 0.8356 | 0.8286 | 0.8135 | 0.7977 | 1.3125 | 1.515 | 0.916 | 0.760 | 0.494 |
| 1-12 UNF-2B | 1.0000 | 0.9535 | 0.9459 | 0.9279 | 0.9098 | 1.5000 | 1.732 | 1.015 | 0.874 | 0.562 |
| $1\frac{1}{8}$-12 UNF-2B | 1.1250 | 1.0787 | 1.0709 | 1.0529 | 1.0348 | 1.6875 | 1.948 | 1.176 | 0.984 | 0.629 |
| $1\frac{1}{4}$-12 UNF-2B | 1.2500 | 1.2039 | 1.1959 | 1.1779 | 1.1598 | 1.8750 | 2.165 | 1.275 | 1.087 | 0.744 |
| $1\frac{3}{8}$-12 UNF-2B* | 1.3750 | 1.3291 | 1.3209 | 1.3029 | 1.2848 | 2.0625 | 2.382 | 1.400 | 1.197 | 0.806 |
| $1\frac{1}{2}$-12 UNF-2B | 1.5000 | 1.4542 | 1.4459 | 1.4279 | 1.4098 | 2.2500 | 2.598 | 1.530 | 1.311 | 0.874 |

*To be dispensed with wherever possible.
For full range and further information see BS 1768.

**Example**

The interpretation of designation $\frac{1}{2}$-20 UNF-2B is as follows: nominal diameter $\frac{1}{2}$ inch; threads per inch 20; ISO unified thread, fine series; thread tolerance classification 2B.

(Dimensions in inches)

**2.32 ISO unified precision external screw threads, fine series (UNF)**

| Designation | Major diameter | | Pitch (effective) diameter | | Minor diameter | | Shank diameter | | Hexagon head (bolt) | | |
|---|---|---|---|---|---|---|---|---|---|---|---|
| | max | min | max | min | max | min | max | min | Max width across flats (A/F) | Max width across corners (A/C) | Max height |
| $\frac{1}{4}$-28 UNF-2A | 0.2490 | 0.2425 | 0.2258 | 0.2225 | 0.2052 | 0.1993 | 0.2500 | 0.2465 | 0.4375 | 0.505 | 0.163 |
| $\frac{5}{16}$-24 UNF-2A | 0.3114 | 0.3042 | 0.2843 | 0.2806 | 0.2603 | 0.2536 | 0.3125 | 0.3090 | 0.5000 | 0.577 | 0.211 |
| $\frac{3}{8}$-24 UNF-2A | 0.3739 | 0.3667 | 0.3468 | 0.3430 | 0.3228 | 0.3160 | 0.3750 | 0.3715 | 0.5625 | 0.650 | 0.243 |
| $\frac{7}{16}$-20 UNF-2A | 0.4362 | 0.4281 | 0.4037 | 0.3995 | 0.3749 | 0.3671 | 0.4375 | 0.4335 | 0.6250 | 0.722 | 0.291 |
| $\frac{1}{2}$-20 UNF-2A | 0.4987 | 0.4906 | 0.4662 | 0.4615 | 0.4374 | 0.4295 | 0.5000 | 0.4960 | 0.7500 | 0.866 | 0.323 |
| $\frac{9}{16}$-18 UNF-2A* | 0.5611 | 0.5524 | 0.5250 | 0.5205 | 0.4929 | 0.4844 | 0.5625 | 0.5585 | 0.8125 | 0.938 | 0.371 |
| $\frac{5}{8}$-18 UNF-2A | 0.6236 | 0.6149 | 0.5875 | 0.5828 | 0.5554 | 0.5467 | 0.6250 | 0.6190 | 0.9375 | 1.083 | 0.403 |
| $\frac{3}{4}$-16 UNF-2A | 0.7485 | 0.7391 | 0.7079 | 0.7029 | 0.6718 | 0.6623 | 0.7500 | 0.7440 | 1.1250 | 1.300 | 0.483 |

(continued)

**2.32** (*continued*)

(Dimensions in inches)

| Designation | Major diameter | | Pitch (effective) diameter | | Minor diameter | | Shank diameter | | Hexagon head (bolt) | | |
|---|---|---|---|---|---|---|---|---|---|---|---|
| | | | | | | | | | Max width across flats (A/F) | Max width across corners (A/C) | Max height |
| | max | min | max | min | max | min | max | min | | | |
| $\frac{7}{8}$-14 UNF-2A | 0.8734 | 0.8631 | 0.8270 | 0.8216 | 0.7858 | 0.7753 | 0.8750 | 0.8670 | 1.3125 | 1.515 | 0.563 |
| 1-12 UNF-2A | 0.9982 | 0.9868 | 0.9441 | 0.9382 | 0.8960 | 0.8841 | 1.0000 | 0.9920 | 1.5000 | 1.732 | 0.627 |
| $1\frac{1}{8}$-12 UNF-2A | 1.1232 | 1.1118 | 1.0691 | 1.0631 | 1.0210 | 1.0090 | 1.1250 | 1.1170 | 1.6875 | 1.948 | 0.718 |
| $1\frac{1}{4}$-12 UNF-2A | 1.2482 | 1.2368 | 1.1941 | 1.1879 | 1.1460 | 1.1338 | 1.2500 | 1.2420 | 1.8750 | 2.165 | 0.813 |
| $1\frac{3}{8}$-12 UNF-2A* | 1.3731 | 1.3617 | 1.3190 | 1.3127 | 1.2709 | 1.2586 | 1.3750 | 1.3650 | 2.0625 | 2.382 | 0.878 |
| $1\frac{1}{2}$-12 UNF-2A | 1.4981 | 1.4867 | 1.4440 | 1.4376 | 1.3959 | 1.3835 | 1.5000 | 1.4900 | 2.2500 | 2.598 | 0.974 |

*To be dispensed with wherever possible.
For full range and further information see BS 1768.

**Example**

The interpretation of designation $\frac{1}{2}$-20 UNF-2A is as follows: nominal diameter $\frac{1}{2}$ inch: threads per inch 20; ISO unified thread, fine series: thread tolerance classification 2A.

## 2.33 ISO unified tapping and clearance drills, fine thread series

| Nominal size in | Tapping drill size | | Clearance drill size | |
|---|---|---|---|---|
| | mm | Letter or in | mm | Letter or in |
| $\frac{1}{4} \times 28$ | 5.50 | $\frac{7}{32}$ | 6.50 | F |
| $\frac{5}{16} \times 24$ | 6.90 | I | 8.00 | O |
| $\frac{3}{8} \times 24$ | 8.50 | R | 9.80 | W |
| $\frac{7}{16} \times 20$ | 9.90 | $\frac{25}{64}$ | 11.30 | $\frac{29}{64}$ |
| $\frac{1}{2} \times 20$ | 11.50 | $\frac{29}{64}$ | 13.00 | $\frac{33}{64}$ |
| $\frac{9}{16} \times 18$ | 12.90 | $\frac{33}{64}$ | 14.75 | $\frac{37}{64}$ |
| $\frac{5}{8} \times 18$ | 14.50 | $\frac{37}{64}$ | 16.50 | $\frac{41}{64}$ |
| $\frac{3}{4} \times 16$ | 17.50 | $\frac{11}{16}$ | 19.50 | $\frac{49}{64}$ |
| $\frac{7}{8} \times 14$ | 20.50 | $\frac{13}{16}$ | 22.75 | $\frac{57}{64}$ |
| $1 \times 12$ | 23.25 | $\frac{59}{64}$ | 25.80 | $1\frac{1}{64}$ |
| $1\frac{1}{8} \times 12$ | 26.50 | $1\frac{3}{64}$ | 29.00 | $1\frac{9}{64}$ |
| $1\frac{1}{4} \times 12$ | 29.50 | $1\frac{11}{64}$ | 32.50 | $1\frac{17}{64}$ |
| $1\frac{3}{8} \times 12$ | 33.00 | $1\frac{19}{64}$ | 35.50 | $1\frac{25}{64}$ |
| $1\frac{1}{2} \times 12$ | 36.00 | $1\frac{27}{64}$ | 38.50 | $1\frac{33}{64}$ |

Note: if number, letter and fractional size drills are not available, see Table 1.5 for nearest metric alternative.

## 2.34 British Association (BA) thread form

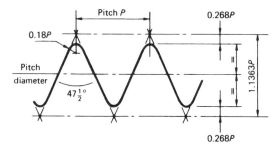

British Association (BA) thread forms are obsolete but are still used in repairs, maintenance and model making.

**BA internal and external screw threads**

(Dimensions in millimetres)

| Designation number | Pitch | Depth of thread | Major diameter | Pitch (effective) diameter | Minor diameter | Crest radius | Root radius |
|---|---|---|---|---|---|---|---|
| 0 | 1.0000 | 0.600 | 6.00 | 5.400 | 4.80 | 0.1808 | 0.1808 |
| 1 | 0.9000 | 0.540 | 5.30 | 4.760 | 4.22 | 0.1627 | 0.1627 |
| 2 | 0.8100 | 0.485 | 4.70 | 4.215 | 3.73 | 0.1465 | 0.1465 |
| 3 | 0.7300 | 0.440 | 4.10 | 3.660 | 3.22 | 0.1320 | 0.1320 |
| 4 | 0.6600 | 0.395 | 3.60 | 3.205 | 2.81 | 0.1193 | 0.1193 |
| 5 | 0.5900 | 0.355 | 3.20 | 2.845 | 2.49 | 0.1067 | 0.1067 |
| 6 | 0.5300 | 0.320 | 2.80 | 2.480 | 2.16 | 0.0958 | 0.0958 |
| 7 | 0.4800 | 0.290 | 2.50 | 2.210 | 1.92 | 0.0868 | 0.0868 |
| 8 | 0.4300 | 0.260 | 2.20 | 1.940 | 1.68 | 0.0778 | 0.0778 |
| 9 | 0.3900 | 0.235 | 1.90 | 1.665 | 1.43 | 0.0705 | 0.0705 |
| 10 | 0.3500 | 0.210 | 1.70 | 1.490 | 1.28 | 0.0633 | 0.0633 |
| 11 | 0.3100 | 0.185 | 1.50 | 1.315 | 1.13 | 0.0561 | 0.0561 |
| 12 | 0.2800 | 0.170 | 1.30 | 1.130 | 0.96 | 0.0506 | 0.0506 |
| 13 | 0.2500 | 0.150 | 1.20 | 1.050 | 0.90 | 0.0452 | 0.0452 |

(continued)

171

(Dimensions in millimetres)

| Designation number | Pitch | Depth of thread | Major diameter | Pitch (effective) diameter | Minor diameter | Crest radius | Root radius |
|---|---|---|---|---|---|---|---|
| 14 | 0.2300 | 0.140 | 1.00 | 0.860 | 0.72 | 0.0416 | 0.0416 |
| 15 | 0.2100 | 0.125 | 0.90 | 0.775 | 0.65 | 0.0380 | 0.0380 |
| 16 | 0.1900 | 0.115 | 0.79 | 0.675 | 0.56 | 0.0344 | 0.0344 |
| 17 | 0.1700 | 0.100 | 0.70 | 0.600 | 0.50 | 0.0307 | 0.0307 |
| 18 | 0.1500 | 0.090 | 0.62 | 0.530 | 0.44 | 0.0271 | 0.0271 |
| 19 | 0.1400 | 0.085 | 0.54 | 0.455 | 0.37 | 0.0253 | 0.0253 |
| 20 | 0.1200 | 0.070 | 0.48 | 0.410 | 0.34 | 0.0217 | 0.0217 |
| 21 | 0.1100 | 0.065 | 0.42 | 0.355 | 0.29 | 0.0199 | 0.0199 |
| 22 | 0.1000 | 0.060 | 0.37 | 0.310 | 0.25 | 0.0181 | 0.0181 |
| 23 | 0.0900 | 0.055 | 0.33 | 0.275 | 0.22 | 0.0163 | 0.0163 |
| 24 | 0.0800 | 0.050 | 0.29 | 0.240 | 0.19 | 0.0145 | 0.0145 |
| 25 | 0.0700 | 0.040 | 0.25 | 0.210 | 0.17 | 0.0127 | 0.0127 |

For further information see BS 57 and BS 93.

## 2.35 BA threads: tapping and clearance drills

| BA no. | Tapping size drill mm | Number or fraction size | Clearance size drill mm | Number or letter |
|---|---|---|---|---|
| 0 | 5.10 | 8 | 6.10 | D |
| 1 | 4.50 | 16 | 5.50 | 2 |
| 2 | 4.00 | 22 | 4.85 | 10 |
| 3 | 3.40 | 29 | 4.25 | 18 |
| 4 | 3.00 | 32 | 3.75 | 24 |
| 5 | 2.65 | 37 | 3.30 | 29 |
| 6 | 2.30 | 43 | 2.90 | 32 |
| 7 | 2.05 | 45/46 | 2.60 | 36 |
| 8 | 1.80 | 50 | 2.25 | 41 |
| 9 | 1.55 | 53 | 1.95 | 45 |
| 10 | 1.40 | 54 | 1.75 | 49 |
| 11 | 1.20 | 56 | 1.60 | 52 |
| 12 | 1.05 | 59 | 1.40 | 54 |
| 13 | 0.98 | 62 | 1.30 | 55 |
| 14 | 0.78 | 68 | 1.10 | 57 |
| 15 | 0.70 | 70 | 0.98 | 60 |
| 16 | 0.60 | 73 | 0.88 | 65 |

## 2.36 Model engineering threads (55° ME)

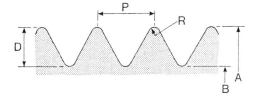

| Dia. | Thread per inch | Out-side dia. A | Core dia. B | Pitch P | Depth D | Radius R |
|---|---|---|---|---|---|---|
| 1/8 | 40 | 0.1250 | 0.093 | 0.025 | 0.016 | 0.003 |
| 5/32 | 40 | 0.1562 | 0.124 | 0.025 | 0.016 | 0.003 |
| 3/16 | 40 | 0.1875 | 0.156 | 0.025 | 0.016 | 0.003 |
| 7/32 | 40 | 0.2187 | 0.187 | 0.025 | 0.016 | 0.003 |
| 1/4 | 40 | 0.2500 | 0.218 | 0.025 | 0.016 | 0.003 |
| 9/32 | 40 | 0.2812 | 0.249 | 0.025 | 0.016 | 0.003 |
| 5/16 | 40 | 0.3125 | 0.281 | 0.025 | 0.016 | 0.003 |
| 5/16 | 32 | 0.3125 | 0.273 | 0.031 | 0.020 | 0.004 |
| 5/16 | 26 | 0.3125 | 0.263 | 0.038 | 0.025 | 0.005 |
| 3/8 | 40 | 0.3750 | 0.343 | 0.025 | 0.016 | 0.003 |
| 3/8 | 32 | 0.3750 | 0.335 | 0.031 | 0.020 | 0.004 |
| 3/8 | 26 | 0.3750 | 0.325 | 0.038 | 0.025 | 0.005 |
| 7/16 | 40 | 0.4375 | 0.406 | 0.025 | 0.016 | 0.003 |
| 7/16 | 32 | 0.4375 | 0.398 | 0.031 | 0.020 | 0.004 |
| 7/16 | 26 | 0.4375 | 0.388 | 0.038 | 0.025 | 0.005 |
| 1/2 | 40 | 0.5000 | 0.468 | 0.025 | 0.016 | 0.003 |
| 1/2 | 32 | 0.5000 | 0.460 | 0.031 | 0.020 | 0.004 |
| 1/2 | 26 | 0.5000 | 0.451 | 0.038 | 0.025 | 0.005 |

## 2.37 Model engineer clearance and trapping drills

| Dia. | Threads per inch | Tapping drill | | Clearance drill | |
|---|---|---|---|---|---|
| | | Imp. | Metric | Imp. | Metric |
| 1/8 | 40 | 3/32 | 2.60 | 30 | 3.30 |
| 5/32 | 40 | 1/8 | 3.40 | 22 | 4.10 |
| 3/16 | 40 | 5/32 | 4.20 | 12 | 5.00 |
| 7/32 | 40 | 3/16 | 5.00 | 2 | 5.70 |
| 1/4 | 40 | 7/32 | 5.80 | F | 6.50 |
| 9/32 | 40 | 1/4 | 6.50 | L | 7.30 |
| 5/16 | 40 | 9/32 | 7.30 | O | 8.00 |
| 5/16 | 32 | J | 7.20 | O | 8.00 |
| 5/16 | 26 | 17/64 | 6.70 | O | 8.00 |
| 3/8 | 40 | 11/32 | 8.80 | V | 9.80 |
| 3/8 | 32 | R | 8.70 | V | 9.80 |
| 3/8 | 26 | 21/64 | 8.40 | V | 9.80 |
| 7/16 | 40 | 13/32 | 10.50 | 29/64 | 11.50 |
| 7/16 | 32 | Y | 10.40 | 29/64 | 11.50 |

| 7/16 | 26 | 25/64 | 10.00 | 29/64 | 11.50 |
| 1/2 | 40 | 15/32 | 12.10 | 33/64 | 13.00 |
| 1/2 | 32 | 15/32 | 11.90 | 33/64 | 13.00 |
| 1/2 | 26 | 29/64 | 11.70 | 33/64 | 13.00 |

*Note:* All tapping drill sizes quoted in these charts are based on the British Standards Institute recomendations and sizes published in the Model Engineering Press.

## 2.38 Friction locking devices

**Lock nut**

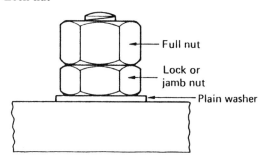

**Stiff nut (insert)**

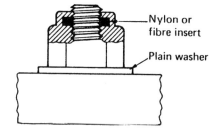

**Stiff nut (slit head)**

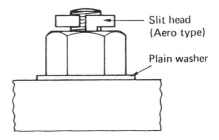

Slit head
(Aero type)

Plain washer

**Stiff nut (slit head)**

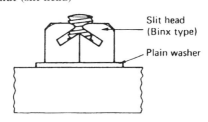

Slit head
(Binx type)

Plain washer

**Serrated (toothed) lock washers**

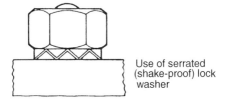

Use of serrated
(shake-proof) lock
washer

*Note:*

Lock washers, see:
Spring washers, Sections 4.1.3–5.
Toothed lock washers, Section 4.1.6.
Serrated lock washers, Section 4.1.7.
Crinkle washers, Section 4.1.8.

## 2.39 Positive locking devices

**Slotted nut**

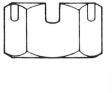

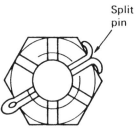

Split pin

**Castle nut**

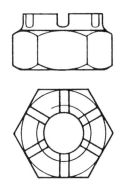

**Tab washer**

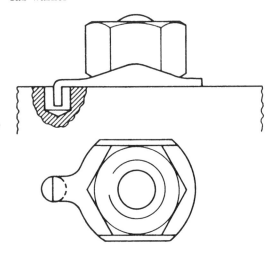

**Lock plate**

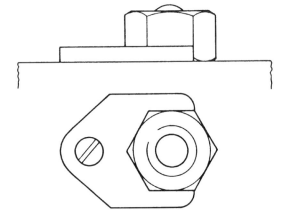

Wiring

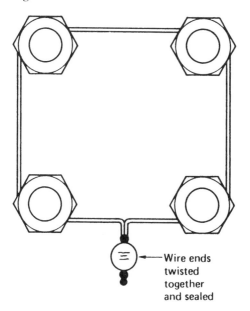

Wire ends
twisted
together
and sealed

# Part 3

# Cutting Tools (HSS) and Abrasive Wheels

## 3.1 Twist drill sizes, metric

| Nominal diameter | Parallel shank jobber series | | Parallel shank stub drills | | Parallel shank long series | | Morse taper (MT) shank two-flute twist and multiflute core drills | | | Oversize Morse taper shank | |
|---|---|---|---|---|---|---|---|---|---|---|---|
| | Flute length | Overall length | Flute length | Overall length | Flute length | Overall length | Flute length | Overall length | MT no. | Overall length | MT no. |
| 0.20 | 2.5 | 19 | | | | | | | | | |
| 0.22 | 2.5 | 19 | | | | | | | | | |
| 0.25 | 3 | 19 | | | | | | | | | |
| 0.28 | 3 | 19 | | | | | | | | | |
| 0.30 | 4 | 19 | | | | | | | | | |
| 0.32 | 4 | 19 | | | | | | | | | |
| 0.35 | 4 | 19 | | | | | | | | | |
| 0.38 | 4 | 19 | | | | | | | | | |
| 0.40 | 5 | 20 | | | | | | | | | |
| 0.42 | 5 | 20 | | | | | | | | | |
| 0.45 | 5 | 20 | | | | | | | | | |
| 0.48 | 5 | 20 | | | | | | | | | |
| 0.50 | 6 | 22 | 3 | 20 | | | | | | | |
| 0.52 | 6 | 22 | | | | | | | | | |

(continued)

183

**3.1** (continued)

(Dimensions in millimetres)

| Nominal diameter | Parallel shank jobber series | | Parallel shank stub drills | | Parallel shank long series | | Morse taper (MT) shank two-flute twist and multiflute core drills | | | Oversize Morse taper shank | |
|---|---|---|---|---|---|---|---|---|---|---|---|
| | Flute length | Overall length | Flute length | Overall length | Flute length | Overall length | Flute length | Overall length | MT no. | Overall length | MT no. |
| 0.55 | 7 | 24 | | | | | | | | | |
| 0.58 | 7 | 24 | | | | | | | | | |
| 0.60 | 7 | 24 | | | | | | | | | |
| 0.62 | 8 | 26 | | | | | | | | | |
| 0.65 | 8 | 26 | | | | | | | | | |
| 0.68 | 9 | 28 | | | | | | | | | |
| 0.70 | 9 | 28 | | | | | | | | | |
| 0.72 | 9 | 28 | | | | | | | | | |
| 0.75 | 9 | 28 | | | | | | | | | |
| 0.78 | 10 | 30 | | | | | | | | | |
| 0.80 | 10 | 30 | 5 | 24 | | | | | | | |
| 0.82 | 10 | 30 | | | | | | | | | |
| 0.85 | 10 | 30 | | | | | | | | | |
| 0.88 | 11 | 32 | | | | | | | | | |
| 0.90 | 11 | 32 | | | | | | | | | |
| 0.92 | 11 | 32 | | | | | | | | | |
| 0.95 | 11 | 32 | | | | | | | | | |
| 0.98 | 12 | 34 | | | | | | | | | |

| | | | | | | |
|---|---|---|---|---|---|---|
| 1.00 | 12 | 34 | 6 | 26 | 33 | 56 |
| 1.05 | 12 | 34 | | | | |
| 1.10 | 14 | 36 | | | 37 | 60 |
| 1.15 | 14 | 36 | | | | |
| 1.20 | 16 | 38 | 8 | 30 | 41 | 65 |
| 1.25 | 16 | 38 | | | | |
| 1.30 | 16 | 38 | | | 41 | 65 |
| 1.35 | 18 | 40 | | | | |
| 1.40 | 18 | 40 | 9 | 32 | 45 | 70 |
| 1.45 | 18 | 40 | | | | |
| 1.50 | 20 | 40 | | | 45 | 70 |
| 1.55 | 20 | 43 | | | | |
| 1.60 | 20 | 43 | 11 | 36 | 50 | 76 |
| 1.65 | 20 | 43 | | | | |
| 1.70 | 22 | 43 | | | 50 | 76 |
| 1.75 | 22 | 46 | | | | |
| 1.80 | 22 | 46 | 12 | 38 | 53 | 80 |
| 1.85 | 22 | 46 | | | | |
| 1.90 | 24 | 46 | | | 53 | 80 |
| 1.95 | 24 | 49 | | | | |
| 2.00 | 24 | 49 | | | 56 | 85 |

(continued)

**3.1** (continued)

(Dimensions in millimetres)

| Nominal diameter | Parallel shank jobber series | | Parallel shank stub drills | | Parallel shank long series | | Morse taper (MT) shank two-flute twist and multiflute core drills | | | Oversize Morse taper shank | |
|---|---|---|---|---|---|---|---|---|---|---|---|
| | Flute length | Overall length | Flute length | Overall length | Flute length | Overall length | Flute length | Overall length | MT no. | Overall length | MT no. |
| 2.05 | 24 | 49 | | | | | | | | | |
| 2.10 | 24 | 49 | | | | | | | | | |
| 2.15 | 27 | 53 | | | | | | | | | |
| 2.20 | 27 | 53 | 13 | 40 | 56 | 85 | | | | | |
| 2.25 | 27 | 53 | | | | | | | | | |
| 2.30 | 27 | 53 | | | 59 | 90 | | | | | |
| 2.35 | 27 | 53 | | | | | | | | | |
| 2.40 | 30 | 57 | | | 59 | 90 | | | | | |
| 2.45 | 30 | 57 | | | 62 | 95 | | | | | |
| 2.50 | 30 | 57 | 14 | 43 | 62 | 95 | | | | | |
| 2.55 | 30 | 57 | | | | | | | | | |
| 2.60 | 30 | 57 | | | 62 | 95 | | | | | |
| 2.65 | 30 | 57 | | | | | | | | | |
| 2.70 | 33 | 61 | | | 66 | 100 | | | | | |

(continued)

| | | | | | | | | | |
|---|---|---|---|---|---|---|---|---|---|
| 2.75 | 33 | 61 | | | | | | | |
| 2.80 | 33 | 61 | 16 | 46 | | | | | |
| 2.85 | 33 | 61 | | | | | | | |
| 2.90 | 33 | 61 | | | | | | | |
| 2.95 | 33 | 61 | 16 | 46 | | | | | |
| 3.00 | 33 | 61 | | | 66 | 100 | 33 | 114 | 1 |
| 3.10 | 36 | 65 | 18 | 49 | 66 | 100 | | | |
| 3.20 | 36 | 65 | | | 66 | 100 | 36 | 117 | 1 |
| 3.30 | 36 | 65 | 20 | 52 | 69 | 106 | | | |
| 3.40 | 39 | 70 | | | 69 | 106 | 39 | 120 | 1 |
| 3.50 | 39 | 70 | 22 | 55 | 69 | 106 | | | |
| 3.60 | 39 | 70 | | | 73 | 112 | | | |
| 3.70 | 39 | 70 | 22 | 55 | 73 | 112 | | | |
| 3.80 | 43 | 75 | | | 73 | 112 | 43 | 123 | 1 |
| 3.90 | 43 | 75 | 22 | 55 | 78 | 119 | | | |
| 4.00 | 43 | 75 | | | 78 | 119 | 43 | 123 | 1 |
| 4.10 | 43 | 75 | | | 78 | 119 | | | |
| 4.20 | 43 | 75 | | | 78 | 126 | 43 | 123 | 1 |
| 4.30 | 47 | 80 | 24 | 58 | 82 | 126 | | | |
| 4.40 | 47 | 80 | | | 82 | 126 | | | |
| 4.50 | 47 | 80 | | | 82 | | | | |
| 4.60 | 47 | 80 | | | 82 | | 47 | 128 | 1 |

**3.1** (continued)

(Dimensions in millimetres)

| Nominal diameter | Parallel shank jobber series | | Parallel shank stub drills | | Parallel shank long series | | Morse taper (MT) shank two-flute twist and multiflute core drills | | | Oversize Morse taper shank | |
|---|---|---|---|---|---|---|---|---|---|---|---|
| | Flute length | Overall length | Flute length | Overall length | Flute length | Overall length | Flute length | Overall length | MT no. | Overall length | MT no. |
| 4.70 | 47 | 80 | | | 82 | 126 | | | | | |
| 4.80 | 52 | 86 | 26 | 62 | 87 | 132 | 52 | 133 | 1 | | |
| 4.90 | 52 | 86 | | | 87 | 132 | | | | | |
| 5.00 | 52 | 86 | 26 | 62 | 87 | 132 | 52 | 133 | 1 | | |
| 5.10 | 52 | 86 | | | 87 | 132 | | | | | |
| 5.20 | 52 | 86 | 26 | 62 | 87 | 132 | 52 | 133 | 1 | | |
| 5.30 | 52 | 86 | | | 87 | 132 | | | | | |
| 5.40 | 57 | 93 | | | 91 | 139 | | | | | |
| 5.50 | 57 | 93 | 28 | 66 | 91 | 139 | 57 | 138 | 1 | | |
| 5.60 | 57 | 93 | | | 91 | 139 | | | | | |
| 5.70 | 57 | 93 | | | 91 | 139 | | | | | |
| 5.80 | 57 | 93 | 28 | 66 | 91 | 139 | 57 | 138 | 1 | | |
| 5.90 | 57 | 93 | | | 91 | 139 | | | | | |
| 6.00 | 57 | 93 | 28 | 66 | 91 | 139 | 57 | 138 | 1 | | |
| 6.10 | 63 | 101 | | | 97 | 148 | | | | | |

| | | | | | | | | | |
|---|---|---|---|---|---|---|---|---|---|
| 6.20 | 63 | 101 | | | 97 | 148 | | | |
| 6.30 | 63 | 101 | 31 | 70 | 97 | 148 | 63 | 144 | 1 |
| 6.40 | 63 | 101 | | | 97 | 148 | | | |
| 6.50 | 63 | 101 | | | 97 | 148 | | | |
| 6.60 | 63 | 101 | 31 | 70 | 97 | 148 | 63 | 144 | 1 |
| 6.70 | 63 | 101 | | | 97 | 148 | | | |
| 6.80 | 69 | 109 | 34 | 74 | 102 | 156 | 69 | 150 | 1 |
| 6.90 | 69 | 109 | | | 102 | 156 | | | |
| 7.00 | 69 | 109 | 34 | 74 | 102 | 156 | 69 | 150 | 1 |
| 7.10 | 69 | 109 | | | 102 | 156 | | | |
| 7.20 | 69 | 109 | 34 | 74 | 102 | 156 | 69 | 150 | 1 |
| 7.30 | 69 | 109 | | | 102 | 156 | | | |
| 7.40 | 69 | 109 | 34 | 74 | 102 | 156 | 69 | 150 | 1 |
| 7.50 | 75 | 117 | | | 109 | 156 | | | |
| 7.60 | 75 | 117 | | | 109 | 165 | | | |
| 7.70 | 75 | 117 | 37 | 79 | 109 | 165 | 75 | 156 | 1 |
| 7.80 | 75 | 117 | | | 109 | 165 | | | |
| 7.90 | 75 | 117 | | | 109 | 165 | | | |

(continued)

**3.1** (continued)

(Dimensions in millimetres)

| Nominal diameter | Parallel shank jobber series | | Parallel shank stub drills | | Parallel shank long series | | Morse taper (MT) shank two-flute twist and multiflute core drills | | | Oversize Morse taper shank | |
|---|---|---|---|---|---|---|---|---|---|---|---|
| | Flute length | Overall length | Flute length | Overall length | Flute length | Overall length | Flute length | Overall length | MT no. | Overall length | MT no. |
| 8.00 | 75 | 117 | 37 | 79 | 109 | 165 | 75 | 156 | 1 | | |
| 8.10 | 75 | 117 | | | 109 | 165 | | | | | |
| 8.20 | 75 | 117 | 37 | 79 | 109 | 165 | 75 | 156 | 1 | | |
| 8.30 | 75 | 117 | | | 109 | 165 | | | | | |
| 8.40 | 75 | 117 | | | 109 | 165 | | | | | |
| 8.50 | 75 | 117 | 37 | 79 | 109 | 165 | 75 | 156 | 1 | | |
| 8.60 | 81 | 125 | | | 115 | 175 | | | | | |
| 8.70 | 81 | 125 | | | 115 | 175 | | | | | |
| 8.80 | 81 | 125 | 40 | 84 | 115 | 175 | 81 | 162 | 1 | | |
| 8.90 | 81 | 125 | | | 115 | 175 | | | | | |
| 9.00 | 81 | 125 | 40 | 84 | 115 | 175 | 81 | 162 | 1 | | |
| 9.10 | 81 | 125 | | | 115 | 175 | | | | | |
| 9.20 | 81 | 125 | 40 | 84 | 115 | 175 | 81 | 162 | 1 | | |
| 9.30 | 81 | 125 | | | 115 | 175 | | | | | |
| 9.40 | 81 | 125 | | | 115 | 175 | | | | | |

| | | | | | | | | | |
|---|---|---|---|---|---|---|---|---|---|
| 9.50 | 81 | 125 | 40 | 84 | 115 | 175 | 81 | 162 | 1 |
| 9.60 | 87 | 133 | | | 121 | 184 | | | |
| 9.70 | 87 | 133 | 43 | 89 | 121 | 184 | 87 | 168 | 1 |
| 9.80 | 87 | 133 | 43 | 89 | 121 | 184 | 87 | 168 | 1 |
| 9.90 | 87 | 133 | 43 | 89 | 121 | 184 | 87 | 168 | 1 |
| 10.00 | 87 | 133 | | | 121 | 184 | | | |
| 10.10 | 87 | 133 | | | 121 | 184 | | | |
| 10.20 | 87 | 133 | | | 121 | 184 | | | |
| 10.30 | 87 | 133 | | | 121 | 184 | | | |
| 10.40 | 87 | 133 | | | 121 | 184 | | | |
| 10.50 | 87 | 133 | 43 | 89 | 121 | 184 | 87 | 168 | 1 |
| 10.60 | 87 | 133 | | | 121 | 184 | | | |
| 10.70 | 94 | 142 | | | 128 | 195 | | | |
| 10.80 | 94 | 142 | 47 | 95 | 128 | 195 | 94 | 175 | 1 |
| 10.90 | 94 | 142 | | | 128 | 195 | | | |
| 11.00 | 94 | 142 | 47 | 95 | 128 | 195 | 94 | 175 | 1 |
| 11.10 | 94 | 142 | | | 128 | 195 | | | |
| 11.20 | 94 | 142 | 47 | 95 | 128 | 195 | 94 | 175 | 1 |
| 11.30 | 94 | 142 | | | 128 | 195 | | | |
| 11.40 | 94 | 142 | | | 128 | 195 | | | |
| 11.50 | 94 | 142 | 47 | 95 | 128 | 195 | 94 | 175 | 1 |
| 11.60 | 94 | 142 | | | 128 | 195 | | | |
| 11.70 | 94 | 142 | 47 | 95 | 128 | 195 | 94 | 175 | 1 |
| 11.80 | 94 | 142 | | | 128 | 195 | | | |
| 11.90 | 101 | 151 | | | 134 | 205 | | | |

(continued)

**3.1** (continued)

(Dimensions in millimetres)

| Nominal diameter | Parallel shank jobber series | | Parallel shank stub drills | | Parallel shank long series | | Morse taper (MT) shank two-flute twist and multiflute core drills | | | Oversize Morse taper shank | |
|---|---|---|---|---|---|---|---|---|---|---|---|
| | Flute length | Overall length | Flute length | Overall length | Flute length | Overall length | Flute length | Overall length | MT no. | Overall length | MT no. |
| 12.00 | 101 | 151 | 51 | 102 | 134 | 205 | 101 | 182 | 1 | 199 | 2 |
| 12.10 | 101 | 151 | | | 134 | 205 | | | | | |
| 12.20 | 101 | 151 | 51 | 102 | 134 | 205 | 101 | 182 | 1 | 199 | 2 |
| 12.30 | 101 | 151 | | | 134 | 205 | | | | | |
| 12.40 | 101 | 151 | | | 134 | 205 | | | | | |
| 12.50 | 101 | 151 | 51 | 102 | 134 | 205 | 101 | 182 | 1 | 199 | 2 |
| 12.60 | 101 | 151 | | | 134 | 205 | | | | | |
| 12.70 | 101 | 151 | | | 134 | 205 | | | | | |
| 12.80 | 101 | 151 | 51 | 102 | 134 | 205 | 101 | 182 | 1 | 199 | 2 |
| 12.90 | 101 | 151 | | | 134 | 205 | | | | | |
| 13.00 | 101 | 151 | 51 | 102 | 134 | 205 | 101 | 182 | 1 | 199 | 2 |
| 13.10 | 101 | 151 | | | 134 | 205 | | | | | |
| 13.20 | 101 | 151 | 51 | 102 | 134 | 205 | 101 | 182 | 1 | 199 | 2 |
| 13.30 | 108 | 160 | | | 140 | 214 | | | | | |
| 13.40 | 108 | 160 | | | 140 | 214 | | | | | |
| 13.50 | 108 | 160 | 54 | 107 | 140 | 214 | 108 | 189 | 1 | 206 | 2 |
| 13.60 | 108 | 160 | | | 140 | 214 | | | | | |
| 13.70 | 108 | 160 | 54 | 107 | 140 | 214 | 108 | 189 | 1 | 206 | 2 |

| | | | | | | | | | | | |
|------|-----|-----|-----|-----|-----|-----|-----|-----|---|-----|---|
| 14.00 | 108 | 160 | 54 | 107 | 140 | 214 | 108 | 189 | 1 | 206 | 2 |
| 14.25 | 114 | 169 | 55 | 111 | 144 | 220 | 114 | 212 | 2 | 206 | 2 |
| 14.50 | 114 | 169 | 56 | 111 | 144 | 220 | 114 | 212 | 2 | 206 | 2 |
| 14.75 | 114 | 169 | 58 | 115 | 144 | 220 | 114 | 212 | 2 | 206 | 2 |
| 15.00 | 114 | 169 | 58 | 115 | 144 | 220 | 114 | 212 | 2 | 206 | 2 |
| 15.25 | 120 | 178 | 60 | 119 | 149 | 227 | 120 | 218 | 2 | 206 | 2 |
| 15.50 | 120 | 178 | | | 149 | 227 | 120 | 218 | 2 | 206 | 2 |
| 15.75 | 120 | 178 | | | 149 | 227 | 120 | 218 | 2 | 206 | 2 |
| 16.00 | 120 | 178 | | | 149 | 235 | 120 | 223 | 2 | 206 | 2 |
| 16.25 | | | | | 154 | 235 | 125 | 223 | 2 | 206 | 2 |
| 16.50 | | | | | 154 | 235 | 125 | 223 | 2 | 206 | 2 |
| 16.75 | | | | | 154 | 241 | 125 | 223 | 2 | 206 | 2 |
| 17.00 | | | | | 158 | 241 | 130 | 228 | 2 | 206 | 2 |
| 17.25 | | | | | 158 | 241 | 130 | 228 | 2 | 206 | 2 |
| 17.50 | | | | | 158 | 247 | 130 | 228 | 2 | 206 | 2 |
| 17.75 | | | | | 158 | 247 | 135 | 233 | 2 | 256 | 3 |
| 18.00 | | | | | 162 | 247 | 135 | 283 | 2 | 256 | 3 |
| 18.25 | | | | | 162 | 247 | 135 | 283 | 2 | 256 | 3 |
| 18.50 | | | | | 162 | 254 | 135 | 233 | 2 | 256 | 3 |
| 18.75 | | | | | 162 | | 140 | 238 | 2 | 261 | 3 |
| 19.00 | | | | | 166 | | | | | | |
| 19.25 | | | | | | | | | | | |

*(continued)*

| Nominal diameter | Parallel shank long series | | Morse taper (MT) shank two-flute twist and multiflute core drills | | | Oversize Morse taper shank | |
|---|---|---|---|---|---|---|---|
| | Flute length | Overall length | Flute length | Overall length | MT no. | Overall length | MT no. |
| 19.50 | 166 | 254 | 140 | 238 | 2 | 261 | 3 |
| 19.75 | 166 | 254 | 140 | 238 | 2 | 261 | 3 |
| 20.00 | 166 | 254 | 140 | 238 | 2 | 261 | 3 |
| 20.25 | 171 | 261 | 145 | 243 | 2 | 266 | 3 |
| 20.50 | 171 | 261 | 145 | 243 | 2 | 266 | 3 |
| 20.75 | 171 | 261 | 145 | 243 | 2 | 266 | 3 |
| 21.00 | 171 | 261 | 145 | 243 | 2 | 266 | 3 |
| 21.25 | 176 | 268 | 150 | 248 | 2 | 271 | 3 |
| 21.50 | 176 | 268 | 150 | 248 | 2 | 271 | 3 |
| 21.75 | 176 | 268 | 150 | 248 | 2 | 271 | 3 |
| 22.00 | 176 | 268 | 150 | 248 | 2 | 271 | 3 |
| 22.25 | 176 | 268 | 150 | 248 | 2 | 271 | 3 |
| 22.50 | 180 | 275 | 155 | 253 | 2 | 276 | 3 |
| 22.75 | 180 | 275 | 155 | 253 | 2 | 276 | 3 |
| 23.00 | 180 | 275 | 155 | 253 | 2 | 276 | 3 |
| 23.25 | 180 | 275 | 155 | 276 | 3 | — | — |
| 23.50 | 180 | 275 | 155 | 276 | 3 | — | — |
| 23.75 | 185 | 282 | 160 | 281 | 3 | — | — |
| 24.00 | 185 | 282 | 160 | 281 | 3 | — | — |
| 24.25 | 185 | 282 | 160 | 281 | 3 | — | — |
| 24.50 | 185 | 282 | 160 | 281 | 3 | — | — |
| 24.75 | 185 | 282 | 160 | 281 | 3 | — | — |
| 25.00 | 185 | 282 | 160 | 281 | 3 | — | — |
| 25.25 | | | 165 | 286 | 3 | — | — |
| 25.50 | | | 165 | 286 | 3 | — | — |
| 25.75 | | | 165 | 286 | 3 | — | — |
| 26.00 | | | 165 | 286 | 3 | — | — |
| 26.25 | | | 165 | 286 | 3 | — | — |
| 26.50 | | | 165 | 286 | 3 | — | — |
| 26.75 | | | 170 | 291 | 3 | 319 | 4 |
| 27.00 | | | 170 | 291 | 3 | 319 | 4 |
| 27.25 | | | 170 | 291 | 3 | 319 | 4 |
| 27.50 | | | 170 | 291 | 3 | 319 | 4 |
| 27.75 | | | 170 | 291 | 3 | 319 | 4 |
| 28.00 | | | 170 | 291 | 3 | 319 | 4 |
| 28.25 | | | 175 | 296 | 3 | 324 | 4 |
| 28.50 | | | 175 | 296 | 3 | 324 | 4 |
| 28.75 | | | 175 | 296 | 3 | 324 | 4 |
| 29.00 | | | 175 | 296 | 3 | 324 | 4 |
| 29.25 | | | 175 | 296 | 3 | 324 | 4 |
| 29.50 | | | 175 | 296 | 3 | 324 | 4 |
| 29.75 | | | 175 | 296 | 3 | 324 | 4 |
| 30.00 | | | 175 | 296 | 3 | 324 | 4 |
| 30.25 | | | 180 | 301 | 3 | 329 | 4 |
| 30.50 | | | 180 | 301 | 3 | 329 | 4 |

| Nominal diameter | Morse taper (MT) shank two-flute twist and multiflute core drills | | | Oversize Morse taper shank | |
|---|---|---|---|---|---|
| | Flute length | Overall length | MT no. | Overall length | MT no. |
| 30.75 | 180 | 301 | 3 | 329 | 4 |
| 31.00 | 180 | 301 | 3 | 329 | 4 |
| 31.25 | 180 | 301 | 3 | 329 | 4 |
| 31.50 | 180 | 301 | 3 | 329 | 4 |
| 31.75 | 185 | 306 | 3 | 334 | 4 |
| 32.00 | 185 | 334 | 4 | — | — |
| 32.50 | 185 | 334 | 4 | — | — |
| 33.00 | 185 | 334 | 4 | — | — |
| 33.50 | 185 | 334 | 4 | — | — |
| 34.00 | 190 | 339 | 4 | — | — |
| 34.50 | 190 | 339 | 4 | — | — |
| 35.00 | 190 | 339 | 4 | — | — |
| 35.50 | 190 | 339 | 4 | — | — |
| 36.00 | 195 | 344 | 4 | — | — |
| 36.50 | 195 | 344 | 4 | — | — |
| 37.00 | 195 | 344 | 4 | — | — |
| 37.50 | 195 | 344 | 4 | — | — |
| 38.00 | 200 | 349 | 4 | — | — |
| 38.50 | 200 | 349 | 4 | — | — |
| 39.00 | 200 | 349 | 4 | — | — |
| 39.50 | 200 | 349 | 4 | — | — |
| 40.00 | 200 | 349 | 4 | — | — |
| 40.50 | 205 | 354 | 4 | 392 | 5 |
| 41.00 | 205 | 354 | 4 | 392 | 5 |
| 41.50 | 205 | 354 | 4 | 392 | 5 |
| 42.00 | 205 | 354 | 4 | 392 | 5 |
| 42.50 | 205 | 354 | 4 | 392 | 5 |
| 43.00 | 210 | 359 | 4 | 397 | 5 |
| 43.50 | 210 | 359 | 4 | 397 | 5 |
| 44.00 | 210 | 359 | 4 | 397 | 5 |
| 44.50 | 210 | 359 | 4 | 397 | 5 |
| 45.00 | 210 | 359 | 4 | 397 | 5 |
| 45.50 | 215 | 364 | 4 | 402 | 5 |
| 46.00 | 215 | 364 | 4 | 402 | 5 |
| 46.50 | 215 | 364 | 4 | 402 | 5 |
| 47.00 | 215 | 364 | 4 | 402 | 5 |
| 47.50 | 215 | 364 | 4 | 402 | 5 |
| 48.00 | 220 | 369 | 4 | 407 | 5 |
| 48.50 | 220 | 369 | 4 | 407 | 5 |
| 49.00 | 220 | 369 | 4 | 407 | 5 |
| 49.50 | 220 | 369 | 4 | 407 | 5 |
| 50.00 | 220 | 369 | 4 | 407 | 5 |
| 50.50 | 225 | 374 | 4 | 412 | 5 |
| 51.00 | 225 | 412 | 5 | — | — |
| 52.00 | 225 | 412 | 5 | — | — |
| 53.00 | 225 | 412 | 5 | — | — |
| 54.00 | 230 | 417 | 5 | — | — |
| 55.00 | 230 | 417 | 5 | — | — |
| 56.00 | 230 | 417 | 5 | — | — |

(*continued*)

(Dimensions in millimetres)

| Nominal diameter | Morse taper (MT) shank two-flute twist and multiflute core drills | | | Oversize Morse taper shank | |
|---|---|---|---|---|---|
| | Flute length | Overall length | MT no. | Overall length | MT no. |
| 57.00 | 235 | 422 | 5 | — | — |
| 58.00 | 235 | 422 | 5 | — | — |
| 59.00 | 235 | 422 | 5 | — | — |
| 60.00 | 235 | 422 | 5 | — | — |
| 61.00 | 240 | 427 | 5 | — | — |
| 62.00 | 240 | 427 | 5 | — | — |
| 63.00 | 240 | 427 | 5 | — | — |
| 64.00 | 245 | 432 | 5 | 499 | 6 |
| 65.00 | 245 | 432 | 5 | 499 | 6 |
| 66.00 | 245 | 432 | 5 | 499 | 6 |
| 67.00 | 245 | 432 | 5 | 499 | 6 |
| 68.00 | 250 | 437 | 5 | 504 | 6 |
| 69.00 | 250 | 437 | 5 | 504 | 6 |
| 70.00 | 250 | 437 | 5 | 504 | 6 |
| 71.00 | 250 | 437 | 5 | 504 | 6 |
| 72.00 | 255 | 442 | 5 | 509 | 6 |
| 73.00 | 255 | 442 | 5 | 509 | 6 |
| 74.00 | 255 | 442 | 5 | 509 | 6 |
| 75.00 | 255 | 442 | 5 | 509 | 6 |
| 76.00 | 260 | 447 | 5 | 514 | 6 |
| 77.00 | 260 | 514 | 6 | — | — |
| 78.00 | 260 | 514 | 6 | — | — |
| 79.00 | 260 | 514 | 6 | — | — |
| 80.00 | 260 | 514 | 6 | — | — |
| 81.00 | 265 | 519 | 6 | — | — |
| 82.00 | 265 | 519 | 6 | — | — |
| 83.00 | 265 | 519 | 6 | — | — |
| 84.00 | 265 | 519 | 6 | — | — |
| 85.00 | 265 | 519 | 6 | — | — |
| 86.00 | 270 | 524 | 6 | — | — |
| 87.00 | 270 | 524 | 6 | — | — |
| 88.00 | 270 | 524 | 6 | — | — |
| 89.00 | 270 | 524 | 6 | — | — |
| 90.00 | 270 | 524 | 6 | — | — |
| 91.00 | 275 | 529 | 6 | — | — |
| 92.00 | 275 | 529 | 6 | — | — |
| 93.00 | 275 | 529 | 6 | — | — |
| 94.00 | 275 | 529 | 6 | — | — |
| 95.00 | 275 | 529 | 6 | — | — |
| 96.00 | 280 | 534 | 6 | — | — |
| 97.00 | 280 | 534 | 6 | — | — |
| 98.00 | 280 | 534 | 6 | — | — |
| 99.00 | 280 | 534 | 6 | — | — |
| 100.00 | 280 | 534 | 6 | — | — |

For further information see BS 328.

## 3.2 Gauge and letter size twist drills

(and alternative New Standard International
Series)

| OLD | | NEW | |
|---|---|---|---|
| Drill No. | Decimal inch | Drill mm | Decimal inch |
| 80 | 0.0135 | 0.35 | 0.0138 |
| 79 | 0.0145 | 0.38 | 0.0150 |
| 78 | 0.0160 | 0.40 | 0.0157 |
| 77 | 0.0180 | 0.45 | 0.0177 |
| 76 | 0.0200 | 0.50 | 0.0197 |
| 75 | 0.0210 | 0.52 | 0.0205 |
| 74 | 0.0225 | 0.58 | 0.0228 |
| 73 | 0.0240 | 0.60 | 0.0236 |
| 72 | 0.0250 | 0.65 | 0.0256 |
| 71 | 0.0260 | 0.65 | 0.0256 |
| 70 | 0.0280 | 0.70 | 0.0276 |
| 69 | 0.0292 | 0.75 | 0.0295 |
| 68 | 0.0310 | $1/32''$ | 0.0312 |
| 67 | 0.0320 | 0.82 | 0.0323 |
| 66 | 0.0330 | 0.85 | 0.0335 |
| 65 | 0.0350 | 0.90 | 0.0354 |
| 64 | 0.0360 | 0.92 | 0.0362 |
| 63 | 0.0370 | 0.95 | 0.0374 |
| 62 | 0.0380 | 0.98 | 0.0386 |
| 61 | 0.0390 | 1.00 | 0.0394 |
| 60 | 0.0400 | 1.00 | 0.0394 |
| 59 | 0.0410 | 1.05 | 0.0413 |
| 58 | 0.0420 | 1.05 | 0.0413 |
| 57 | 0.0430 | 1.10 | 0.0433 |
| 56 | 0.0465 | $3/64''$ | 0.0469 |
| 55 | 0.0520 | 1.30 | 0.0512 |
| 54 | 0.0550 | 1.40 | 0.0551 |
| 53 | 0.0595 | 1.50 | 0.0591 |
| 52 | 0.0635 | 1.60 | 0.0630 |
| 51 | 0.0670 | 1.70 | 0.0669 |
| 50 | 0.0700 | 1.80 | 0.0709 |
| 49 | 0.0730 | 1.85 | 0.0728 |
| 48 | 0.0760 | 1.95 | 0.0768 |
| 47 | 0.0785 | 2.00 | 0.0787 |
| 46 | 0.0810 | 2.05 | 0.0807 |

*(continued)*

| OLD | | NEW | |
|---|---|---|---|
| *Drill No.* | *Decimal inch* | *Drill mm* | *Decimal inch* |
| 45 | 0.0820 | 2.10 | 0.0827 |
| 44 | 0.0860 | 2.20 | 0.0866 |
| 43 | 0.0890 | 2.25 | 0.0886 |
| 42 | 0.0935 | $3/_{32}''$ | 0.0938 |
| 41 | 0.0960 | 2.45 | 0.0965 |
| 40 | 0.0980 | 2.50 | 0.0984 |
| 39 | 0.0995 | 2.55 | 0.1004 |
| 38 | 0.1015 | 2.60 | 0.1024 |
| 37 | 0.1040 | 2.65 | 0.1043 |
| 36 | 0.1065 | 2.70 | 0.1063 |
| 35 | 0.1100 | 2.80 | 0.1102 |
| 34 | 0.1110 | 2.80 | 0.1102 |
| 33 | 0.1130 | 2.85 | 0.1122 |
| 32 | 0.1160 | 2.95 | 0.1161 |
| 31 | 0.1200 | 3.00 | 0.1181 |
| 30 | 0.1285 | 3.30 | 0.1299 |
| 29 | 0.1360 | 3.50 | 0.1378 |
| 28 | 0.1405 | $9/_{64}''$ | 0.1406 |
| 27 | 0.1440 | 3.70 | 0.1457 |
| 26 | 0.1470 | 3.70 | 0.1457 |
| 25 | 0.1495 | 3.80 | 0.1496 |
| 24 | 0.1520 | 3.90 | 0.1535 |
| 23 | 0.1540 | 3.90 | 0.1535 |
| 22 | 0.1570 | 4.00 | 0.1575 |
| 21 | 0.1590 | 4.00 | 0.1575 |
| 20 | 0.1610 | 4.10 | 0.1614 |
| 19 | 0.1660 | 4.20 | 0.1654 |
| 18 | 0.1695 | 4.30 | 0.1693 |
| 17 | 0.1730 | 4.40 | 0.1732 |
| 16 | 0.1770 | 4.50 | 0.1772 |
| 15 | 0.1800 | 4.60 | 0.1811 |
| 14 | 0.1820 | 4.60 | 0.1811 |
| 13 | 0.1850 | 4.70 | 0.1850 |
| 12 | 0.1890 | 4.80 | 0.1890 |
| 11 | 0.1910 | 4.90 | 0.1929 |
| 10 | 0.1935 | 4.90 | 0.1929 |
| 9 | 0.1960 | 5.00 | 0.1968 |
| 8 | 0.1990 | 5.10 | 0.2008 |
| 7 | 0.2010 | 5.10 | 0.2008 |
| 6 | 0.2040 | 5.20 | 0.2047 |

| | | | |
|---|---|---|---|
| 5 | 0.2055 | 5.20 | 0.2047 |
| 4 | 0.2090 | 5.30 | 0.2087 |
| 3 | 0.2130 | 5.40 | 0.2126 |
| 2 | 0.2210 | 5.60 | 0.2205 |
| 1 | 0.2280 | 5.80 | 0.2283 |
| A | 0.2340 | $15\!/_{64}''$ | 0.2344 |
| B | 0.2380 | 6.00 | 0.2362 |
| C | 0.2420 | 6.10 | 0.2402 |
| D | 0.2460 | 6.20 | 0.2441 |
| E | 0.2500 | $1\!/_4''$ | 0.2500 |
| F | 0.2570 | 6.50 | 0.2559 |
| G | 0.2610 | 6.60 | 0.2598 |
| H | 0.2660 | $17\!/_{64}''$ | 0.2656 |
| I | 0.2720 | 6.90 | 0.2717 |
| J | 0.2770 | 7.00 | 0.2756 |
| K | 0.2810 | $9\!/_{32}''$ | 0.2812 |
| L | 0.2900 | 7.40 | 0.2913 |
| M | 0.2950 | 7.50 | 0.2953 |
| N | 0.3020 | 7.70 | 0.3031 |
| O | 0.3160 | 8.00 | 0.3150 |
| P | 0.3230 | 8.20 | 0.3228 |
| Q | 0.3320 | 8.40 | 0.3307 |
| R | 0.3390 | 8.60 | 0.3386 |
| S | 0.3480 | 8.80 | 0.3465 |
| T | 0.3580 | 9.10 | 0.3583 |
| U | 0.3680 | 9.30 | 0.3661 |
| V | 0.3770 | $3\!/_8''$ | 0.3750 |
| W | 0.3860 | 9.80 | 0.3858 |
| X | 0.3970 | 10.10 | 0.3976 |
| Y | 0.4040 | 10.30 | 0.4055 |
| Z | 0.4130 | 10.50 | 0.4134 |

Drill gauge and letters sizes are now obsolete and should not be used in new designs – ref. BS 328.

### 3.3 Hand reamer (normal lead)

Taper lead (1°) = 1 1/2 × diameter or 20 mm whichever is the smaller

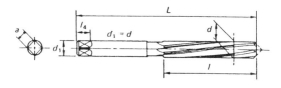

(Dimensions in millimetres)

| Preferred cutting diameters* | Cutting edge length | Overall length | Driving square | |
|---|---|---|---|---|
| d | l | L | a(h12) | $l_4$ |
| 1.5 | 20 | 41 | 1.12 | 4 |
| **1.6** | 21 | 44 | 1.25 | |
| **1.8** | 23 | 47 | 1.40 | |
| **2.0** | 25 | 50 | 1.60 | |
| 2.2 | 27 | 54 | 1.80 | |
| 2.5 | 29 | 58 | 2.00 | |
| **2.8** / **3.0** | 31 | 62 | 2.24 | 5 |
| **3.5** | 35 | 71 | 2.80 | |
| **4.0** | 38 | 76 | 3.15 | 6 |
| **4.5** | 41 | 81 | 3.55 | |
| **5.0** | 44 | 87 | 4.00 | 7 |
| **5.5** / **6.0** | 47 | 93 | 4.50 | |
| **7.0** | 54 | 107 | 5.60 | 8 |
| **8.0** | 58 | 115 | 6.30 | 9 |
| **9.0** | 62 | 124 | 7.10 | 10 |
| **10.0** | 66 | 133 | 8.00 | 11 |
| **11.0** | 71 | 142 | 9.00 | 12 |
| **12.0** / 13.0 | 76 | 152 | 10.00 | 13 |
| **14.0** / 15.0 | 81 | 163 | 11.20 | 14 |
| **16.0** / 17.0 | 87 | 175 | 12.50 | 16 |

| | | | | |
|---|---|---|---|---|
| **18.0** 19.0 | 93 | 188 | 14.00 | 18 |
| **20.0** 21.0 | 100 | 201 | 16.00 | 20 |
| **22** 23 | 107 | 215 | 18.00 | 22 |
| 24 **25** 26 | 115 | 231 | 20.00 | 24 |
| 27 **28** 30 | 124 | 247 | 22.40 | 26 |
| **32** | 133 | 265 | 25.00 | 28 |
| 34 35 **36** | 142 | 284 | 28.00 | 31 |
| 38 **40** 42 | 152 | 305 | 31.50 | 34 |
| 44 **45** 46 | 163 | 326 | 35.50 | 38 |
| 48 **50** 52 | 174 | 347 | 40.00 | 42 |
| 55 **56** 58 60 | 184 | 367 | 45.00 | 46 |
| 62 **63** **67** | 194 | 387 | 50.00 | 51 |
| **71** | 203 | 406 | 56.00 | 56 |

*The diameters in bold type should be used whenever possible.

This table is based on a table from ISO 236/1, except that the latter uses the symbols $l$ for $L$ and $l_1$ for $l$. For full range and further information see BS 328: Pt 4: 1983.

## 3.4 Long flute machine reamers

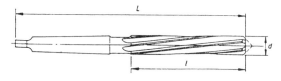

(Dimensions in millimetres)

| Preferred cutting diameters* $d$ | Cutting edge length $l$ | Overall length $L$ | Morse taper shank |
|:---:|:---:|:---:|:---:|
| **7** | 54 | 134 | |
| **8** | 58 | 138 | |
| **9** | 62 | 142 | |
| **10** | 66 | 146 | |
| **11** | 71 | 151 | no. 1 |
| **12** 13 | 76 | 156 | |
| **14** 15 | 81 | 161 181 | |
| **16** 17 | 87 | 187 | |
| **18** 19 | 93 | 193 | no. 2 |
| **20** 21 | 100 | 200 | |
| **22** 23 | 207 | 207 | |
| 24 **25** 26 | 115 | 242 | |
| 27 **28** 30 | 124 | 251 | no. 3 |

| | | | |
|---|---|---|---|
| **32** | 133 | 293 | |
| 34 | | | |
| 35 | 142 | 302 | |
| **36** | | | |
| 38 | | | |
| **40** | 152 | 312 | no. 4 |
| 42 | | | |
| 44 | | | |
| **45** | 163 | 323 | |
| 46 | | | |
| 48 | | 334 | |
| **50** | 174 | | |
| 52 | | 371 | |
| 55 | | | |
| **56** | 184 | 381 | |
| 58 | | | |
| 60 | | | no. 5 |
| 62 | | | |
| **63** | 194 | 391 | |
| **67** | | | |
| **71** | 203 | 400 | |

*The diameters in bold type should be used whenever possible.

This table is based on a table from ISO 236/II, except that the latter uses the symbols $l$ for $L$ and $l_1$ for $l$.

For tool definitions, full range and further information see BS 328: 4: 1983.

## 3.5 Machine chucking reamers with Morse taper shanks

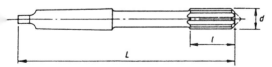

## Machine chucking reamers with Morse taper shanks, dimensions

(Dimensions in millimetres)

| Preferred cutting diameters* d | Cutting edge length l | Overall length L | Morse taper shank |
|---|---|---|---|
| 5.5 }<br>6 } | 26 | 138 | |
| 7 | 31 | 150 | |
| 8 | 33 | 156 | |
| 9 | 36 | 162 | |
| 10 | 38 | 168 | no. 1 |
| 11 | 41 | 175 | |
| 12 }<br>13 } | 44 | 182 | |
| 14 | 47 | 189 | |
| 15 | 50 | 204 | |
| 16 | 52 | 210 | |
| 17 | 54 | 214 | |
| 18 | 56 | 219 | no. 2 |
| 19 | 58 | 223 | |
| 20 | 60 | 228 | |
| 22 | 64 | 237 | |
| 24 }<br>25 } | 68 | 268 | |
| 26 | 70 | 273 | no. 3 |
| 28 | 71 | 277 | |
| 30 | 73 | 281 | |
| 32 | 77 | 317 | |
| 34 }<br>35 } | 78 | 321 | |
| 36 | 79 | 325 | |
| 38 }<br>40 } | 81 | 329 | |
| 42 | 82 | 333 | no. 4 |
| 44 }<br>45 } | 83 | 336 | |
| 46 | 84 | 340 | |
| 48 }<br>50 } | 86 | 344 | |

*The diameters in bold type should be used whenever possible.
This table is based on a table from ISO 521. For tool
definitions, full range and further information see
BS 328: Pt 4: 1983.

## 3.6 Shell reamers with taper bore

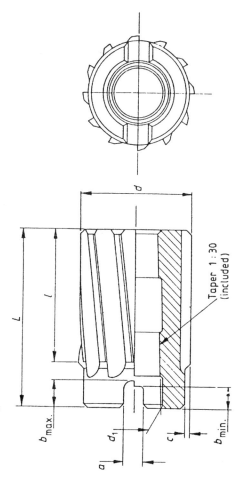

Taper 1:30 (included)

(continued)

205

**Arbor for shell reamer with taper bore**

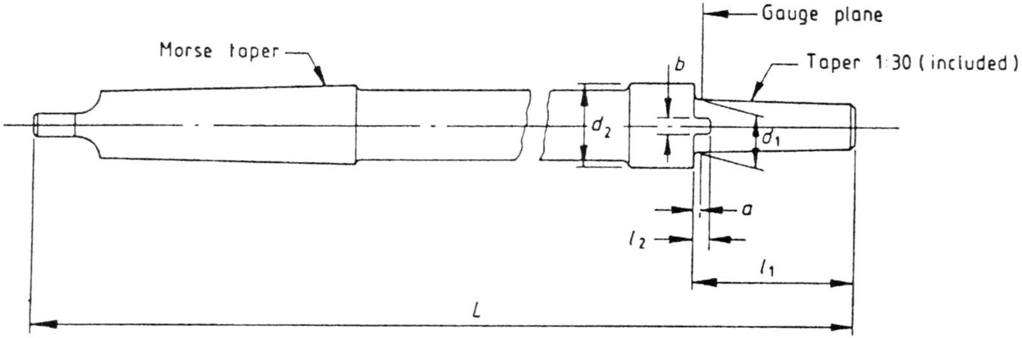

Gauge plane

Taper 1:30 (included)

$d_1$

$a$

$l_1$

$b$

$l_2$

$d_2$

$L$

Morse taper

(Dimensions in millimetres)

| Reamer diameter d | | | Diameter of large end of taper bore $d_1$ | Width of driving slot a (H13)* | Depth of driving slot b | | Relief depth C max. | Cutting edge length l | Overall length L |
|---|---|---|---|---|---|---|---|---|---|
| Over | Up to and including | Preferred sizes | | | min. | max. | | | |
| 19.9 | 23.6 | – | 10 | 4.3 | 5.4 | 7.0 | 1.0 | 28 | 40 |
| 23.6 | 30.0 | 25 26 27 28 30 | 13 | 4.3 | 5.4 | 7.0 | 1.0 | 32 | 45 |
| 30.0 | 35.5 | 32 34 35 | 16 | 5.4 | 6.2 | 8.3 | 1.5 | 36 | 50 |
| 35.5 | 42.5 | 36 38 40 42 | 19 | 6.4 | 7.8 | 10.2 | 1.5 | 40 | 56 |
| 42.5 | 50.8 | 45 47 48 50 | 22 | 7.4 | 8.6 | 11.3 | 1.5 | 45 | 63 |

(continued)

**3.6** (continued)

(Dimensions in millimetres)

| Reamer diameter d | | | Diameter of large end of taper bore $d_1$ | Width of driving slot a (H13)* | Depth of driving slot b | | Relief depth C max. | Cutting edge length l | Overall length L |
|---|---|---|---|---|---|---|---|---|---|
| Over | Up to and including | Preferred sizes | | | min. | max. | | | |
| 50.8 | 60.0 | 52 55 58 60 | 27 | 8.4 | 9.3 | 12.5 | 2.0 | 50 | 71 |
| 60.0 | 71.0 | 62 65 70 | 32 | 10.4 | 10.5 | 14.5 | 2.0 | 56 | 80 |
| 71.0 | 85.0 | 72 75 80 85 | 40 | 12.4 | 11.2 | 16.2 | 2.5 | 63 | 90 |
| 85.0 | 101.6 | 90 95 100 | 50 | 14.4 | 13.1 | 18.7 | 2.5 | 71 | 100 |

*For values of the tolerance H13, see BS 328: Pt 4 Appendix B.
The dimensions shown in this table are in accordance with ISO 2402, except that the latter does not include preferred diameters.

## 3.7 Hand taper pin reamer

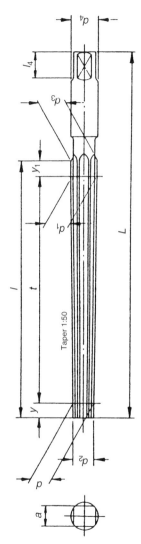

*(continued)*

**3.7** (*continued*)                                    (Dimensions in millimetres)

| $d$ nom. | $d_1$ | $t$ | $y$ | $y_1$ | $d_2$ | $d_3$ | $l$ | $d_4$ (h11)* | $L$ | $a$ (h12)* | $l_4$ |
|---|---|---|---|---|---|---|---|---|---|---|---|
| 0.6 | 0.76 | 8 | 5 | 7 | 0.5 | 0.90 | 20 | | 38 | † | † |
| 0.8 | 1.04 | 12 | 5 | 7 | 0.7 | 1.18 | 24 | | 42 | 0.90 | 4 |
| 1.0 | 1.32 | 16 | 5 | 7 | 0.9 | 1.46 | 28 | | 46 | 1.12 | 4 |
| 1.2 | 1.60 | 20 | 5 | 7 | 1.1 | 1.74 | 32 | | 50 | 1.40 | 4 |
| 1.5 | 2.00 | 25 | 5 | 7 | 1.4 | 2.14 | 37 | $d_4 = d_3$ | 57 | 1.80 | 4 |
| 2.0 | 2.70 | 35 | 5 | 8 | 1.9 | 2.86 | 48 | | 68 | 2.24 | 5 |
| 2.5 | 3.20 | 35 | 5 | 8 | 2.4 | 3.36 | 48 | ———— | 68 | 2.80 | 5 |
| 3.0 | 3.90 | 45 | 5 | 8 | 2.9 | 4.06 | 58 | 4.0 | 80 | 3.15 | 6 |
| 4.0 | 5.10 | 55 | 5 | 8 | 3.9 | 5.26 | 68 | 5.0 | 93 | 4.00 | 7 |
| 5.0 | 6.20 | 60 | 5 | 8 | 4.9 | 6.36 | 73 | 6.3 | 100 | 5.00 | 8 |
| 6.0 | 7.80 | 90 | 5 | 10 | 5.9 | 8.00 | 105 | 8.0 | 135 | 6.30 | 9 |
| 8.0 | 10.60 | 130 | 5 | 10 | 7.9 | 10.80 | 145 | 10.0 | 180 | 8.00 | 11 |
| 10.0 | 13.20 | 160 | 5 | 10 | 9.9 | 13.40 | 175 | 12.5 | 215 | 10.00 | 13 |
| 12.0 | 15.60 | 180 | 10 | 20 | 11.8 | 16.00 | 210 | 14.0 | 255 | 11.20 | 14 |
| 16.0 | 20.00 | 200 | 10 | 20 | 15.8 | 20.40 | 230 | 18.0 | 280 | 14.00 | 18 |
| 20.0 | 24.40 | 220 | 10 | 20 | 19.8 | 24.80 | 250 | 22.4 | 310 | 18.00 | 22 |
| 25.0 | 29.80 | 240 | 15 | 45 | 24.7 | 30.70 | 300 | 28.0 | 370 | 22.40 | 26 |
| 30.0 | 35.20 | 260 | 15 | 45 | 29.7 | 36.10 | 320 | 31.5 | 400 | 25.00 | 28 |
| 40.0 | 45.60 | 280 | 15 | 45 | 39.7 | 46.50 | 340 | 40.0 | 430 | 31.50 | 34 |
| 50.0 | 56.00 | 300 | 15 | 45 | 49.7 | 56.90 | 360 | 50.0 | 460 | 40.00 | 42 |

*For the values of the tolerances h11 and h12 see BS 328: Pt 4: 1983 Appendix B.

†This shank size is smaller than the size range for which a size of driving square is specified in ISO 237. A parallel shank should be used without a square.

This table is in accordance with ISO 3465, except that in the latter, for values of $d$ equal to or less than 2.5 mm, $d_4$ has a constant value equal to 3.15 mm. The values of $a$ and $l_4$ are in accordance with ISO 237. For further information see BS 328: Pt 4: 1983.

## 3.8 Counterbores with parallel shanks and integral pilots

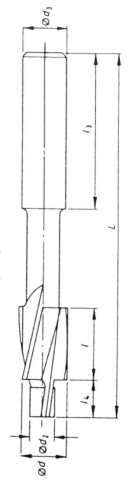

**General dimensions**

(Dimensions in millimetres)

| Cutting diameter $d$ $(z9)^*$ | | Pilot diameter $d_2$ | Shank diameter $d_3$ $(h9)^*$ | Overall length $L$ | Cutting length $l$ | Shank length $l_3$ | Pilot length (approx.) $l_4$ |
|---|---|---|---|---|---|---|---|
| over | to | | | | | | |
| $2.00^\dagger$ | 3.15 | For all cutting diameters: $d/3$ min. | $d_3 = d$ | 45 | 7 | | $d_2$ |
| 3.15 | 5.00 | Limits of tolerance on selected pilot diameter: $e8^*$ | $d_3 = d$ | 56 | 10 | | $d_2$ |
| 5.00 | 8.00 | The selected pilot diameter is to be specified, when ordering, to suit the pilot hole diameter | 5.0 | 71 | 14 | 31.5 | $d_2$ |
| 8.00 | 12.50 | | 8.0 | 80 | 18 | 35.5 | $d_2$ |
| 12.50 | 20.00 | | 12.5 | 100 | 22 | 40.0 | $d_2$ |

*For values of the tolerances $z9$, $e8$ and $h9$ see Tables 11, 8 and 10 in BS 328: Pt 5: 1983 Appendix A.

$^\dagger$Includes 2 mm.

This table is in accordance with ISO 4206 except that the latter uses $l_1$ for $L$, $l_2$ for $l$ and $d_1$ for $d$.

**Diameters** (Dimensions in millimetres)

| Preferred cutting diameters $d$ (z9) | Pilot diameter $d_2$ (38) | Cap screw size | Cap screw head diameter |
|---|---|---|---|
| 6.0 | 2.5<br>3.2<br>3.4* | M3 | 5.5 |
| 8.0 | 3.3<br>4.3<br>4.5* | M4 | 7.0 |
| 10.0 | 4.2<br>5.3<br>5.5* | M5 | 8.5 |
| 11.0 | 5.0<br>6.4<br>6.6* | M6 | 10.0 |
| 15.0 | 6.8<br>8.4<br>9.0* | M8 | 13.0 |
| 18.0 | 8.5<br>10.5<br>11.0* | M10 | 16.0 |
| 20.0 | 10.2<br>13.0<br>14.0* | M12 | 18.0 |

*These are the preferred pilot diameters, being the diameters of clearance holes for the sizes of cap screw indicated. For further information see BS 328: Pt 5: 1983.

## 3.9 Counterbores with Morse taper shanks and detachable pilots

Morse taper shank (BS 1660)

**General dimensions**

(Dimensions in millimetres)

| Cutting diameter d (z9)* | | Pilot diameter d₂ (e8)* | | Diameter of hole for pilot d₃ (H8) | Set screw size d₄ | Overall length L | Cutting length l | Pilot shank L₃ | Set screw position l₄ | Morse taper shank no. |
|---|---|---|---|---|---|---|---|---|---|---|
| over | to | over | to | | | | | | | |
| 12.5 | 16.0 | 5.0 | 14.0 | 4 | M3 | 132 | 22 | 30 | 16 | 2 |
| 16.0 | 20.0 | 6.3 | 18.0 | 5 | M4 | 140 | 25 | 38 | 19 | 2 |
| 20.0 | 25.0 | 8.0 | 22.4 | 6 | M5 | 150 | 30 | 46 | 23 | 2 |
| 25.0 | 31.5 | 10.0 | 28.0 | 8 | M6 | 180 | 35 | 54 | 27 | 3 |
| 31.5 | 40.0 | 12.5 | 35.5 | 10 | M8 | 190 | 40 | 64 | 32 | 3 |
| 40.0 | 50.0 | 16.0 | 45.0 | 12 | M8 | 236 | 50 | 76 | 42 | 4 |
| 50.0 | 63.0 | 20.0 | 56.0 | 16 | M10 | 250 | 63 | 88 | 53 | 4 |

*For values of the tolerances z9, e8 and H8 see Tables 11, 8 and 12 in BS 328: Pt 5: 1983 Appendix A.
This table is in accordance with ISO 4207 except that the latter uses $l_1$ for $l$, $l_2$ for $l$ and $d_1$ for $d$.

**Diameters**               (Dimensions in millimetres)

| Preferred cutting diameters $d$ (z9) | Pilot diameter $d_2$ (e8) | Pilot shank diameter $d_3$ (f7) | Cap screw size | Cap screw head diameter |
|---|---|---|---|---|
| 15.0 | 6.8<br>8.4<br>9.0* | 4.0 | M8 | 13.0 |
| 18.0 | 8.5<br>10.2<br>10.5 | 5.0 | M10 | 16.0 |
| 20.0 | 11.0*<br>13.0<br>14.0* | | M12 | 18.0 |
| 24.0 | 12.0<br>15.0<br>16.0* | 6.0 | M14 | 21.0 |
| 26.0 | 14.0<br>15.5<br>17.0 | 8.0 | M16 | 24.0 |
| 30.0 | 18.0<br>19.0<br>20.0* | | M18 | 27.0 |
| 33.0 | 17.5<br>19.5 | 10.0 | M20 | 30.0 |
| 36.0 | 21.0<br>22.0*<br>23.0 | | M22 | 33.0 |
| 40.0 | 24.0*<br>25.0<br>26.0* | | M24 | 36.0 |

*These are the preferred pilot diameters, being the diameters of clearance holes for the sizes of cap screw indicated.

For further information see BS 328: Pt 5.

# 3.10 Detachable pilots for counterbores

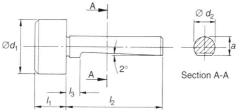

(Dimensions in millimetres)

| Pilot shank diameter $d_2$ (f7)* | Pilot diameter $d_1$ (e8)* over | $d_1$ (e8)* to | $a_{-0.1}^{0}$ | Pilot length $l_1$ | Pilot shank length $l_2$ | $l_3$ |
|---|---|---|---|---|---|---|
| 4 | 5.0 | 6.3 | 3.6 | 5 | 20 | 3 |
| 4 | 6.3 | 8.0 | 3.6 | 6 | 20 | 3 |
| 4 | 8.0 | 10.0 | 3.6 | 7 | 20 | 3 |
| 4 | 10.0 | 12.5 | 3.6 | 8 | 20 | 4 |
| 4 | 12.5 | 14.0 | 3.6 | 10 | 20 | 4 |
| 5 | 6.3 | 8.0 | 4.6 | 6 | 23 | 3 |
| 5 | 8.0 | 10.0 | 4.6 | 7 | 23 | 3 |
| 5 | 10.0 | 12.5 | 4.6 | 8 | 23 | 4 |
| 5 | 12.5 | 16.0 | 4.6 | 10 | 23 | 4 |
| 5 | 16.0 | 18.0 | 4.6 | 12 | 23 | 4 |
| 6 | 8.0 | 10.0 | 5.5 | 7 | 28 | 4 |
| 6 | 10.0 | 12.5 | 5.5 | 8 | 28 | 4 |
| 6 | 12.5 | 16.0 | 5.5 | 10 | 28 | 4 |
| 6 | 16.0 | 20.0 | 5.5 | 12 | 28 | 5 |
| 6 | 20.0 | 22.4 | 5.5 | 15 | 28 | 5 |
| 8 | 10.0 | 12.5 | 7.5 | 8 | 32 | 4 |
| 8 | 12.5 | 16.0 | 7.5 | 10 | 32 | 4 |
| 8 | 16.0 | 20.0 | 7.5 | 12 | 32 | 5 |
| 8 | 20.0 | 25.0 | 7.5 | 15 | 32 | 5 |
| 8 | 25.0 | 28.0 | 7.5 | 18 | 32 | 5 |
| 10 | 12.5 | 16.0 | 9.1 | 10 | 40 | 5 |
| 10 | 16.0 | 20.0 | 9.1 | 12 | 40 | 5 |
| 10 | 20.0 | 25.0 | 9.1 | 15 | 40 | 5 |
| 10 | 25.0 | 31.5 | 9.1 | 18 | 40 | 6 |
| 10 | 31.5 | 35.5 | 9.1 | 22 | 40 | 6 |
| 12 | 16.0 | 20.0 | 11.3 | 12 | 50 | 5 |
| 12 | 20.0 | 25.0 | 11.3 | 15 | 50 | 5 |
| 12 | 25.0 | 31.5 | 11.3 | 18 | 50 | 6 |
| 12 | 31.5 | 40.0 | 11.3 | 22 | 50 | 6 |
| 12 | 40.0 | 45.0 | 11.3 | 27 | 50 | 6 |
| 16 | 20.0 | 25.0 | 15.2 | 15 | 60 | 6 |
| 16 | 25.0 | 31.5 | 15.2 | 18 | 60 | 6 |
| 16 | 31.5 | 40.0 | 15.2 | 22 | 60 | 6 |
| 16 | 40.0 | 50.0 | 15.2 | 27 | 60 | 6 |
| 16 | 50.0 | 56.0 | 15.2 | 30 | 60 | 6 |

'or values of the tolerances f7 and e8 see Tables 8 and 9 in BS 328:
5: 1983 Appendix A.
is table is in accordance with ISO 4208. For further information
: BS 328: Pt 5: 1983.

## 3.11 Countersinks with parallel shanks

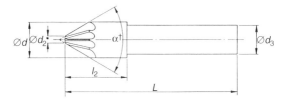

(Dimensions in millimetres)

| Nominal size $d$ | Small diameter* $d_2$ | Overall length $L$[†] | | Body length $l_2$[†] | | Shank diameter $d_3$(h9)[‡] |
|---|---|---|---|---|---|---|
| | | $\alpha = 60°$ | $\alpha = 90°$ and $120°$ | $\alpha = 60°$ | $\alpha = 90°$ and $120°$ | |
| 8.0 | 1.6 | 48 | 44 | 16 | 12 | 8 |
| 10.0 | 2.0 | 50 | 46 | 18 | 14 | 8 |
| 12.5 | 2.5 | 52 | 48 | 20 | 16 | 8 |
| 16.0 | 3.2 | 60 | 56 | 24 | 20 | 10 |
| 20.0 | 4.0 | 64 | 60 | 28 | 24 | 10 |
| 25.0 | 7.0 | 69 | 65 | 33 | 29 | 10 |

*Front end design optional.
[†]Tolerance on $\alpha$ is $^{0}_{-1}$ degrees.
[‡]For values of the tolerance h9 see Table 10 in BS 328 : Pt 5 : 1983 Appendix A.
This table is in accordance with ISO 3294, except that the latter uses $l_1$ for $L$ and $d_1$ for $d$.
For further information see BS 328: Pt 5: 1983.

## 3.12 Countersinks with Morse taper shanks

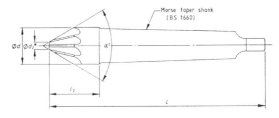

| Nominal size d | Small diameter* $d_2$ | Overall length $L^\dagger$ | | Body length $l_2^\dagger$ | | Morse taper shank no. |
|---|---|---|---|---|---|---|
| | | $\alpha = 60°$ | $\alpha = 90°$ and $120°$ | $\alpha = 60°$ | $\alpha = 90°$ and $120°$ | |
| 16.0 | 3.2 | 97 | 93 | 24 | 20 | 1 |
| 20.0 | 4.0 | 120 | 116 | 28 | 24 | 2 |
| 25.0 | 7.0 | 125 | 121 | 33 | 29 | 2 |
| 31.5 | 9.0 | 132 | 124 | 40 | 32 | 2 |
| 40.0 | 12.5 | 160 | 150 | 45 | 35 | 3 |
| 50.0 | 16.0 | 165 | 153 | 50 | 38 | 3 |
| 63.0 | 20.0 | 200 | 185 | 58 | 43 | 4 |
| 80.0 | 25.0 | 215 | 196 | 73 | 54 | 4 |

*Front end design optional.
$\dagger$Tolerance on $\alpha$ is $_{-1}^{0}$ degrees.
This table is in accordance with ISO 3293, except that the latter uses $l_1$ for L and $d_1$ for d.
For further information see BS 328: Pt 5: 1983.

## 3.13 Single point cutting tools: butt welded high-speed steel

### Light turning and facing tool

No. 1 Right hand as drawn
No. 2 Left hand opposite to drawing

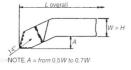

NOTE. A = from 0.5W to 0.7W

### Straight nosed roughing tool

No. 3 Right hand as drawn
No. 4 Left hand opposite to drawing

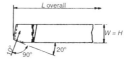

| *Preferred sizes* (mm) | | | *Preferred sizes* (mm) | | |
|---|---|---|---|---|---|
| H | W | L | H | W | L |
| 12 | 12 | 100 | 12 | 12 | 100 |
| 16 | 16 | 110 | 16 | 16 | 110 |
| 20 | 20 | 125 | 20 | 20 | 125 |
| 25 | 16 | 200 | 25 | 16 | 200 |
| 32 | 20 | 250 | 32 | 20 | 250 |
| 40 | 25 | 315 | 40 | 25 | 315 |
| 20 | 16) | 140 | (20 | 16) | 200 |
| 25 | 20) | 200 | (25 | 20) | 200 |

## Knife tool or side-cutting tool

No. 7  Right hand as drawn
No. 8  Left hand opposite to drawing

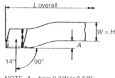

NOTE. A = from 0.3W to 0.5W

## External screw-cutting tool

No. 13  As drawn

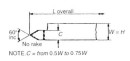

NOTE. C = from 0.5W to 0.75W

| *Preferred sizes* (mm) | | |
|---|---|---|
| H | W | L |
| 12 | 12 | 100 |
| 16 | 16 | 110 |
| 20 | 20 | 125 |
| 25 | 16 | 200 |
| 32 | 20 | 250 |
| 40 | 25 | 315 |
| (20 | 16) | 140 |
| (25 | 20) | 200 |

| *Preferred sizes* (mm) | | |
|---|---|---|
| H | W | L |
| 12 | 12 | 100 |
| 16 | 16 | 110 |
| 20 | 20 | 125 |
| 25 | 16 | 200 |
| 32 | 20 | 250 |
| 40 | 25 | 315 |
| (20 | 16) | 140 |
| (25 | 20) | 200 |

## Parting-off tool

No. 16RH  Right hand as drawn
No. 16LH  Left hand opposite to drawing

NOTE. B = from 1.2H to 1.4H
C = from 0.2H to 0.4H

## Facing tool

No. 19  Right hand as drawn
No. 20  Left hand opposite to drawing

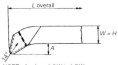

NOTE. A = from 0.5W to 0.7W

| Preferred sizes (mm) | | |
|---|---|---|
| H | W | L |
| 12 | 12 | 100 |
| 16 | 16 | 110 |
| 20 | 20 | 125 |
| 25 | 16 | 200 |
| 32 | 20 | 250 |
| 40 | 25 | 315 |
| (20 | 16) | 140 |
| (25 | 20) | 200 |

**Round nosed planing or shaping tool**

No. 17 Cuts either right hand or left hand

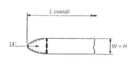

| Preferred sizes (mm) | | |
|---|---|---|
| H | W | L |
| 12 | 12 | 100 |
| 16 | 16 | 110 |
| 20 | 20 | 125 |
| 25 | 25 | 200 |
| 32 | 32 | 315 |
| 40 | 40 | 315 |
| 25 | 16 | 200 |
| 32 | 20 | 250 |
| 40 | 25 | 315 |
| 20 | 16) | 140 |
| 25 | 20) | 200 |
| 50 | 40) | 400 |

| Preferred sizes (mm) | | |
|---|---|---|
| H | W | L |
| 12 | 12 | 100 |
| 16 | 16 | 110 |
| 20 | 20 | 125 |
| 25 | 16 | 200 |
| 32 | 20 | 250 |
| 40 | 25 | 315 |
| (20 | 16) | 140 |
| (25 | 20) | 200 |

**Right-angle recessing tool**

No. 25 Right hand as drawn
No. 26 Left hand opposite to drawing

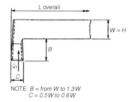

NOTE. B = from W to 1.3 W
C = 0.5 W to 0.6 W

| Preferred sizes (mm) | | |
|---|---|---|
| H | W | L |
| 12 | 12 | 100 |
| 16 | 16 | 110 |
| (20 | 16) | 140 |
| (25 | 20) | 200 |

## Right-angle parting-off tool

No. 27 Right hand as drawn
No. 28 Left hand opposite to drawing

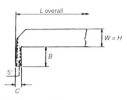

NOTE. *B* = from 1.0*H* to 1.2*H*
      *C* = from 0.2*H* to 0.4*H*

| *Preferred sizes* (mm) | | |
|---|---|---|
| *H* | *W* | *L* |
| 12 | 12 | 100 |
| 16 | 16 | 110 |
| (20 | 16) | 140 |
| (25 | 20) | 200 |

## Cranked turning or recessing tool

No. 39 Right hand as drawn
No. 40 Left hand opposite to drawing

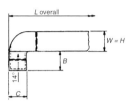

NOTE. *B* = from 0.9*W* to 1.2*W*
      *C* = *W*

| *Preferred sizes* (mm) | | |
|---|---|---|
| *H* | *W* | *L* |
| 12 | 12 | 100 |
| 16 | 16 | 110 |
| 20 | 20 | 125 |
| (20 | 16) | 140 |
| (25 | 20) | 200 |

## Hardened blank

No. 47

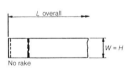

## Hardened blank

No. 62

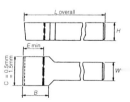

| Preferred sizes (mm) | | |
| --- | --- | --- |
| H | W | L |
| 12 | 12 | 100 |
| 16 | 16 | 110 |
| 20 | 20 | 125 |
| 25 | 25 | 200 |
| 32 | 32 | 315 |
| 40 | 40 | 315 |
| 25 | 16 | 200 |
| 32 | 20 | 250 |
| 40 | 25 | 315 |
| (20 | 16) | 140 |
| (25 | 20) | 200 |
| (50 | 40) | 400 |

| Preferred sizes (mm) | | | | | |
| --- | --- | --- | --- | --- | --- |
| H | W | C | B | E | L |
| 16 | 16 | 20 | 25 | 16 | 140 |
| 16 | 16 | 25 | 25 | 16 | 140 |
| 20 | 20 | 25 | 25 | 18 | 140 |
| 20 | 20 | 32 | 32 | 25 | 200 |
| 25 | 25 | 40 | 36 | 28 | 200 |

## Boring tool

No. 50    Square nose
No. 50A   V-nose for internal screw cutting
No. 50B   Round nose

## Swan-necked finishing tool

No. 52   Cuts either right hand or left hand

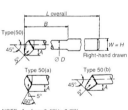

NOTE. A = from 0.4W to 0.6W
B = from 0.4W to 0.6W of the overall length of tool
D = from 0.6W to 1.0W

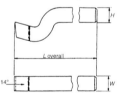

NOTE. The cutting edge is on or below the level of the base of the tool.

| Preferred sizes (mm) | | |
| --- | --- | --- |
| H | W | L |
| 12 | 12 | 160 |
| 16 | 16 | 200 |
| 20 | 16) | 200 |
| 25 | 20) | 250 |

| Preferred sizes (mm) | | |
| --- | --- | --- |
| H | W | L |
| 40 | 25 | 355 |
| (20 | 16) | 200 |
| (25 | 20) | 250 |

For further details, including non-preferred sizes, nomenclature and shank sections, see BS 1296: Pts 1 to 4 inclusive.

## 3.14 Tool bits: ground high-speed steel

### Round section tool bits

(Dimensions in millimetres)

| diameter (h12)* | $L_{-3}^{+0}$ | | | | |
|---|---|---|---|---|---|
| | 63 | 80 | 100 | 160 | 180 |
| 4 | × | × | × | − | − |
| 5 | × | × | × | − | − |
| 6 | × | × | × | × | − |
| 8 | − | × | × | × | − |
| 10 | − | × | × | × | × |
| 12 | − | − | × | × | × |
| 16 | − | − | × | × | × |
| 18 | − | − | − | − | × |

*For tolerance sizes see BS 4500.
For further information see BS 1296.

### Square section tool bits

(Dimensions in millimetres)

| breadth (h13)* | height (h13)* | $L_{-3}^{+0}$ | | | | |
|---|---|---|---|---|---|---|
| | | 63 | 80 | 100 | 160 | 180 |
| 4 | 4 | × | − | − | − | − |
| 5 | 5 | × | − | − | − | − |
| 6 | 6 | × | × | × | × | × |
| 8 | 8 | × | × | × | × | × |
| 10 | 10 | × | × | × | × | × |
| 12 | 12 | × | × | × | × | × |
| 16 | 16 | − | − | × | × | × |
| 20 | 20 | − | − | − | × | × |
| 25 | 25 | − | − | − | − | × |

*For tolerance sizes see BS 4500.
For further information see BS 1296.

**Rectangular section tool bits**

(Dimensions in millimetres)

| breadth (h13)* | height (h13)* | $L_{-3}^{+0}$ | | |
|---|---|---|---|---|
| | | 100 | 160 | 200 |
| 4 | 6 | × | − | − |
| 4 | 8 | × | − | − |
| 5 | 8 | × | − | − |
| 5 | 10 | × | − | − |
| 6 | 10 | − | × | × |
| 6 | 12 | − | × | × |
| 8 | 12 | − | × | × |
| 8 | 16 | − | × | × |
| 10 | 16 | − | × | × |
| 10 | 20 | − | × | × |
| 12 | 20 | − | × | × |
| 12 | 25 | − | − | × |
| 16 | 25 | − | − | × |
| 16 | − | − | − | − |

*For tolerance sizes see BS 4500.
For further information see BS 1296.

## 3.15 Milling cutters

### 3.15.1 Cylindrical cutters

**Light duty cylindrical cutter**

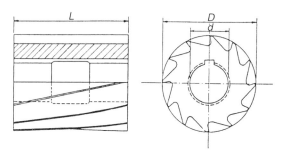

**High power cylindrical cutter**

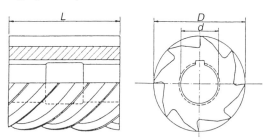

The dimensions of light duty or high power cylindrical cutters are as given in the table. These cutters are normally supplied with left-hand helix as shown in the figures. The cutters have keyways in accordance with BS 122: Pt 3 Clause 3.2.

For further information, see BS 122: Pt 3: 1987.

(Dimensions in millimetres)

| Diameter of cutter $D$ (js16)* | Diameter of bore $d$ (H7)* | Lengths $L$ (js15)* |
|---|---|---|
| 50 | 22 | 40, 63, 80 |
| 63 | 27 | 50, 70 |
| 80 | 32 | 63, 80, 100, 125 |
| 100 | 40 | 70, 100, 125, 160 |
| 125 | 50 | 125, 200 |

*For tolerances see BS 4500: Pt 1.

### 3.15.2 High helix cylindrical cutters

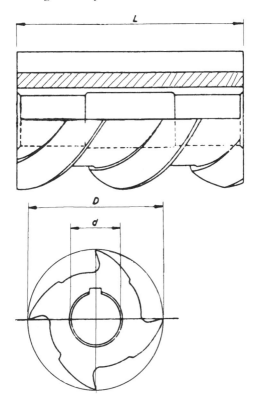

The dimensions of high helix cylindrical cutters are as given in the table. These cutters are normally supplied with left-hand helix as shown in the figure. The cutters have keyways in accordance with BS 122: Pt 3 Clause 3.2.

For further information see BS 122: Pt 3: 1987.

(Dimensions in millimetres)

| Diameter of cutter D (js16)* | Diameter of bore d (H7)* | Lengths L (js15)* |
|---|---|---|
| 80 | 32 | 70, 100, 160 |
| 100 | 40 | 70, 100, 160 |
| 125 | 50 | 70, 100, 160 |

*For tolerances see BS 4500: Pt 1.

### 3.15.3 Side and face cutters

| Light duty side and face cutter | High power side and face cutter |
|---|---|
| 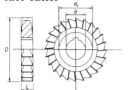 | 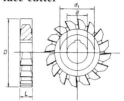 |

(Dimensions in millimetres)

| Diameter of cutter D (js16)* | Diameter of boss $d_1$ (min.) | Diameter of bore d (H7)* | Width of cutting edges and boss† L (k11)* |
|---|---|---|---|
| 50 | 27 | 16 | 6, 8, 10 |
| 63 | 34 | 22 | 6, 8, 10, 12, (14), 16 |
| 80 | 41 | 27 | 6, 8, 10, 12, (14), 16, (18), 20 |
| 100 | 47 | 32 | 6, 8, 10, 12, (14), 16, (18), 20, (22), 25 |
| 125 | 47 | 32 | 8, 10, 12, (14), 16, (18), 20, (22), 25, (28) |
| 160 | 55 | 40 | 10, 12, (14), 16, (18), 20, (22), 25, (28), 32 |
| 200 | 55 | 40 | 12, (14), 16, (18), 20, (22), 25, (28), 32, (36), 40 |

*For tolerances see BS 4500: Pt 1.

†Dimensions in parentheses are least preferred.

The dimensions of light duty or high power side and face cutters shall be as given in the table. The cutters shall have keyways in accordance with BS 122: Pt 3 Clause 3.2. Each side of a cutter may be ground to provide $\frac{1}{4}^{\circ}$ side clearance, when the clearance shall be amplified by recessing.

For further information see BS 122: Pt 3: 1987.

### 3.15.4 Staggered tooth side and face cutters

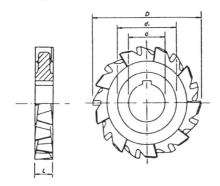

The dimensions of staggered tooth side and face cutters shall be as given in the table. The cutters shall have keyways in accordance with BS 122: Pt 3: Clause 3.2. Each side of a cutter may be ground to provide $\frac{1}{4}^{\circ}$ side clearance, when the clearance shall be amplified by recessing.

For further information see BS 122: Pt 3: 1987.

(Dimensions in millimetres)

| Diameter of cutter $D$ (js16)* | Diameter of boss $d_1$ (min.) | Diameter of bore $d$ (H7)* | Width† $L$ (k11)* |
|---|---|---|---|
| 63 | 34 | 22 | 6, 8, 10, 12, (14), 16 |
| 80 | 41 | 27 | 6, 8, 10, 12, (14), 16, (18), 20 |
| 100 | 47 | 32 | 6, 8, 10, 12, (14), 16, (18), 20, (22), 25 |
| 125 | 47 | 32 | 8, 10, 12, (14), 16, (18), 20, (22), 25, (28) |
| 160 | 55 | 40 | 10, 12, (14), 16, (18), 20, (22), 25, (28), 32 |
| 200 | 55 | 40 | 12, (14), 16, (18), 20, (22), 25, (28), 32, (36), 40 |

*For tolerances see BS 4500: Pt 1.
†Dimensions in parentheses are least preferred.

### 3.15.5 Slotting cutters

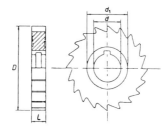

The dimensions of slotting cutters are as given in the table. The cutters have each side ground to provide side clearance and the clearance shall, where the width and diameter of the cutter permit, be amplified by recessing. The cutters have keyways in accordance with BS 122: Pt 3 Clause 3.2. The use of these cutters for cutting keyways is not recommended as the width of the slot produced cannot be guaranteed. If, however, the cutters are required for cutting keyways to BS 46: Pt 1, the special tolerances should be agreed between the purchaser and the manufacturer.

For further information see BS 122: Pt 3: 1987.

(Dimensions in millimetres)

| Diameter of cutter $D$ (js16)* | Diameter of boss $d_1$ (min.) | Diameter of bore $d$ (H7)* | Width[†] $L$ (js4)* |
|---|---|---|---|
| 50 | 27 | 16 | 6, 7, 8, 10 |
| 63 | 34 | 22 | 6, 7, 8, 10, 12, 14 |
| 80 | 41 | 27 | 6, 7, 8, 10, 12, 14, 16, 18 |
| 100 | 47 | 32 | 6, 7, 8, 10, 12, 14, 16, 18, 20, (22), 25 |
| 125 | 47 | 32 | 8, 10, 12, 14, 16, 18, 20, (22), 25 |
| 160 | 55 | 40 | 10, 12, 14, 16, 18, 20, (22), 25, (28), 32 |
| 200 | 55 | 40 | 12, 14, 16, 18, 20, (22), 25, (28), 32, (36), 40 |

*For tolerances see BS 4500: Pt 1.
[†]Dimensions in parentheses are least preferred.

### 3.15.6 Metal slitting saws without side chip clearance: fine teeth

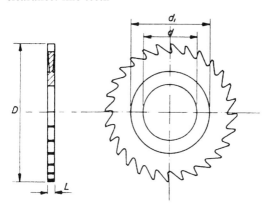

The dimensions of metal slitting saws without side chip clearance are as given in the tables in this section for fine teeth and in Section 3.15.7 for coarse teeth. The value for the tooth pitch in relation to the number of teeth of a saw of a given diameter is expressed as an approximate rounded figure. The saws have side clearance either up to the bore or up to the boss.

For further information see BS 122: Pt 3: 1987.

(Dimensions in millimetres)

| Diameter of saw D (js16)* | 20 | 25 | 32 | 40 | 50 | 63 | 80 | 100 | 125 | 160 | 200 | 250 | 315 |
|---|---|---|---|---|---|---|---|---|---|---|---|---|---|
| Diameter of boss $d_1$ (min.) | – | – | – | – | – | – | 34 | 34 | 34 | 47 | 63 | 63 | 80 |
| Diameter of bore d (H7)* | 5 | 8 | 8 | 10 | 13 | 16 | 22 | 22 | 22 | 32 | 32 | 32 | 40 |

Number of teeth

| Width L (js11)* | Tooth pitch | 20 | 25 | 32 | 40 | 50 | 63 | 80 | 100 | 125 | 160 | 200 | 250 | 315 |
|---|---|---|---|---|---|---|---|---|---|---|---|---|---|---|
| 0.2 | 0.8 | 80 | 100 | 128 | | | | | | | | | | |
| 0.25 | 0.8 | 80 | 100 | 128 | | | | | | | | | | |
| 0.3 | 1.0 | 64 | 80 | 100 | 128 | | | | | | | | | |
| 0.4 | 1.0 | 64 | 80 | 100 | 128 | | | | | | | | | |
| 0.5 | 1.0 | 64 | 80 | 100 | 128 | | | | | | | | | |
| 0.6 | 1.25 | 48 | 64 | 80 | 100 | 128 | 160 | | | | | | | |
| 0.8 | 1.25 | 48 | 64 | 80 | 100 | 128 | 160 | | | | | | | |
| 1.0 | 1.25 | 48 | 64 | 80 | 100 | 128 | 160 | | | | | | | |

| | 1.6 | 2.0 | 2.5 | 3.15 | 4.0 | 5.0 | 6.3 | | |
|-----|-----|-----|-----|------|-----|-----|-----|-----|-----|
| 1.2 | 40 | 32 | 40 | | | | | | |
| 1.6 | | 40 | 40 | 48 | 48 | | | | |
| 2.0 | | | 48 | 40 | 64 | 64 | | 100 | |
| 2.5 | | | | 40 | 64 | 80 | 100 | 128 | 160 |
| 3.0 | | | | | 48 | 64 | 80 | 100 | 160 |
| 4.0 | | | | | | | 64 | 100 | 128 |
| 5.0 | | | | | | | | 80 | 128 |
| 6.0 | | | | | | | | | 160 |

*For tolerances see BS 4500: Pt 1.

### 3.15.7 Metal slitting saws without side chip clearance: coarse teeth

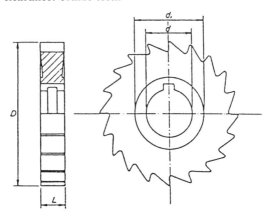

The value of the tooth pitch in relation to the number of teeth of a saw of a given diameter is expressed as an approximate rounded figure. The saws have side clearance either up to the bore or up to the boss.

| Diameter of saw D (js16)* | 32 | 40 | 50 | 63 | 80 | 100 | 125 | 160 | 200 | 250 | 315 |
|---|---|---|---|---|---|---|---|---|---|---|---|
| Diameter of boss $d_1$ (min.) | — | — | — | — | 34 | 34 | 34 | 47 | 63 | 63 | 80 |
| Diameter of bore $d$ (H7) | 8 | 10 | 13 | 16 | 22 | 22 | 22 | 32 | 32 | 32 | 40 |

| Width L (js 11)* | Tooth pitch | 32 | 40 | 50 | 63 | 80 | 100 | 125 | 160 | 200 | 250 | 315 |
|---|---|---|---|---|---|---|---|---|---|---|---|---|
| 0.3 | | | | 48 | 64 | | | | | | | |
| 0.4 | 2.5 | | 40 | | 64 | | | | | | | |
| 0.5 | | | | 48 | | | | | | | | |
| 0.6 | | | 40 | | | 64 | | | | | | |
| 0.8 | 3.15 | 32 | | | 48 | | | | | | | |
| 1.0 | | | | 40 | | | 64 | 80 | | | | |
| 1.2 | | | 32 | | | 48 | | | 80 | | | |
| 1.6 | 4.0 | 24 | | | | 40 | | | 64 | | | |
| 2.0 | | | | | 32 | | 48 | | | 80 | 100 | |
| 2.5 | | | | 24 | | | 40 | | 64 | | | 100 |
| 3.0 | 5.0 | | 20 | | | 32 | | | 48 | | 80 | |
| 4.0 | | | | 20 | 24 | | 40 | | | 64 | | |
| 5.0 | | | | | | 24 | 32 | | 48 | | | 80 |
| | | | | | | | | | 40 | | | |
| 6.0 | | | | | | | 32 | | | 48 | 64 | |
| | 6.3 | | | | 8.0 | | 10.0 | | 12.5 | | | |

*For tolerances see BS 4500: Pt 1.

### 3.15.8 Metal slitting saws with side chip clearance

#### Type A: staggered teeth

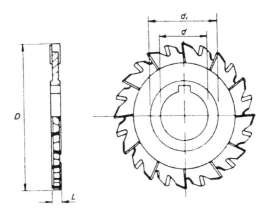

#### Type B: straight teeth

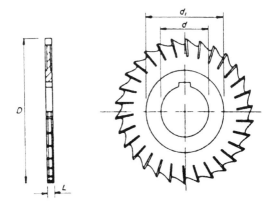

(Dimensions in millimetres)

| Diameter of saw $D$ (js16)* | 63 | 80 | 100 | 125 | 160 | 200 | 250 |
|---|---|---|---|---|---|---|---|
| Diameter of boss $d_1$ (min.) | — | 34 | 34 | 34 | 47 | 63 | 63 |
| Diameter of bore $d$ (H7)* | 16 | 22 | 22 | 22 | 32 | 32 | 32 |

*Type A: staggered teeth*

| Width<br>L (js10)* | Number of teeth (and pitch) | | | | | | |
|---|---|---|---|---|---|---|---|
| 4.0<br>5.0<br>6.0 | 28<br>(7.1) | 32<br>(7.8) | 36<br>(8.7) | 40<br>(9.8) | 44<br>(11.4) | 52<br>(12.0) | 64<br>(12.3) |

*Type B: straight teeth*

| Width<br>L (js10)* | Number of teeth (and pitch) | | | | | | |
|---|---|---|---|---|---|---|---|
| 1.6<br>2.0<br>2.5<br>3.0 | 32<br>(6.2) | 36<br>(7.0) | 40<br>(7.8) | 44<br>(8.9) | 48<br>(10.5) | 56<br>(11.2) | 68<br>(11.5) |

*For tolerances see BS 4500: Pt 1.
For further information see BS 122: Pt 3: 1987.

### 3.15.9 Convex milling cutters

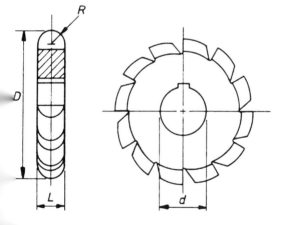

### Convex milling cutters

The dimensions of convex milling cutters are as given in the table. The cutters have keyways in accordance with BS 122: Pt 3 Clause 3.2.

For further information see BS 122: Pt 3: 1987.

| $R$ (k11)* | $D$ (js16)* | $d$ (H7)* | $L$ |
|---|---|---|---|
| 1 | | | 2 |
| 1.25 | | | 2.5 |
| 1.6 | 50 | 16 | 3.2 |
| 2 | | | 4 |
| 2.5 | | | 5 |
| 3.15 or 3 | | | 6.3 or 6 |
| 4 | 63 | 22 | 8 |
| 5 | | | 10 |
| 6.3 or 6 | 80 | 27* | 12.6 or 12 |
| 8 | | | 16 |
| 10 | 100 | | 20 |
| 12.5 or 12 | | 32 | 25 or 24 |
| 16 | 125 | | 32 |
| 20 | | | 40 |

*For tolerances see BS 4500: Pt 1.

### 3.15.10 Concave milling cutters

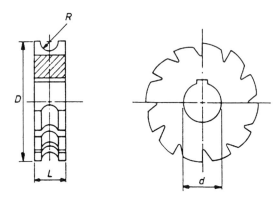

**Concave milling cutters**

The dimensions of concave milling cutters are as given in the table. For these cutters, radius $R$ is struck from the outside diameter of the cutter and chamfers have been eliminated from the intersection of the profile and the outside diameter. The cutters have keyways in accordance with BS 122: Pt 3 Clause 3.2.

For further information see BS 122: Pt 3: 1987.

| R<br>(N11)* | D<br>(js16)* | d<br>(h7)* | L |
|---|---|---|---|
| | | | (Dimensions in millimetres) |

| R<br>(N11)* | D<br>(js16)* | d<br>(h7)* | L |
|---|---|---|---|
| 1 | | | 6 |
| 1.25 | 50 | 16 | |
| 1.6 | | | 8 |
| 2 | | | 9 |
| 2.5 | | | 10 |
| 3.15 or 3 | 63 | 22 | 12 |
| 4 | | | 16 |
| 5 | | | 20 |
| 6.3 or 6 | 80 | 27 | 24 |
| 8 | | | 32 |
| 10 | 100 | | 36 |
| 12.6 or 12 | | 32 | 40 |
| 16 | 125 | | 50 |
| 20 | | | 60 |

*For tolerances see BS 4500: Pt 1.

### 3.15.11 Corner rounding concave milling cutters

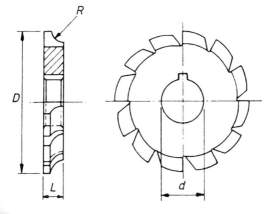

he dimensions of corner rounding concave milling
utters are as given in the table. For these cutters,
adius R is struck from the outside diameter of the
utter and chamfers have been eliminated from the
tersection of the profile and the outside diameter.
he cutters have keyways in accordance with BS 122:
3 Clause 3.2.

For further information see BS 122: Pt 3: 1987.

(Dimensions in millimetres)

| R (N11)* | D (js16)* | d (H7)* | L |
|---|---|---|---|
| 1 | | | 4 |
| 1.25 | 50 | 16 | |
| 1.6 | | | |
| 2 | | | 5 |
| 2.5 | | | |
| 3.15 or 3 | 63 | 22 | 6 |
| 4 | | | 8 |
| 5 | | | 10 |
| 6.3 or 6 | 80 | 27 | 12 |
| 8 | | | 16 |
| 10 | 100 | | 18 |
| 12.5 or 12 | | 32 | 20 |
| 16 | 125 | | 24 |
| 20 | | | 28 |

*For tolerances see BS 4500: Pt 1.

### 3.15.12 Double equal angle milling cutters

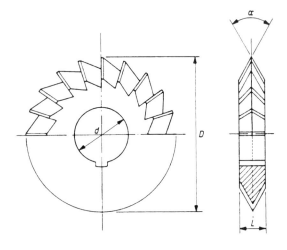

The dimensions of double equal angle milling cutters are as given in the table. The cutters have keyways in accordance with BS 122: Pt 3 Clause 3.2.

For further information see BS 122: Pt 3: 1987.

| D (js16)* mm | d (H7)* mm | α(±15')* degrees | L (js16)* mm |
|---|---|---|---|
| 50 | 16 | 45 | 8 |
| | | 60 | 10 |
| | | 90 | 14 |
| 63 | 22 | 45 | 10 |
| | | 60 | 14 |
| | | 90 | 20 |
| 80 | 27 | 45 | 12 |
| | | 60 | 18 |
| | | 90 | 22 |
| 100 | 32 | 45 | 18 |
| | | 60 | 25 |
| | | 90 | 32 |

*For tolerances see BS 4500: Pt 1.

### 3.15.13 T-slot cutters with Morse taper shanks

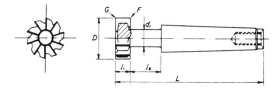

The dimensions of T-slot cutters with Morse taper shanks are as given in the table. However, the cutters may be made with square corners throughout, as an alternative to the radiused corners G and F shown. It is recommended that the corners be radiused when considerations of strength preclude the use of slots with square corners. The cutter teeth shall be either straight (as shown) or staggered. The cutters shall be designated by the nominal size of the slot.

For further information see BS 122: Pt 3: 1987.

(Dimensions in millimetres)

| T-slot | | | | | T-slot cutter | | | |
|---|---|---|---|---|---|---|---|---|
| Nominal size of slot* | Diameter of cutter $D$ (h12)† | Width of cutter $l_1$ (h12)† | Diameter of neck $d_1$ (max.) | Length of neck $l_2$ | No. of Morse taper shank, tapped | Overall length $L$ | Radius $G$ (max.) | Radius $F$ (max.) |
| 10 | 18 | 8 | 8 | $17^{+1}_{-0}$ | 1 | 82 | 1.0 | 0.6 |
| 12 | 21 | 9 | 10 | $20^{+1}_{-0}$ | 2 | 98 | 1.0 | 0.6 |
| 14 | 25 | 11 | 12 | $23^{+1}_{-0}$ | 2 | 103 | 1.6 | 0.6 |
| 18 | 32 | 14 | 15 | $28^{+1}_{-0}$ | 2 | 111 | 1.6 | 1.0 |
| 22 | 40 | 18 | 19 | $34^{+1}_{-0}$ | 3 | 138 | 2.5 | 1.0 |
| 28 | 50 | 22 | 25 | $42^{+1}_{-0}$ | 4 | 173 | 2.5 | 1.0 |
| 36 | 60 | 28 | 30 | $51^{+1}_{-0}$ | 4 | 188 | 2.5 | 1.0 |
| 42 | 72 | 35 | 36 | $58^{+1}_{-0}$ | 5 | 229 | 4.0 | 1.6 |
| 48 | 85 | 40 | 42 | $64^{+1}_{-0}$ | 5 | 240 | 6.0 | 2.0 |
| 54 | 95 | 44 | 44 | $71^{+1}_{-0}$ | 5 | 251 | 6.0 | 2.0 |

*See BS 2485.

### 3.15.14 Shell end mills

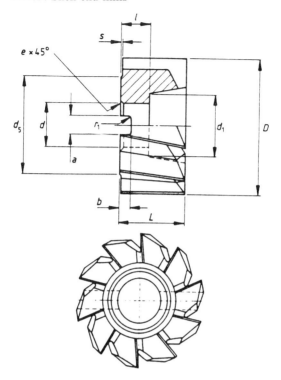

The dimensions of shell end milling cutters are as given in the table. However, the boss $d_5$ and the side clearance $S$ are optional. The milling cutters shall not be reversible on their arbors and the direction of rotation shall be specified by describing them as either right-hand or left-hand cutters. Milling cutters for right-hand rotation with right-hand helical cutting edges are normally supplied.

For further information see BS 122: Pt 3: 1987.

(Dimensions in millimetres)

| Diameter of cutter D (js16)* | Length of cutter L (H15)* | Diameter of bore d (H7)* | Length of bore l (H14)* | Diameter of boss $d_1$ | Width of driving slot $a_1$ (min.) | Depth of driving slot $b_1$ (min.) | Maximum radius of driving slot $r_1$ | Diameter of boss $d_5$ | Chamfer on bore e | Shoulder S |
|---|---|---|---|---|---|---|---|---|---|---|
| 40 | 32 | 16 | 18 | 22 | 8.4 | 5.6 | 1.0 | 33 | $0.6^{+0.2}_{-0}$ | 0.5 |
| 50 | 36 | 22 | 20 | 30 | 10.4 | 6.3 | 1.2 | 41 | $0.6^{+0.2}_{-0}$ | 0.5 |
| 63 | 40 | 27 | 22 | 38 | 12.4 | 7.0 | 1.2 | 49 | $0.8^{+0.2}_{-0}$ | 0.5 |
| 80 | 45 | 27 | 22 | 38 | 12.4 | 7.0 | 1.2 | 49 | $0.8^{+0.2}_{-0}$ | 0.5 |
| 100 | 50 | 32 | 25 | 45 | 14.4 | 8.0 | 1.6 | 59 | $0.8^{+0.2}_{-0}$ | 0.5 |
| 125 | 56 | 40 | 28 | 56 | 16.4 | 9.0 | 2.0 | 71 | $1.0^{+0.3}_{-0}$ | 0.5 |
| 160 | 63 | 50 | 31 | 67 | 18.4 | 10.0 | 2.0 | 91 | $1.0^{+0.3}_{-0}$ | 0.5 |

*For tolerances see BS 4500: Pt 1.

## 3.15.15 Arbors for shell end mills

**Spigot**

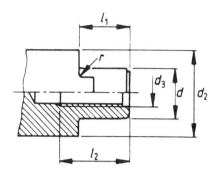

    The dimensions of spigots, tenons and retaining bolts for shell end mills are as given in the tables.

    For further information see BS 122: Pt 3: 1987.

(Dimensions in millimetres)

| $d$ (h6)* | $l_1$ (max.) | $d_2$ | $d_3$ | $l_2$ (min.) | $r$ (max.) |
|---|---|---|---|---|---|
| 16 | $17^{+0}_{-1}$ | 32 | M8 | 22 | 0.6 |
| 22 | $19^{+2}_{-1}$ | 40 | M10 | 28 | 0.6 |
| 27 | $21^{+0}_{-1}$ | 48 | M12 | 32 | 0.8 |
| 32 | $24^{+0}_{-1}$ | 58 | M16 | 36 | 0.8 |
| 40 | $27^{+0}_{-1}$ | 70 | M20 | 45 | 1.0 |
| 50 | $30^{+0}_{-1}$ | 90 | M24 | 50 | 1.0 |

*For tolerances see BS 4500: Pt 1.

**Tenon**

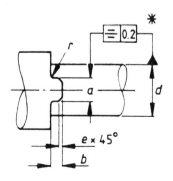

| d | Arbor a | b | r (max.) | Chamfer e |
|---|---|---|---|---|
| 16 | 8 | 5.0 | 0.6 | $0.6^{+0.2}_{-0}$ |
| 22 | 10 | 5.6 | | |
| 27 | 12 | 6.3 | 0.8 | $0.8^{+0.2}_{-0}$ |
| 32 | 14 | 7.0 | | |
| 40 | 16 | 8.0 | 1.0 | $1.0^{+0.3}_{-0}$ |
| 50 | 18 | 9.0 | | |

*See BS 308: Pt 3.

## Retaining bolt

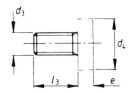

(Dimensions in millimetres)

| Spigot diameter (nominal) | $d_3$ | $l_3$ | $d_4$ (max.)* | e |
|---|---|---|---|---|
| 16 | M8 | $16^{+3}_{-0}$ | 20 | 6 |
| 22 | M10 | $18^{+3}_{-0}$ | 28 | 7 |
| 27 | M12 | $22^{+3}_{-0}$ | 35 | 8 |
| 32 | M16 | $26^{+3}_{-0}$ | 42 | 9 |
| 40 | M20 | $30^{+3}_{-0}$ | 52 | 10 |
| 50 | M24 | $36^{+3}_{-0}$ | 63 | 10 |

*The shape of the head of the bolt is not specified.

## 3.15.16 Screwed shank end mills: normal series

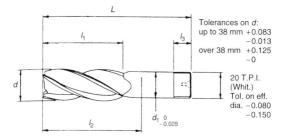

(Dimensions in millimetres)

| Cutter diameter | Cut length | Shank diameter | Nominal length below chuck | Overall length | Thread length |
|---|---|---|---|---|---|
| $d$ | $l_1$ | $d_1$ | $l_2$ | $L$ | $l_3$ |
| 2.5 | 6.5 | 6 | 13.5 | 51 | |
| 3 | 9.5 | 6 | 16.5 | 54 | |
| 3.5 | | | | | |
| 4 | 12.5 | 6 | 19.5 | 57 | |
| 4.5 | | | | | |
| 5 | | | | | |
| 5.5 | 16 | 6 | 23 | 60.5 | |
| 6 | | | | | |
| 6.5 | 16 | 10 | 22.5 | 60.5 | |
| 7 | 15 | | | | |
| 7.5 | 18 | 10 | 25.5 | 63.5 | |
| 8 | | | | | |
| 8.5 | | | | | 9.5 |
| 9 | 21 | 10 | 28.5 | 66.5 | |
| 9.5 | | | | | |
| 10 | | | | | |
| 10.5 | 19 | 12 | 28.5 | 66.5 | |
| 11 | | | | | |
| 11.5 | 22.5 | | | | |
| 12 | 24 | 12 | 32 | 70 | |
| 13 | 24.5 | | | | |

(*continued*)

| Cutter diameter | Cut length | Shank diameter | Nominal length below chuck | Overall length | Thread length |
|---|---|---|---|---|---|
| $d$ | $l_1$ | $d_1$ | $l_2$ | $L$ | $l_3$ |
| 14 | 28.5 | 12 | 35 | 73 | |
| 15<br>16 | 26.5 | 16 | 38 | 77 | |
| 17<br>18 | 32<br>35 | 16 | 41 | 80 | 9.5 |
| 19<br>20 | 38 | 16 | 44.5 | 83.5 | |
| 21 | 38 | 25 | 42.5 | 95 | |
| 22<br>23<br>24 | 41.5 | 25 | 46 | 98.5 | |
| 25<br>26 | 44.5<br>43 | 25 | 49 | 101.5 | |
| 28<br>30 | 46 | 25 | 52 | 104.5 | |
| 32<br>34 | 49 | 25 | 55.5 | 108 | |
| 35<br>36 | 52.5 | 25 | 58.5 | 111 | 15 |
| 38 | 55.5 | 25 | 62 | 114.5 | |
| 40<br>42 | 58.5<br>60.5 | 25 | 65 | 117.5 | |
| 44<br>45 | 63.5 | 25 | 68 | 120.5 | |
| 32<br>33<br>34<br>35 | 51 | 32 | 58.5 | 112.5 | |
| 36<br>38 | 54 | 32 | 62 | 116 | |
| 40<br>42 | 55.5<br>54 | 32 | 63.5<br>62 | 117.5<br>116 | 15 |
| 44<br>45 | 57 | 32 | 65 | 119 | |
| 50 | 65 | 32 | 73 | 127 | |

For further information see BS 122: Pt 4: 1980.

## 3.15.17 Screwed shank slot drills: normal series

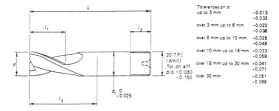

Tolerances on d:

| | |
|---|---|
| up to 3 mm | -0.013 / -0.033 |
| over 3 mm up to 6 mm | -0.020 / -0.038 |
| over 6 mm up to 10 mm | -0.025 / -0.046 |
| over 10 mm up to 18 mm | -0.033 / -0.058 |
| over 18 mm up to 30 mm | -0.041 / -0.071 |
| over 30 mm | -0.051 / -0.089 |

(Dimensions in millimetres)

| Cutter diameter | Cut length | Shank diameter | Nominal length below chuck | Overall length | Thread length |
|---|---|---|---|---|---|
| $d$ | $l_1$ | $d_1$ | $l_2$ | $L$ | $l_3$ |
| 1.5 | 2.5 | 6 | 11 | 48.5 | |
| 2 | 3 | 6 | 11.5 | 49 | |
| 2.5 | 4.5 | 6 | 13.5 | 51 | |
| 3 | 7 | | | | |
| 3.5 | 7.5 | | | | |
| 4 | 9.5 | 6 | 15 | 52.5 | |
| 4.5 | | | | | |
| 5 | | | | | |
| 5.5 | 11 | 6 | 18 | 55.5 | 9.5 |
| 6 | | | 19 | 56.5 | |
| 6.5 | | | | | |
| 7 | 11 | 10 | 20.5 | 58.5 | |
| 7.5 | | | | | |
| 8 | 12.5 | 10 | 21.5 | 59.5 | |
| 8.5 | | | | | |
| 9 | 14.5 | 10 | 22.5 | 60.5 | |
| 9.5 | | | | | |
| 10 | | | | | |

(*continued*)

**3.15.17** (continued)        (Dimensions in millimetres)

| Cutter diameter | Cut length | Shank diameter | Nominal length below chuck | Overall length | Thread length |
|---|---|---|---|---|---|
| d | $l_1$ | $d_1$ | $l_2$ | L | $l_3$ |
| 10.5 | | | | | |
| 11 | 17.5 | 12 | 27 | 65 | |
| 11.5 | | | | | |
| 12 | 19 | 12 | 28.5 | 66.5 | |
| 13 | | | | | |
| 14 | 22 | 12 | 30.5 | 68.5 | 9.5 |
| 15 | 22 | 16 | 33 | 72 | |
| 16 | | | | | |
| 17 | 24 | 16 | 35 | 74 | |
| 18 | | | | | |
| 19 | 25.5 | 16 | 38 | 77 | |
| 20 | | | | | |
| 21 | 25.5 | 25 | 46.0 | 98.5 | |
| 22 | 25.5 | 25 | 47.5 | 100 | |
| 23 | 25.5 | 25 | 49 | 101.5 | |
| 24 | 25.5 | 25 | 50.5 | 103 | |
| 25 | 27 | 25 | 42.5 | 95 | |
| 26 | | | | | |
| 27 | 28.5 | 25 | 41 | 93.5 | |
| 28 | 30 | 25 | 42.5 | 95 | |
| 29 | 30 | 25 | 41 | 93.5 | |
| 30 | | | | | |
| 32 | 38 | 25 | 49 | 101.5 | |
| 34 | | | | | |
| 35 | 39.5 | 25 | 50.5 | 103 | |
| 36 | | | | | |
| 38 | 43 | 25 | 54 | 106.5 | 15 |
| 40 | 46 | 25 | 58.5 | 111 | |
| 42 | 47.5 | 25 | 60 | 112.5 | |
| 44 | 51 | 25 | 63.5 | 116 | |
| 45 | | | | | |
| 32 | 35 | 32 | 63.5 | 117.5 | |
| 34 | 35 | 32 | 65 | 119 | |
| 35 | 39.5 | 32 | 57 | 111 | |
| 36 | | | | | |
| 38 | 43 | 32 | 60.5 | 114.5 | |
| 40 | 46 | 32 | 63.5 | 117.5 | |
| 42 | 47.5 | | | | |
| 44 | 47.5 | 32 | 65 | 119 | |
| 45 | | | | | |
| 50 | 51 | 32 | 63.5 | 117.5 | |

For further information see BS 122: Pt 4: 1980.

## 3.15.18 Screwed shank slot drills, ball nosed: normal series

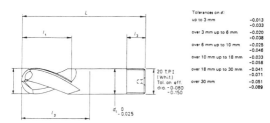

Tolerances on d:
| | |
|---|---|
| up to 3 mm | −0.013 / −0.033 |
| over 3 mm up to 6 mm | −0.020 / −0.038 |
| over 6 mm up to 10 mm | −0.025 / −0.046 |
| over 10 mm up to 18 mm | −0.033 / −0.058 |
| over 18 mm up to 30 mm | −0.041 / −0.071 |
| over 30 mm | −0.051 / −0.089 |

20 T.P.I
[Whit.]
Tol. on eff.
dia. −0.080 / −0.150

$d_{1} \, {}^{0}_{-0.025}$

(Dimensions in millimetres)

| Cutter diameter | Cut length | Shank diameter | Nominal length below chuck | Overall length | Thread length |
|---|---|---|---|---|---|
| $d$ | $l_1$ | $d_1$ | $l_2$ | $L$ | $l_3$ |
| 2 | 3 | 6 | 11.5 | 49 | |
| 2.5 | 4.5 | 6 | 13.5 | 51 | |
| 3 | 7 | | | | |
| 4 | 9.5 | 6 | 15 | 52.5 | |
| 5 | | | | | |
| 6 | 11 | 6 | 19 | 56.5 | |
| 7 | 11 | 10 | 20.5 | 58.5 | |
| 8 | 12.5 | 10 | 21.5 | 59.5 | |
| 9 | 14.5 | 10 | 20.5 | 58.5 | |
| 10 | 14.5 | 10 | 22.5 | 60.5 | 9.5 |
| 11 | 17.5 | 12 | 27 | 65 | |
| 12 | 19 | 12 | 28.5 | 66.5 | |
| 13 | | | | | |
| 14 | 22 | 12 | 30.5 | 68.5 | |
| 15 | 22 | 16 | 33 | 72 | |
| 16 | | | | | |
| 17 | 24 | 16 | 34 | 73 | |
| 18 | 24 | 16 | 35 | 74 | |
| 19 | 25.5 | 16 | 38 | 77 | |
| 20 | | | | | |
| 22 | 25.5 | 25 | 47.5 | 100 | |
| 24 | 25.5 | 25 | 50.5 | 103 | |
| 25 | 28.5 | 25 | 44.5 | 97 | |
| 26 | | | | | 15 |
| 28 | 30 | 25 | 42.5 | 95 | |
| 30 | 30 | 25 | 41 | 93.5 | |
| 32 | 36.5 | 25 | 47.5 | 100 | |

For further information see BS 122: Pt 4: 1980.

## 3.16 British Standard centre drill (60°)

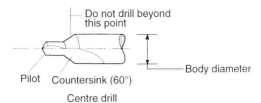

Do not drill beyond this point

Body diameter

Pilot  Countersink (60°)

Centre drill

(Dimensions in inches)

| Size | Body diameter | Pilot diameter | Length overall |
|------|---------------|----------------|----------------|
| BS1 | ⅛ | ³⁄₆₄ | 1 ½ |
| BS2 | ³⁄₁₆ | ¹⁄₁₆ | 1 ¾ |
| BS3 | ¼ | ³⁄₃₂ | 2 |
| BS4 | ⁵⁄₁₆ | ⅛ | 2 ¼ |
| BS5 | ⁷⁄₁₆ | ³⁄₁₆ | 2 ½ |
| BS6 | ⅝ | ¼ | 3 |
| BS7 | ¾ | ⁵⁄₁₆ | 3 ½ |

For further information see BS 328.

Metric          (Dimensions in mm)

| Body diameter | Pilot diameter | Length overall |
|---------------|----------------|----------------|
| 3.15 | 1.00 | 31.50 |
| 4.00 | 1.60 | 35.50 |
| 5.00 | 2.00 | 40.00 |
| 6.30 | 2.50 | 45.00 |
| 8.00 | 3.15 | 50.00 |
| 10.00 | 4.00 | 56.00 |

For further information see BS 328:
Part 2: DIN333.

## 3.17 Engineers' files – popular stock sizes

Available grades of cut: rough*, bastard, second cut, smooth, deadsmooth*.

### Hand

Parallel in width, cut on both sides and one edge.

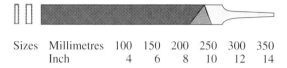

| Sizes | Millimetres | 100 | 150 | 200 | 250 | 300 | 350 |
|-------|-------------|-----|-----|-----|-----|-----|-----|
|       | Inch        | 4   | 6   | 8   | 10  | 12  | 14  |

### Flat

Tapered in width, cut on both sides and edges.

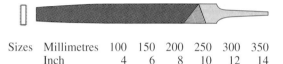

| Sizes | Millimetres | 100 | 150 | 200 | 250 | 300 | 350 |
|-------|-------------|-----|-----|-----|-----|-----|-----|
|       | Inch        | 4   | 6   | 8   | 10  | 12  | 14  |

### Round

For circular openings and concave surfaces, tapers slightly towards the point.

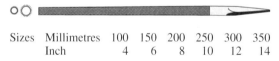

| Sizes | Millimetres | 100 | 150 | 200 | 250 | 300 | 350 |
|-------|-------------|-----|-----|-----|-----|-----|-----|
|       | Inch        | 4   | 6   | 8   | 10  | 12  | 14  |

### Half round

For filing both flat and concave surfaces, parallel in width and thickness, but tapers slightly towards the point.

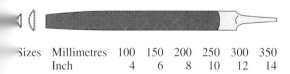

| Sizes | Millimetres | 100 | 150 | 200 | 250 | 300 | 350 |
|-------|-------------|-----|-----|-----|-----|-----|-----|
|       | Inch        | 4   | 6   | 8   | 10  | 12  | 14  |

---

* These cuts are to special order only.
  For further information see: BS 498.

### Square

For slots and keyways, tapers slightly to a point. Double cut on all four sides.

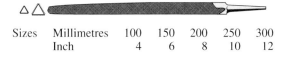

| Sizes | Millimetres | 100 | 150 | 200 | 250 | 300 |
|-------|-------------|-----|-----|-----|-----|-----|
|       | Inch        | 4   | 6   | 8   | 10  | 12  |

### Threesquare

For filing out sharp corners and internal angles. Double cut on all three sides. Tapers slightly towards point.

| Sizes | Millimetres | 100 | 150 | 200 | 250 | 300 |
|-------|-------------|-----|-----|-----|-----|-----|
|       | Inch        | 4   | 6   | 8   | 10  | 12  |

### Warding

Uniform in thickness and tapered in width to a narrow point. Double cut on sides and single cut on edges. Made of a special flexible steel which will not snap under normal use.

| Sizes | Millimetres | 100 | 150 | 200 |
|-------|-------------|-----|-----|-----|
|       | Inch        | 4   | 6   | 8   |

### Knife

Shaped like a wedge or knife with one thick edge tapering to a thin edge. For filing all work having acute angles. Double cut sides and single cut on thin edge.

| Sizes | Millimetres | 150 | 200 |
|-------|-------------|-----|-----|
|       | Inch        | 6   | 8   |

### Pillar

Parallel in width and thickness, cut on both sides.

| Sizes | Millimetres | 150 | 200 |
|-------|-------------|-----|-----|
|       | Inch        | 6   | 8   |

## Safety

The tang of a file must always be protected with a properly fitted handle.

## 3.18 Miscellaneous files

### 3.18.1 Needle (Swiss) files

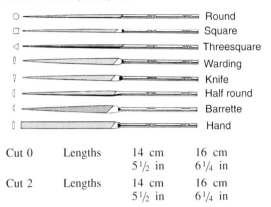

| | | | | |
|---|---|---|---|---|
| Cut 0 | Lengths | 14 cm | 16 cm | |
| | | 5½ in | 6¼ in | |
| Cut 2 | Lengths | 14 cm | 16 cm | |
| | | 5½ in | 6¼ in | |

### 3.18.2 Milled tooth files

Due to the undercut and generous radius at the root of each tooth, milled tooth files clear themselves whilst in use. So instead of a build-up of irregular swarf that will eventually clog the file, the metal is simply removed quickly and efficiently in the form of spirals. This is both faster and safer. Milled tooth files are particularly efficient when used to cut soft materials such as aluminium and thermoplastics. They can also be used on harder materials such as cast irons and steels.

### Dreadnought files (curved tooth, tanged)

| Hand | millimetre | 200 | 250 | 300 | 350 | } standard cut |
|---|---|---|---|---|---|---|
| | inch | 8 | 10 | 12 | 14 | |
| Half round | millimetre | | 250 | 300 | | } standard cut |
| | inch | | 10 | 12 | | |
| Flat | millimetre | | 250 | 300 | 350 | } standard cut |
| | inch | | 10 | 12 | 14 | |

## Millenicut files (straight tooth, tanged)

| | | | | | |
|---|---|---|---|---|---|
| *Hand* | millimetre | 200 | 250 | 300 | 350 } standard cut |
| | inch | 8 | 10 | 12 | 14 |

| | | | | | |
|---|---|---|---|---|---|
| *Half round* | millimetre | 200 | 250 | 300 } standard cut |
| | inch | 8 | 10 | 12 |

## 3.19 Hacksaw blades (high-speed steel – all hard)

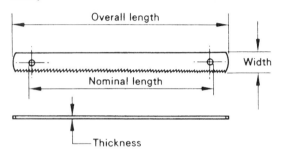

Sizes (Inch)

| Length (nominal) | Width | Thickness | Cut TPI* |
|---|---|---|---|
| 10 | 1/2 | 0.025 | 18 |
| 10 | 1/2 | 0.025 | 24 |
| 12 | 1/2 | 0.025 | 14 |
| 12 | 1/2 | 0.025 | 18 |
| 12 | 1/2 | 0.025 | 24 |
| 12 | 1/2 | 0.025 | 32 |
| 12 | 5/8 | 0.032 | 14 |
| 12 | 5/8 | 0.032 | 18 |
| 12 | 5/8 | 0.032 | 24 |
| 12 | 1 | 0.050 | 10 |
| 12 | 1 | 0.050 | 14 |
| 14 | 1 | 0.050 | 10 |
| 14 | 1 | 0.050 | 14 |
| 14 | 1 1/4 | 0.062 | 6 |
| 14 | 1 1/4 | 0.062 | 10 |
| 14 | 1 1/2 | 0.075 | 6 |
| 16 | 1 1/4 | 0.062 | 6 |
| 16 | 1 1/4 | 0.062 | 10 |
| 16 | 1 1/2 | 0.075 | 4 |
| 16 | 1 1/2 | 0.075 | 6 |

| | | | |
|---|---|---|---|
| 17 | 1 | 0.050 | 10 |
| 17 | 1 | 0.050 | 14 |
| 17 | 1 $1/4$ | 0.062 | 6 |
| 17 | 1 $1/4$ | 0.062 | 10 |
| 18 | 1 $1/4$ | 0.062 | 6 |
| 21 | 1 $1/2$ | 0.075 | 6 |
| 21 | 1 $3/4$ | 0.088 | 6 |
| 24 | 1 $3/4$ | 0.088 | 6 |
| 24 | 2 | 0.100 | 4 |
| 24 | 2 | 0.100 | 6 |

Sizes (millimetres)

| Length (*nominal*) | Width | Thickness | Cut TPI* |
|---|---|---|---|
| 250 | 13 | 0.65 | 18 |
| 250 | 13 | 0.65 | 24 |
| 300 | 13 | 0.65 | 14 |
| 300 | 13 | 0.65 | 18 |
| 300 | 13 | 0.65 | 24 |
| 300 | 13 | 0.65 | 32 |
| 300 | 16 | 0.80 | 14 |
| 300 | 16 | 0.80 | 18 |
| 300 | 16 | 0.80 | 24 |
| 300 | 25 | 1.25 | 10 |
| 300 | 25 | 1.25 | 14 |
| 350 | 25 | 1.25 | 10 |
| 350 | 25 | 1.25 | 14 |
| 350 | 32 | 1.6 | 6 |
| 350 | 32 | 1.6 | 10 |
| 350 | 40 | 2.0 | 6 |
| 400 | 32 | 1.6 | 6 |
| 400 | 32 | 1.6 | 10 |
| 400 | 40 | 2.0 | 4 |
| 400 | 40 | 2.0 | 6 |
| 425 | 25 | 1.25 | 10 |
| 425 | 25 | 1.25 | 4 |
| 425 | 32 | 1.6 | 6 |
| 425 | 32 | 1.6 | 10 |
| 450 | 32 | 1.6 | 6 |
| 450 | 32 | 1.6 | 10 |
| 450 | 40 | 2.0 | 6 |
| 525 | 40 | 2.0 | 6 |
| 525 | 45 | 2.25 | 6 |
| 600 | 45 | 2.25 | 6 |
| 600 | 50 | 2.5 | 4 |
| 600 | 50 | 2.5 | 6 |

*TPI = Teeth per inch despite
other dimensions being metric.

257

## 3.20 Bonded abrasives

### 3.20.1 Example of the complete marking of an abrasive wheel

| Order of marking | 0 | 1 | 2 | 3 | 4 | 5 | 6 |
|---|---|---|---|---|---|---|---|
| | Type of abrasive* | Nature of abrasive | Grain size | Grade | Structure* | Nature of bond | Type of bond etc.* |
| Example | 51 | A | 36 3 | L | 5 | V | 23 |

Aluminium abrasives — A
Silicon carbides — C

Spacing from the closest to the most open

| | |
|---|---|
| 0 | 8 |
| 1 | 9 |
| 2 | 10 |
| 3 | 11 |
| 4 | 12 |
| 5 | 13 |
| 6 | 14 |
| 7 | etc. |

| Coarse | Medium | Fine | Very fine |
|---|---|---|---|
| 8 | 30 | 70 | 220 |
| 10 | 36 | 80 | 240 |
| 12 | 46 | 90 | 280 |
| 14 | 54 | 100 | 320 |
| 16 | 60 | 120 | 400 |
| 20 | | 150 | 500 |
| 24 | | 180 | 600 |

| | |
|---|---|
| V | Vitrified |
| B | Resinoid (synthetic resins) |
| BF | Resinoid (synthetic resins) reinforced |
| R | Rubber |
| RF | Rubber reinforced |
| E | Shellac |
| S | Silicate |
| Mg | Magnesia |

Additional number for grain size mixtures (optional): 1 2 3 etc.

| Soft | Medium | Hard |
|---|---|---|
| A B C D E F G H I J K | L M N O P Q R S T U V W X Y Z | |

*Symbols at positions 0 and 6 are the manufacturer's own choice.

## 3.20.2 Classification of wheel and product shapes by type numbers

For further information see BS 4481: Pt 1: 1981.

**Type 1  Straight wheels**

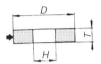

**Type 2  Cylinder wheels**

**Type 3  Taper one side (for use only with straight flanges)**

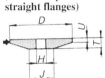

**Type 4  Taper sided portable (for use with tapered flanges)**

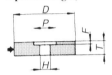

**Type 5  Recessed one side**

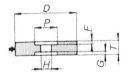

**Type 6  Straight cup wheel**

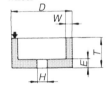

**Type 7  Double recessed wheel**

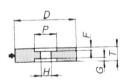

**Type 9  Double cup wheels**

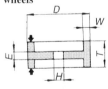

**Type 11  Taper cup wheels**

**Type 12  Dish wheels**

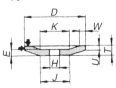

*(continued)*

**Type 13  Saucer wheels**

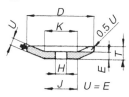

**Type 16  Cone**

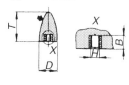

**Type 17  Cone**

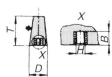

**Type 18 Plug**

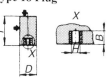

**Type 18R  Plug**

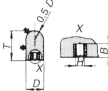

**Type 19  Cone**

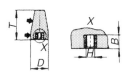

**Type 20  Relieved one side**

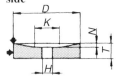

**Type 21  Relieved two sides**

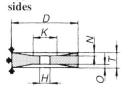

**Type 22 Relieved o/s recess o/s**

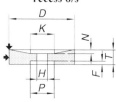

**Type 23 Relieved o/s recess same side**

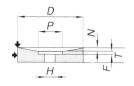

**Type 24 Relieved o/s recessed b/s**

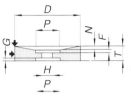

**Type 25 Relieved b/s recessed o/s**

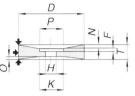

**Type 26 Relieved and recessed b/s**

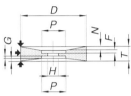

**Type 27 Depressed centre**

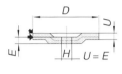

**Type 28 Coolie hat**

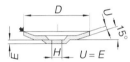

## Symbolisation of dimension

| Symbols | Designation | Typed of wheels concerned |
|---|---|---|
| A | Small base of trapezoidal segment | 31 |
| B | Width of segment, stick or brick | 31-54-90 |
|  | Length of threaded insert | 16 to 19 |
| C | Thickness of segment, stick or brick | 31-54-90 |
| D | Outside diameter | All types of wheels |
| E | Back thickness of cup or dish wheels | 6-9-11-12-13 |
|  | Thickness at hole of relieved wheels with recess | 20 to 28 |
| F | Depth of first recess | 5-7-22 to 26 |
| G | Depth of second recess | 7-24-26 |
| H | Diameter of insert thread | All types of wheels with cut 2-37-51 |
| J | Small outside diameter of tapered wheel, of taper cup, of dish or saucer wheels, outside diameter of hub | 3-11-12-13-39-39 |
| K | Back diameter of taper cup, of dish and saucer wheels, inner diameter relief | 11-12-13-20-21-22-25 |
| L | Spindle length of mounted wheels | 31-52-54-90 |
|  | Length of segments, sticks or bricks |  |
| N | Depth of first relief | 20 to 26 |
| O | Depth of second relief | 21-25-26 |
| P | Recess diameter | 5-7-22 to 26 |
| T | Overall thickness | All types of wheels |
| U | Thickness of grinding face when smaller than T for wheels used on their periphery | 3-12-13-27-28-38-39 |
| W | Width of grinding face for wheels used laterally | 2-6-9-11-12-37 |

Profile elements: U no grinding face
                V profile angle
                X other profile element

▼ Symbolises the grinding face of bonded abrasive products.

See also: BS 4481: Pt 2: for abrasive wheel sizes.
          BS 4481: Pt 3: for abrasive wheel balancing.

### 3.20.3 Maximum permissible peripheral speeds of abrasive wheels

The maximum speeds listed in this table are not necessarily the recommended speeds of operation for optimum grinding efficiency. For higher speeds and further information see BS 4481: Pt 1: 1981.

**3.20.3** (continued)

| Machine classification and grinding operation | | Type of wheel (Section 3.20.2) | Max speed m/s | Special conditions |
|---|---|---|---|---|
| | External cylindrical | 1, 5, 7 20–26 | 35 35 | |
| | Tool room (universal) | 1, 5, 7 | 43 | |
| | (Crankshaft) | 1, 5, 7 | 43 | |
| | Camshaft | 1 | 60 | |
| | Thread | 5, 7 | 45 | |
| | Thread | 1 | 45 | Thicker than 35 mm |
| 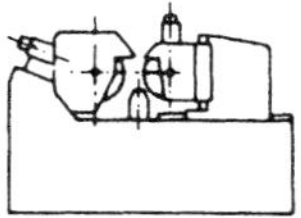 | Centreless | 1, 5, 7 | 35 | |
| | Control wheels | 1, 5, 7 | 12 | |

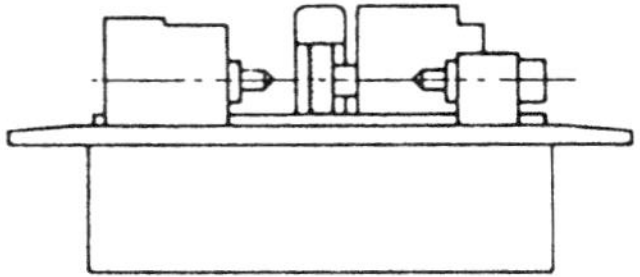

264

| | | | Without overhang |
|---|---|---|---|
| Internal | 1, 5 52 | 35 50 | |
| Surface | | | |
| Horizontal spindle, reciprocating table | 1, 5, 7 | 35 | |
| Horizontal spindle, rotary table | 1, 5, 7 | 35 | |

**3.20.3** (continued)

| Machine classification and grinding operation | | Type of wheel (Section 3.20.2) | Max speed m/s | Special conditions |
|---|---|---|---|---|
| Surface | | | | |
| Vertical spindle, reciprocating table | | 2, 37 | 25 | Inorganic bonds |
| | | | 30 | Organic bonds |
| | | 6 | 30 | |
| | | 35, 36 | 32 | |
| Vertical spindle, rotary table | | 2, 37 | 25 | Inorganic bonds |
| | | | 30 | Organic bonds |
| | | 6 | 30 | |
| | | 35, 36 | 32 | |
| Duplex | | 2, 37 | 25 | Inorganic bonds |
| | | | 30 | Organic bonds |
| | | 6 | 30 | |
| | | 35, 36 | 32 | |

| | | | |
|---|---|---|---|
| Off-hand grinding and fettling | | | |
| Bench | 1, 5, 7 | 35 | Organic bond only |
| Floor stand | 1 | 50 | |
| Side grinding | 6, 35, 36 | 32 | |
| Billet and slab | | | |
| Mechanical control | 1 | 63 | Special high density organic bond |
| Swing frame, manual control | 1 | 50 | Organic bond only |

(continued)

**3.20.3** (continued)

| Machine classification and grinding operation | | Type of wheel (Section 3.20.2) | Max speed m/s | Special conditions |
|---|---|---|---|---|
| | Cutting off | 1 | 80 | Reinforced organic bond only |
| | Cutting off (fully guarded) | 1 | 80 | Organic bonds only |

| Machine type | Operation | Wheel shapes | Speed | Notes |
|---|---|---|---|---|
| Portable, right angle | Grinding | 6, 11 | 50 | Organic bonds only |
| | | 27 | 80 | Reinforced organic bonds only |
| | Cutting off | 1, 27 | 80 | Reinforced Organic bonds only |
| Portable, vertical spindle grinder | | 6, 11 | 50 | Organic bonds only |
| Portable, straight grinder | | 1, 4 | 50 | Organic bonds only |
| | | 16, 17, 18, 18R, 19 | 50 | |
| | | 52 | 50 | Without overhang |
| Tool and cutter | Grinding and sharpening | 1, 5, 7 | 35 | |
| | | 6, 11, 12, 13 | 32 | |

# Part 4

# Miscellaneous

# 4.1 Washers

## 4.1.1 Plain washers, bright: metric series

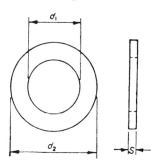

(Dimensions in millimetres)

| Designation (thread diameter)* | Internal diameter $d_1$ | | External diameter $d_2$ | | Thickness S | | | |
|---|---|---|---|---|---|---|---|---|
| | | | | | Thick (normal) | | Thin | |
| | max. | min. | max. | min. | max. | min. | max. | min. |
| M1.0 | 1.25 | 1.1 | 2.5 | 2.3 | 0.4 | 0.2 | — | — |
| M1.2 | 1.45 | 1.3 | 3.0 | 2.8 | 0.4 | 0.2 | — | — |
| M1.4) | 1.65 | 1.5 | 3.0 | 2.8 | 0.4 | 0.2 | — | — |
| M1.6 | 1.85 | 1.7 | 4.0 | 3.7 | 0.4 | 0.2 | — | — |
| M2.0 | 2.35 | 2.2 | 5.0 | 4.7 | 0.4 | 0.2 | — | — |
| M2.2) | 2.55 | 2.4 | 5.0 | 4.7 | 0.6 | 0.4 | — | — |
| M2.5 | 2.85 | 2.7 | 6.5 | 6.2 | 0.6 | 0.4 | — | — |
| M3 | 3.4 | 3.2 | 7.0 | 6.7 | 0.6 | 0.4 | — | — |
| M3.5) | 3.9 | 3.7 | 7.0 | 6.7 | 0.6 | 0.4 | — | — |
| M4 | 4.5 | 4.3 | 9.0 | 8.7 | 0.9 | 0.7 | — | — |
| M4.5) | 5.0 | 4.8 | 9.0 | 8.7 | 0.9 | 0.7 | — | — |
| M5 | 5.5 | 5.3 | 10.0 | 9.7 | 1.1 | 0.9 | — | — |
| M6 | 6.7 | 6.4 | 12.5 | 12.1 | 1.8 | 1.4 | 0.9 | 0.7 |
| M7) | 7.7 | 7.4 | 14.0 | 13.6 | 1.8 | 1.4 | 0.9 | 0.7 |
| M8 | 8.7 | 8.4 | 17.0 | 16.6 | 1.8 | 1.4 | 1.1 | 0.9 |
| M10 | 10.9 | 10.5 | 21.0 | 20.5 | 2.2 | 1.8 | 1.45 | 1.05 |
| M12 | 13.4 | 13.0 | 24.0 | 23.5 | 2.7 | 2.3 | 1.8 | 1.4 |
| M14) | 15.4 | 15.0 | 28.0 | 27.5 | 2.7 | 2.3 | 1.8 | 1.4 |
| M16 | 17.4 | 17.0 | 30.0 | 29.5 | 3.3 | 2.7 | 2.2 | 1.8 |
| M18) | 19.5 | 19.0 | 34.0 | 33.2 | 3.3 | 2.7 | 2.2 | 1.8 |
| M20 | 21.5 | 21.0 | 37.0 | 36.2 | 3.3 | 2.7 | 2.2 | 1.8 |
| M22) | 23.5 | 23.0 | 39.0 | 38.2 | 3.3 | 2.7 | 2.2 | 1.8 |
| M24 | 25.5 | 25.0 | 44.0 | 43.2 | 4.3 | 3.7 | 2.7 | 2.3 |
| M27) | 28.5 | 28.0 | 50.0 | 49.2 | 4.3 | 3.7 | 2.7 | 2.3 |
| M30 | 31.6 | 31.0 | 56.0 | 55.0 | 4.3 | 3.7 | 2.7 | 2.3 |
| M33) | 34.6 | 34.0 | 60.0 | 59.0 | 5.6 | 4.4 | 3.3 | 2.7 |
| M36 | 37.6 | 37.0 | 66.0 | 65.0 | 5.6 | 4.4 | 3.3 | 2.7 |
| M39) | 40.6 | 40.0 | 72.0 | 71.0 | 6.6 | 5.4 | 3.3 | 2.7 |

on-preferred sizes in parentheses ( ).
full information see BS 4320.

## 4.1.2 Plain washers, black: metric series

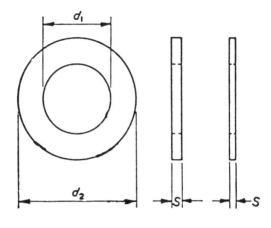

(Dimensions in millimetres)

| Designation (thread diameter)* | Internal diameter $d_1$ | | External diameter $d_2$ | | Thickness $S$ | |
|---|---|---|---|---|---|---|
| | max. | min. | max. | min. | max. | min |
| M5 | 5.8 | 5.5 | 10.0 | 9.2 | 1.2 | 0.8 |
| M6 | 7.0 | 6.6 | 12.5 | 11.7 | 1.9 | 1.3 |
| (M7) | 8.0 | 7.6 | 14.0 | 13.2 | 1.9 | 1.3 |
| M8 | 9.4 | 9.0 | 17.0 | 16.2 | 1.9 | 1.3 |
| M10 | 11.5 | 11.0 | 21.0 | 20.2 | 2.3 | 1.7 |
| M12 | 14.5 | 14.0 | 24.0 | 23.2 | 2.8 | 2.2 |
| (M14) | 16.5 | 16.0 | 28.0 | 27.2 | 2.8 | 2.2 |
| M16 | 18.5 | 18.0 | 30.0 | 29.2 | 3.6 | 2.4 |
| (M18) | 20.6 | 20.0 | 34.0 | 33.8 | 3.6 | 2.4 |
| M20 | 22.6 | 22.0 | 37.0 | 35.8 | 3.6 | 2. |
| (M22) | 24.6 | 24.0 | 39.0 | 37.8 | 3.6 | 2. |
| M24 | 26.6 | 26.0 | 44.0 | 42.8 | 4.6 | 3. |
| (M27) | 30.6 | 30.0 | 50.0 | 48.8 | 4.6 | 3. |
| M30 | 33.8 | 33.0 | 56.0 | 54.5 | 4.6 | 3. |
| (M33) | 36.8 | 36.0 | 60.0 | 58.5 | 6.0 | 4 |
| M36 | 39.8 | 39.0 | 66.0 | 64.5 | 6.0 | 4 |
| (M39) | 42.8 | 42.0 | 72.0 | 70.5 | 7.0 | 5 |
| M42 | 45.8 | 45.0 | 78.0 | 76.5 | 8.2 | 5 |
| (M45) | 48.8 | 48.0 | 85.0 | 83.0 | 8.2 | 5 |
| M48 | 53.0 | 52.0 | 92.0 | 90.0 | 9.2 | 6 |
| (M52) | 57.0 | 56.0 | 98.0 | 96.0 | 9.2 | 6 |
| M56 | 63.0 | 62.0 | 105.0 | 103.0 | 10.2 | 7 |
| (M60) | 67.0 | 66.0 | 110.0 | 108.0 | 10.2 | 7 |
| M64 | 71.0 | 70.0 | 115.0 | 113.0 | 10.2 | 7 |
| (M68) | 75.0 | 74.0 | 120.0 | 118.0 | 11.2 | 8 |

*Non-preferred sizes in parentheses ( ).
For full information see BS 4320.

## 4.1.3 Single coil square section spring washers: metric series, type A

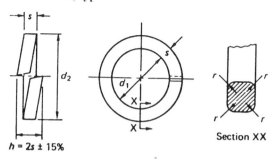

$h = 2s \pm 15\%$

Section XX

(Dimensions in millimetres)

| Nominal size and thread diameter* d | Inside diameter $d_1$ | | Thickness and width s | Outside diameter $d_2$ | Radius r |
|---|---|---|---|---|---|
| | max. | min. | | max. | max. |
| M3 | 3.3 | 3.1 | $1 \pm 0.1$ | 5.5 | 0.3 |
| (M3.5) | 3.8 | 3.6 | $1 \pm 0.1$ | 6.0 | 0.3 |
| M4 | 4.35 | 4.1 | $1.2 \pm 0.1$ | 6.95 | 0.4 |
| M5 | 5.35 | 5.1 | $1.5 \pm 0.1$ | 8.55 | 0.5 |
| M6 | 6.4 | 6.1 | $1.5 \pm 0.1$ | 9.6 | 0.5 |
| M8 | 8.55 | 8.2 | $2 \pm 0.1$ | 12.75 | 0.65 |
| M10 | 10.6 | 10.2 | $2.5 \pm 0.15$ | 15.9 | 0.8 |
| M12 | 12.6 | 12.2 | $2.5 \pm 0.15$ | 17.9 | 0.8 |
| (M14) | 14.7 | 14.2 | $3 \pm 0.2$ | 21.1 | 1.0 |
| M16 | 16.9 | 16.3 | $3.5 \pm 0.2$ | 24.3 | 1.15 |
| (M18) | 19.0 | 18.3 | $3.5 \pm 0.2$ | 26.4 | 1.15 |
| M20 | 21.1 | 20.3 | $4.5 \pm 0.2$ | 30.5 | 1.5 |
| (M22) | 23.3 | 22.4 | $4.5 \pm 0.2$ | 32.7 | 1.5 |
| M24 | 25.3 | 24.4 | $5 \pm 0.2$ | 35.7 | 1.65 |
| (M27) | 28.5 | 27.5 | $5 \pm 0.2$ | 38.9 | 1.65 |
| M30 | 31.5 | 30.5 | $6 \pm 0.2$ | 43.9 | 2.0 |
| (M33) | 34.6 | 33.5 | $6 \pm 0.2$ | 47.0 | 2.0 |
| M36 | 37.6 | 36.5 | $7 \pm 0.25$ | 52.1 | 2.3 |
| (M39) | 40.8 | 39.6 | $7 \pm 0.25$ | 55.3 | 2.3 |
| M42 | 43.8 | 42.6 | $8 \pm 0.25$ | 60.3 | 2.65 |
| (M45) | 46.8 | 45.6 | $8 \pm 0.25$ | 63.3 | 2.65 |
| M48 | 50.0 | 48.8 | $8 \pm 0.25$ | 66.5 | 2.65 |

Sizes shown in parentheses are non-preferred and are not usually stock sizes.

or further information see BS 4464.

## 4.1.4 Single coil rectangular section spring washers: metric series, types B and BP

Type BP     Type B

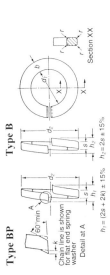

Chain line is shown
for flat end spring
washer

Detail at A

$h_1 = (2s + 2k) \pm 15\%$     $h_2 = 2s \pm 15\%$

Section XX

(Dimensions in millimetres)

| Nominal size and thread diameter* $d$ | Inside diameter $d_1$ | | Width $b$ | Thickness $s$ | Outside diameter $d_2$ max. | Radius $r$ max. | $k$ (type BP only) |
|---|---|---|---|---|---|---|---|
| | max. | min. | | | | | |
| M1.6 | 1.9 | 1.7 | $0.7 \pm 0.1$ | $0.4 \pm 0.1$ | 3.5 | 0.15 | — |
| M2 | 2.3 | 2.1 | $0.9 \pm 0.1$ | $0.5 \pm 0.1$ | 4.3 | 0.15 | — |
| (M2.2) | 2.5 | 2.3 | $1.0 \pm 0.1$ | $0.6 \pm 0.1$ | 4.7 | 0.2 | — |
| M2.5 | 2.8 | 2.6 | $1.0 \pm 0.1$ | $0.6 \pm 0.1$ | 5.0 | 0.2 | — |
| M3 | 3.3 | 3.1 | $1.3 \pm 0.1$ | $0.8 \pm 0.1$ | 6.1 | 0.25 | 0.15 |
| | | | $1.3 \pm 0.1$ | $0.8 \pm 0.1$ | 6.6 | 0.25 | 0.15 |
| | | | | | 7.5 | 0.3 | |

| | | | | | | |
|---|---|---|---|---|---|---|
| M5 | 5.35 | 1.8 ± 0.1 | 5.1 | 1.2 ± 0.1 | 9.15 | 0.4 | 0.15 |
| M6 | 6.4 | 2.5 ± 0.15 | 6.1 | 1.6 ± 0.1 | 11.7 | 0.5 | 0.2 |
| M8 | 8.55 | 3.0 ± 0.15 | 8.2 | 2.0 ± 0.1 | 14.85 | 0.65 | 0.3 |
| M10 | 10.6 | 3.5 ± 0.2 | 10.2 | 2.2 ± 0.15 | 18.0 | 0.7 | 0.3 |
| M12 | 12.6 | 4.0 ± 0.2 | 12.2 | 2.5 ± 0.15 | 21.0 | 0.8 | 0.4 |
| (M14) | 14.7 | 4.5 ± 0.2 | 14.2 | 3.0 ± 0.15 | 24.1 | 1.0 | 0.4 |
| M16 | 16.9 | 5.0 ± 0.2 | 16.3 | 3.5 ± 0.2 | 27.3 | 1.15 | 0.4 |
| (M18) | 19.0 | 5.0 ± 0.2 | 18.3 | 3.5 ± 0.2 | 29.4 | 1.15 | 0.4 |
| M20 | 21.1 | 6 ± 0.2 | 20.3 | 4 ± 0.2 | 33.5 | 1.3 | 0.4 |
| (M22) | 23.3 | 6 ± 0.2 | 22.4 | 4 ± 0.2 | 35.7 | 1.3 | 0.4 |
| M24 | 25.3 | 7 ± 0.25 | 24.4 | 5 ± 0.2 | 39.8 | 1.65 | 0.5 |
| (M27) | 28.5 | 7 ± 0.25 | 27.5 | 5 ± 0.2 | 43.0 | 1.65 | 0.5 |
| M30 | 31.5 | 8 ± 0.25 | 30.5 | 6 ± 0.25 | 48.0 | 2.0 | 0.8 |
| (M33) | 34.6 | 10 ± 0.25 | 33.5 | 6 ± 0.25 | 55.1 | 2.0 | 0.8 |
| M36 | 37.6 | 10 ± 0.25 | 36.5 | 6 ± 0.25 | 58.1 | 2.0 | 0.8 |
| (M39) | 40.8 | 12 ± 0.25 | 39.6 | 6 ± 0.25 | 61.3 | 2.3 | 0.8 |
| M42 | 43.8 | 12 ± 0.25 | 42.6 | 7 ± 0.25 | 68.3 | 2.3 | 0.8 |
| (M45) | 46.8 | 12 ± 0.25 | 45.6 | 7 ± 0.25 | 71.3 | 2.3 | 0.8 |
| M48 | 50.0 | 14 ± 0.25 | 48.8 | 7 ± 0.25 | 74.5 | 2.3 | 0.8 |
| (M52) | 54.1 | 14 ± 0.25 | 52.8 | 8 ± 0.25 | 82.6 | 2.65 | 1.0 |
| M56 | 58.1 | 14 ± 0.25 | 56.8 | 8 ± 0.25 | 86.6 | 2.65 | 1.0 |
| (M60) | 62.3 | 14 ± 0.25 | 60.9 | 8 ± 0.25 | 90.8 | 2.65 | 1.0 |
| M64 | 66.3 | 14 ± 0.25 | 64.9 | 8 ± 0.25 | 93.8 | 2.65 | 1.0 |
| (M68) | 70.5 | 14 ± 0.25 | 69.0 | 8 ± 0.25 | 99.0 | 2.65 | 1.0 |

*Sizes shown in parentheses are non-preferred, and are not usually stock sizes.
For further information see BS 4464.

## 4.1.5 Double coil rectangular section spring washers: metric series, type D

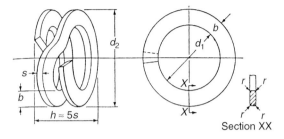

Section XX

(Dimensions in millimetres)

| Nominal size and thread diameter* d | Inside diameter $d_1$ max. | Inside diameter $d_1$ max. | Width b | Thickness s | Outside diameter $d_2$ max. | Radius r max. |
|---|---|---|---|---|---|---|
| M2 | 2.4 | 2.1 | 0.9 ± 0.1 | 0.5 ± 0.05 | 4.4 | 0.15 |
| (M2.2) | 2.6 | 2.3 | 1 ± 0.1 | 0.6 ± 0.05 | 4.8 | 0.2 |
| M2.5 | 2.9 | 2.6 | 1.2 ± 0.1 | 0.7 ± 0.1 | 5.5 | 0.23 |
| M3.0 | 3.6 | 3.3 | 1.2 ± 0.1 | 0.8 ± 0.1 | 6.2 | 0.25 |
| (M3.5) | 4.1 | 3.8 | 1.6 ± 0.1 | 0.8 ± 0.1 | 7.5 | 0.25 |
| M4 | 4.6 | 4.3 | 1.6 ± 0.1 | 0.8 ± 0.1 | 8.0 | 0.25 |
| M5 | 5.6 | 5.3 | 2 ± 0.1 | 0.9 ± 0.1 | 9.8 | 0.3 |
| M6 | 6.6 | 6.3 | 3 ± 0.15 | 1 ± 0.1 | 12.9 | 0.33 |
| M8 | 8.8 | 8.4 | 3 ± 0.15 | 1.2 ± 0.1 | 15.1 | 0.4 |
| M10 | 10.8 | 10.4 | 3.5 ± 0.20 | 1.2 ± 0.1 | 18.2 | 0.4 |
| M12 | 12.8 | 12.4 | 3.5 ± 0.2 | 1.6 ± 0.1 | 20.2 | 0.5 |
| (M14) | 15.0 | 14.5 | 5 ± 0.2 | 1.6 ± 0.1 | 25.4 | 0.5 |
| M16 | 17.0 | 16.5 | 5 ± 0.2 | 2 ± 0.1 | 27.4 | 0.65 |
| (M18) | 19.0 | 18.5 | 5 ± 0.2 | 2 ± 0.1 | 29.4 | 0.65 |
| M20 | 21.5 | 20.8 | 5 ± 0.2 | 2 ± 0.1 | 31.9 | 0.65 |
| (M22) | 23.5 | 22.8 | 6 ± 0.2 | 2.5 ± 0.15 | 35.9 | 0.8 |
| M24 | 26.0 | 25.0 | 6.5 ± 0.2 | 3.25 ± 0.15 | 39.4 | 1.1 |
| (M27) | 29.5 | 28.0 | 7 ± 0.25 | 3.25 ± 0.15 | 44.0 | 1.1 |
| M30 | 33.0 | 31.5 | 8 ± 0.25 | 3.25 ± 0.15 | 49.5 | 1.1 |
| (M33) | 36.0 | 34.5 | 8 ± 0.25 | 3.25 ± 0.15 | 52.5 | 1.1 |
| M36 | 40.0 | 38.0 | 10 ± 0.25 | 3.25 ± 0.15 | 60.5 | 1.1 |
| (M39) | 43.0 | 41.0 | 10 ± 0.25 | 3.25 ± 0.15 | 63.5 | 1.1 |
| M42 | 46.0 | 44.0 | 10 ± 0.25 | 4.5 ± 0.2 | 66.5 | 1.5 |
| M48 | 52.0 | 50.0 | 10 ± 0.25 | 4.5 ± 0.2 | 72.5 | 1.5 |
| M56 | 60.0 | 58.0 | 12 ± 0.25 | 4.5 ± 0.2 | 84.5 | 1.5 |
| M64 | 70.0 | 67.0 | 12 ± 0.25 | 4.5 ± 0.2 | 94.5 | 1.5 |

* Sizes shown in parentheses are non-preferred, and are not usually stock sizes.

*Note:* The free height of double coil washers before compression is normally approximately five times the thickness but, if required, washers with other free heights may be obtained by arrangement between the purchaser and the manufacturer.

For further information see BS 4464.

### 4.1.6 Toothed lock washers, metric

**Type A externally toothed**

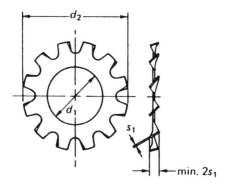

**Type J internally toothed**

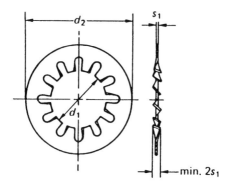

**Type V countersunk**

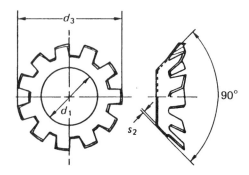

Details left unspecified are to be designed as appropriate.

Designation of a toothed lock washer type J with hole diameter $d_1 = 6.4$ mm of spring steel, surface phosphated for protection against rusting (phr): toothed lock washer J 6.4 DIN 6797 – phr.

If toothed lock washers are required for left-hand threaded bolts, the designation reads: toothed lock washer J 6.4 left DIN 6797 – phr.

(Dimensions in millimetres)

| $d_1$ (H13) | $d_2$ (h14) | $d_3$ ≈ | $s_1$ | $s_2$ | Number of teeth min. | | Weight (7.85 kg/dm³) kg/1000 pieces ≈ | | | For thread diameter |
|---|---|---|---|---|---|---|---|---|---|---|
| | | | | | A and J | V | A | J | V | |
| 1.7 | 3.6 | — | 0.3 | — | 6 | — | 0.01 | — | — | 1.6 |
| 1.8 | 3.8 | — | 0.3 | — | 6 | — | 0.015 | — | — | 1.7 |
| 1.9 | 4 | — | 0.3 | — | 6 | — | 0.02 | 0.03 | — | 1.8 |
| 2.2 | 4.5 | 4.2 | 0.3 | 0.2 | 6 | 6 | 0.025 | 0.04 | 0.02 | 2 |
| 2.5 | 5 | — | 0.4 | 0.2 | 6 | 6 | 0.03 | 0.025 | — | 2.3 |
| 2.7 | 5.5 | 5.1 | 0.4 | 0.2 | 6 | 6 | 0.04 | 0.045 | 0.025 | 2.5 |
| 2.8 | 5.5 | — | 0.4 | 0.2 | 6 | 6 | 0.04 | 0.045 | — | 2.6 |
| 3.2 | 6 | 6 | 0.4 | 0.2 | 6 | 6 | 0.045 | 0.045 | 0.025 | 3 |
| 3.7 | 7 | 7 | 0.5 | 0.25 | 6 | 6 | 0.075 | 0.085 | 0.04 | 3.5 |
| 4.3 | 8 | 8 | 0.5 | 0.25 | 8 | 8 | 0.095 | 0.1 | 0.05 | 4 |
| 5.1* | 9 | — | 0.5 | — | 8 | — | 0.14 | 0.15 | — | 5 |
| 5.3 | 10 | 9.8 | 0.6 | 0.3 | 8 | 8 | 0.18 | 0.2 | 0.12 | 5 |
| 6.4 | 11 | 11.8 | 0.7 | 0.4 | 8 | 10 | 0.22 | 0.25 | 0.2 | 6 |
| 7.4 | 12.5 | — | 0.8 | — | 8 | — | 0.3 | 0.35 | — | 7 |

| | | | | | | | | | | |
|---|---|---|---|---|---|---|---|---|---|---|
| 8.2* | 14 | — | 0.8 | — | 8 | — | 0.4 | 0.45 | — | 8 |
| 8.4 | 15 | 15.3 | 0.8 | 0.4 | 8 | 10 | 0.45 | 0.55 | 0.4 | 8 |
| 10.5 | 18 | 19 | 0.9 | 0.5 | 9 | 10 | 0.8 | 0.9 | 0.7 | 10 |
| 12.5 | 20.5 | 23 | 1.0 | 0.5 | 10 | 10 | 1.1 | 1.3 | 1.2 | 12 |
| 14.5 | 24 | 26.2 | 1.0 | 0.6 | 10 | 12 | 1.7 | 2.0 | 1.4 | 14 |
| 16.5 | 26 | 30.2 | 1.2 | 0.6 | 12 | 12 | 2.1 | 2.5 | 1.4 | 16 |
| 19 | 30 | — | 1.4 | — | 12 | — | 3.5 | 3.7 | — | 18 |
| 21 | 33 | — | 1.4 | — | 12 | — | 3.8 | 4.1 | — | 20 |
| 23 | 36 | — | 1.5 | — | 14 | — | 5 | 6.0 | — | 22 |
| 25 | 38 | — | 1.5 | — | 14 | — | 6 | 6.5 | — | 24 |
| 28 | 44 | — | 1.6 | — | 14 | — | 8 | 8.5 | — | 27 |
| 31 | 48 | — | 1.6 | — | 14 | — | 9 | 9.5 | — | 30 |

*Only for hexagon head bolts.
For further details see DIN 6797.

### 4.1.7 Serrated lock washers, metric

Details left unspecified are to be designed as appropriate.

Designation of a serrated lock washer type J with hole diameter $d_1 = 6.4$ mm in spring steel, surface phosphated for protection against rusting (phr): serrated lock washer J 6.4 DIN 6798 – phr.

If serrated lock washers are required for left-hand threaded bolts, the designation reads: serrated lock washer J 6.4 left DIN 6798 – phr.

**Type A serrated externally**

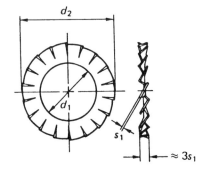

**Type J serrated internally**

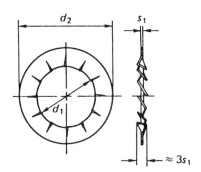

**Type V countersunk**

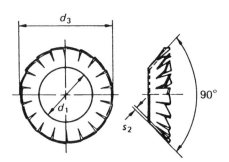

(Dimensions in millimetres)

| $d_1$ (H13) | $d_2$ (h14) | $d_3$ ≈ | $s_1$ | $s_2$ | Number of teeth min. | | | Weight (7.85 kg/dm³) ≈ kg/1000 pieces ≈ | | For thread diameter |
|---|---|---|---|---|---|---|---|---|---|---|
| | | | | | A | J | V | A and J | V | |
| 1.7 | 3.6 | — | 0.3 | — | 9 | 7 | — | 0.02 | — | 1.6 |
| 1.8 | 3.8 | — | 0.3 | — | 9 | 7 | — | 0.02 | — | 1.7 |
| 1.9 | 4 | — | 0.3 | — | 9 | 7 | — | 0.025 | — | — |
| 2.2 | 4.5 | 4.2 | 0.3 | 0.2 | 9 | 7 | 10 | 0.03 | 0.025 | 2 |
| 2.5 | 5 | — | 0.4 | 0.2 | 9 | 7 | 10 | 0.04 | — | 2.3 |
| 2.7 | 5.5 | 5.1 | 0.4 | 0.2 | 9 | 7 | 10 | 0.045 | 0.03 | 2.5 |
| 2.8 | 5.5 | — | 0.4 | 0.2 | 9 | 7 | 10 | 0.05 | — | 2.6 |
| 3.2 | 6 | 6 | 0.4 | 0.2 | 9 | 7 | 12 | 0.06 | 0.04 | 3 |
| 3.7 | 7 | 7 | 0.5 | 0.25 | 10 | 8 | 12 | 0.11 | 0.075 | 3.5 |
| 4.3 | 8 | 8 | 0.5 | 0.25 | 11 | 8 | 14 | 0.14 | 0.1 | 4 |
| 5.1* | 9 | — | 0.5 | — | 11 | 8 | | 0.22 | — | 5 |
| 5.3 | 10 | 9.8 | 0.6 | 0.3 | 11 | 8 | 14 | 0.28 | 0.2 | 5 |
| 6.4 | 11 | 11.8 | 0.7 | 0.4 | 12 | 9 | 16 | 0.36 | 0.3 | 6 |
| 7.4 | 12.5 | — | 0.8 | — | 14 | 10 | — | 0.5 | — | 7 |
| 8.2* | 14 | — | 0.8 | — | 14 | 10 | — | 0.75 | — | 8 |

(continued)

**4.1.7** (continued)

(Dimensions in millimetres)

| $d_1$ (H13) | $d_2$ (h14) | $d_3$ ≈ | $s_1$ | $s_2$ | Number of teeth min. | | | Weight (7.85 kg/dm³) kg/1000 pieces ≈ | | For thread diameter |
|---|---|---|---|---|---|---|---|---|---|---|
| | | | | | A | J | V | A and J | V | |
| 8.4 | 15 | 15.3 | 0.8 | 0.4 | 14 | 10 | 18 | 0.8 | 0.5 | 8 |
| 10.5 | 18 | 19 | 0.9 | 0.5 | 16 | 12 | 20 | 1.25 | 1 | 10 |
| 12.5 | 20.5 | 23 | 1 | 0.5 | 16 | 12 | 26 | 1.7 | 1.5 | 12 |
| 14.5 | 24 | 26.2 | 1 | 0.6 | 18 | 14 | 28 | 2.4 | 2 | 14 |
| 16.5 | 26 | 30.2 | 1.2 | 0.6 | 18 | 14 | 30 | 3 | 2.4 | 16 |
| 19 | 30 | — | 1.4 | — | 18 | 14 | — | 5 | — | 18 |
| 21 | 33 | — | 1.4 | — | 20 | 16 | — | 6 | — | 20 |
| 23 | 36 | — | 1.5 | — | 20 | 16 | — | 7.5 | — | 22 |
| 25 | 38 | — | 1.5 | — | 20 | 16 | — | 8 | — | 24 |
| 28 | 44 | — | 1.6 | — | 22 | 18 | — | 12 | — | 27 |
| 31 | 48 | — | 1.6 | — | 22 | 18 | — | 14 | — | 30 |

*Only for hexagon head bolts.
For further details see DIN 6797.

## 4.1.8 ISO metric crinkle washers: general engineering

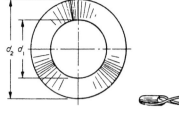

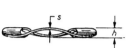

(Dimensions in millimetres)

| Nominal (thread) diameter* | Inside diameter $d_1$ | | Outside diameter $d_2$ | | Height $h$ | | Thickness $s$ |
|---|---|---|---|---|---|---|---|
| | max. | min. | max. | min. | max. | min. | |
| M1.6 | 1.8 | 1.7 | 3.7 | 3.52 | 0.51 | 0.36 | 0.16 |
| M2 | 2.3 | 2.2 | 4.6 | 4.42 | 0.53 | 0.38 | 0.16 |
| M2.5 | 2.8 | 2.7 | 5.8 | 5.62 | 0.53 | 0.38 | 0.16 |
| M3 | 3.32 | 3.2 | 6.4 | 6.18 | 0.61 | 0.46 | 0.16 |
| M4 | 4.42 | 4.3 | 8.1 | 7.88 | 0.84 | 0.69 | 0.28 |
| M5 | 5.42 | 5.3 | 9.2 | 8.98 | 0.89 | 0.74 | 0.30 |
| M6 | 6.55 | 6.4 | 11.5 | 11.23 | 1.14 | 0.99 | 0.40 |
| M8 | 8.55 | 8.4 | 15.0 | 14.73 | 1.40 | 1.25 | 0.40 |
| M10 | 10.68 | 10.5 | 19.6 | 19.27 | 1.70 | 1.55 | 0.55 |
| M12 | 13.18 | 13.0 | 22.0 | 21.67 | 1.90 | 1.65 | 0.55 |
| (M14) | 15.18 | 15.0 | 25.5 | 25.17 | 2.06 | 1.80 | 0.55 |
| M16 | 17.18 | 17.0 | 27.8 | 27.47 | 2.41 | 2.16 | 0.70 |
| (M18) | 19.21 | 19.0 | 31.3 | 30.91 | 2.41 | 2.16 | 0.70 |
| M20 | 21.21 | 21.0 | 34.7 | 34.31 | 2.66 | 2.16 | 0.70 |

*Second choice sizes in parentheses ( ).
For full range and further information see BS 4463.

## 4.2 T-slot profiles

Remove sharp corner.
0.3 × 45° max.

(Dimensions in millimetres)

| Designations of T-slot | Width of throat $A_1$ Nominal | Width of throat $A_1$ Ordinary (H12) | Width of throat $A_1$ For use as tenon (HB) | Width of recess $B_1$ min. | Width of recess $B_1$ max. | Depth of recess $c_1$ min. | Depth of recess $c_1$ max. | Overall depth of T-slot $H$ min. | Overall depth of T-slot $H$ max. | Chamfer × 45° or radius $K$ max. | Chamfer × 45° $F$ max. | Chamfer × 45° $G$ max. | Pitch P (avoid pitch values in brackets as they lead to weakness) |
|---|---|---|---|---|---|---|---|---|---|---|---|---|---|
| M4 | 5 | +0.12 / 0 | +0.018 / 0 | 10 | 11 | 3.5 | 4.5 | 8 | 10 | 1.0 | 0.6 | 1.0 | 20 25 32 |
| M5 | 6 | +0.12 / 0 | +0.018 / 0 | 11 | 12.5 | 5 | 6 | 11 | 13 | 1.0 | 0.6 | 1.0 | 25 32 40 |
| M6 | 8 | +0.15 / 0 | +0.022 / 0 | 14.5 | 16 | 7 | 8 | 15 | 18 | 1.0 | 0.6 | 1.0 | 32 40 50 |
| M8 | 10 | +0.15 / 0 | +0.022 / 0 | 16 | 18 | 7 | 8 | 17 | 21 | 1.0 | 0.6 | 1.0 | 40 50 63 |
| M10 | 12 | +0.18 / 0 | +0.027 / 0 | 19 | 21 | 8 | 9 | 20 | 25 | 1.0 | 0.6 | 1.6 | (40) 50 63 80 |
| M12 | 14 | +0.18 / 0 | +0.027 / 0 | 23 | 25 | 9 | 11 | 23 | 28 | 1.6 | 1.0 | 1.6 | (50) 63 80 100 |
| M16 | 18 | +0.18 / 0 | +0.027 / 0 | 30 | 32 | 12 | 14 | 30 | 36 | 1.6 | 1.0 | 1.6 | (63) 80 100 125 |
| M20 | 22 | +0.21 / 0 | +0.033 / 0 | 37 | 40 | 16 | 18 | 38 | 45 | 2.5 | 1.0 | 2.5 | (80) 100 125 160 |
| M24 | 28 | +0.21 / 0 | +0.033 / 0 | 46 | 50 | 20 | 22 | 48 | 56 | 2.5 | 1.0 | 2.5 | 100 125 160 200 |
| M30 | 36 | +0.25 / 0 | +0.039 / 0 | 56 | 60 | 25 | 28 | 61 | 71 | 2.5 | 1.0 | 2.5 | 125 160 200 250 |
| M36 | 42 | +0.25 / 0 | +0.039 / 0 | 68 | 72 | 32 | 35 | 74 | 85 | 2.5 | 1.6 | 4.0 | 160 200 250 320 |
| M42 | 48 | +0.30 / 0 | +0.046 / 0 | 80 | 85 | 36 | 40 | 84 | 95 | 2.5 | 2.0 | 4.0 | 200 250 320 400 |
| M48 | 54 | +0.30 / 0 | +0.046 / 0 | 90 | 95 | 40 | 44 | 94 | 106 | 2.5 | 2.0 | 6.0 | 250 320 400 500 |

### 4.2.1 Tolerance on pitches *p* of T-slots

| Pitch mm | Tolerance mm |
|----------|--------------|
| 20 to 25 | ±0.2 |
| 32 to 100 | ±0.3 |
| 125 to 250 | ±0.5 |
| 320 to 500 | ±0.8 |

For further information see BS 2485.

## 4.3 Dimensions of T-bolts and T-nuts

### 4.3.1 T-nut

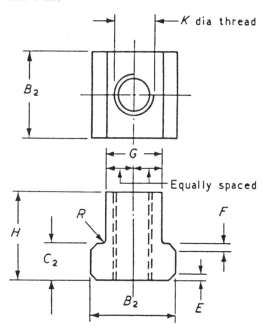

**4.3.2 T-bolt**

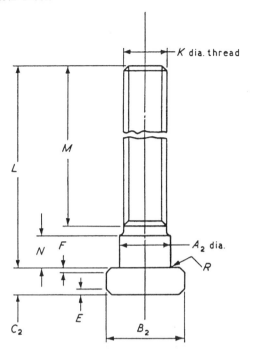

## 4.3.3 Dimensions of T-bolts and T-nuts

| Designation of T-bolt and diameter of thread K | Width of head (square) $B_2$ nom. | Width of head (square) $B_2$ tol. | Depth of head $C_2$ tol. ±0.25 | Length of shoulder N | Diameter of bolt $A_2$ | Chamfer E max. | Radius R max. | Width of T-nut shank G nom. | Width of T-nut shank G tol. | Height of T-nut H max. | Chamfer F max. |
|---|---|---|---|---|---|---|---|---|---|---|---|
| M4 | 9 | | 2.5 | — | 4 | | | 5 | | 6 | |
| M5 | 10 | | 4 | — | 5 | | | 6 | −0.3 | 8 | |
| M6 | 13 | 0 / −0.5 | 6 | — | 6 | 1.0 | | 8 | | 10 | |
| M8 | 15 | | 6 | — | 8 | | | 10 | −0.5 | 12 | |
| M10 | 18 | | 7 | — | 10 | | 1.0 | 12 | | 16 | |
| M12 | 22 | | 8 | — | 12 | | | 14 | −0.3 | 19 | 0.3 × 45° |
| M16 | 28 | | 10 | — | 16 | | | 18 | | 25 | |
| M20 | 34 | | 14 | — | 20 | 1.5 | | 22 | −0.6 | 30 | |
| M24 | 43 | | 18 | 20 | 26 | | | 28 | | 40 | |
| M30 | 53 | | 23 | 20 | 33 | | | 36 | −0.4 / −0.7 | 45 | |
| M36 | 64 | | 28 | 25 | 39 | | | 42 | | 55 | |
| M42 | 75 | | 32 | 25 | 46 | 2.5 | | 48 | −0.4 / −0.8 | 65 | |
| M48 | 85 | | 36 | 30 | 52 | | | 54 | | 75 | |

| Diameter of thread K | Recommended length of bolt stem L | | | | | | | | | | | | | Length of threaded portion of bolt stem M |
|---|---|---|---|---|---|---|---|---|---|---|---|---|---|---|
| M4 | 30 | 40 | 50 | 60 | 70 | 80 | 100 | | | | | | | For L ≤ 100 M = 0.5 L |
| M5 | 30 | 40 | 50 | 60 | 70 | 80 | 100 | | | | | | | |
| M6 | | | | 60 | 70 | 80 | 100 | | | | | | | |
| M8 | | | | 60 | 70 | 80 | 100 | | | | | | | |
| M10 | | | | 60 | 70 | 80 | 100 | 125 | 160 | 180 | | | | For L > 100 M = 0.3 L |
| M12 | | | | | 70 | 80 | 100 | 125 | 160 | 180 | | | | |
| M16 | | | | | 70 | 80 | 100 | 125 | 160 | 180 | 200 | 250 | 300 | |
| M20 | | | | | 70 | 80 | 100 | 125 | 160 | 180 | 200 | 250 | 300 | |
| M24 M30 M36 M42 M48 | | | | | | | 100 | 125 | 160 | 180 | 200 | 250 | 300 | |

For further information see BS 2485.

## 4.4 Dimensions of tenons for T-slots

(Dimensions in millimetres)

| Designation of T-slot | Width of tenon shank $B_2$ | | Overall width of tenon $W$ | | Depth of head of tenon $Q$ | Overall height of tenon $P$ | Length of tenon $L$ | Radius $R$ | Fixing hole | | | |
|---|---|---|---|---|---|---|---|---|---|---|---|---|
| | nom. | tol. (h7) | nom. | tol. (h7) | | | | max. | To suit socket head cap screw to BS 4168: Pt 1 | Clearance hole diameter $A$ to BS 4186 medium fit | Counterbores diameter $B$ tolerance H13 | Counterbore depth $D$ tolerance +0.2 |
| M4 | 5 | 0 / −0.012 | 16 | 0 / −0.018 | 5 | 10 | 25 | 0.6 | M2 | 2.4 | 4.3 | 2.5 |
| M5 | 6 | 0 / −0.015 | 16 | 0 / −0.018 | 5 | 10 | 25 | 0.6 | M3 | 3.4 | 6.0 | 3.5 |
| M6 | 8 | 0 / −0.015 | 16 | 0 / −0.018 | 5 | 10 | 25 | 0.6 | M3 | 3.4 | 6.0 | 3.5 |
| M8 | 10 | 0 / −0.018 | 16 | 0 / −0.018 | 5 | 10 | 25 | 0.6 | M6 | 6.6 | 11.0 | 6.5 |
| M10 | 12 | 0 / −0.018 | 16 | 0 / −0.018 | 5 | 10 | 25 | 0.6 | M6 | 6.6 | 11.0 | 6.5 |
| M12 | 14 | 0 / −0.018 | 30 | 0 / −0.021 | 5.5 | 12 | 25 | 0.6 | M6 | 6.6 | 11.0 | 6.5 |
| M16 | 18 | 0 / −0.018 | 30 | 0 / −0.021 | 5.5 | 12 | 30 | 0.6 | M6 | 6.6 | 11.0 | 6.5 |

| Thread size | Width across flats | Tolerance | Width across corners | Tolerance | | | Thread pitch | Nominal head size | Height |
|---|---|---|---|---|---|---|---|---|---|
| M20 | 22 | 0 / −0.021 | | | | | 1.0 | | |
| M24 | 28 | | 50 | 0 | 15 | 30 | | | 9.0 |
| M30 | 36 | 0 / −0.025 | | −0.025 | 25 | 40 | | M8 | 14.0 |
| M36 | 42 | | 70 | 0 | | | 1.5 | | |
| M42 | 48 | | | −0.030 | 40 | 60 | | | |
| M48 | 54 | 0 / −0.030 | | | | | | | 8.5 |

For further information see BS 2485: 1987.

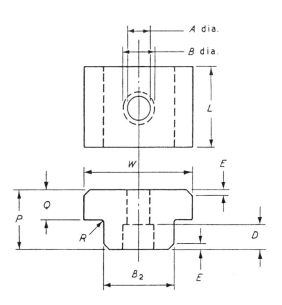

## 4.5 Taper pins, unhardened

**Dimensions**

**Type A** (ground pins): Surface finish $R_a = 0.8\,\mu\mathrm{m}$
**Type B** (turned pins): Surface finish $R_a = 3.2\,\mu\mathrm{m}$

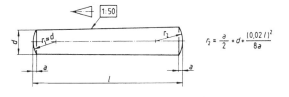

$$r_2 = \frac{a}{2} + d + \frac{(0.02\,l)^2}{8a}$$

(Dimensions in millimetres)

| d | a | | l²⁾ (nom.) → | | | | | | | | | | | | | | | | | | | |
|---|---|---|---|---|---|---|---|---|---|---|---|---|---|---|---|---|---|---|---|---|---|---|---|
| **nom.** h10¹⁾ ≈ | **min.** | **max.** | **0.6** | **0.8** | **1** | **1.2** | **1.5** | **2** | **2.5** | **3** | **4** | **5** | **6** | **8** | **10** | **12** | **16** | **20** | **25** | **30** | **40** | **50** |
| | | | 0.08 | 0.1 | 0.12 | 0.16 | 0.2 | 0.25 | 0.3 | 0.4 | 0.5 | 0.63 | 0.8 | 1 | 1.2 | 1.6 | 2 | 2.5 | 3 | 4 | 5 | 6.3 |
| **2** | 1.75 | 2.25 | | | | | | | | | | | | | | | | | | | | |
| **3** | 2.75 | 3.25 | | | | | | | | | | | | | | | | | | | | |
| **4** | 3.75 | 4.25 | | | | | | | | | | | | | | | | | | | | |
| **5** | 4.75 | 5.25 | | | | | | | | | | | | | | | | | | | | |
| **6** | 5.75 | 6.25 | | | | | | | | | | | | | | | | | | | | |
| **8** | 7.75 | 8.25 | | | | | | | | | | | | | | | | | | | | |
| **10** | 9.75 | 10.25 | | | | | | | | | | | | | | | | | | | | |
| **12** | 11.5 | 12.5 | | | | | | | | | | | | | | | | | | | | |
| **14** | 13.5 | 14.5 | | | | | | | | | | | | | | | | | | | | |
| **16** | 15.5 | 16.5 | | | | | | | | | | | | | | | | | | | | |
| **18** | 17.5 | 18.5 | | | | | | | | | | | | | | | | | | | | |
| **20** | 19.5 | 20.5 | | | | | | | | | | | | | | | | | | | | |
| **22** | 21.5 | 22.5 | | | | | | | | | | | | | | | | | | | | |
| **24** | 23.5 | 24.5 | | | | | | | | | | | | | | | | | | | | |
| **26** | 25.5 | 26.5 | | | | | | | | | | | | | | | | | | | | |
| **28** | 27.5 | 28.5 | | | | | | | | | | | | | | | | | | | | |
| **30** | 29.5 | 30.5 | | | | | | | | | | | | | | | | | | | | |

Range of commercial

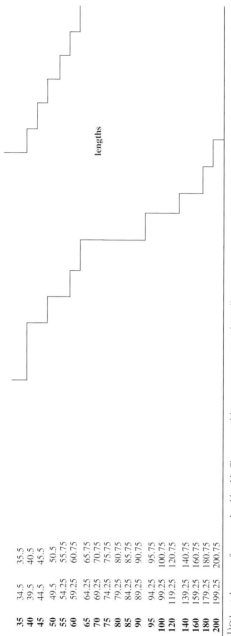

lengths

| | | |
|---|---|---|
| **35** | 34.5 | 35.5 |
| **40** | 39.5 | 40.5 |
| **45** | 44.5 | 45.5 |
| **50** | 49.5 | 50.5 |
| **55** | 54.25 | 55.75 |
| **60** | 59.25 | 60.75 |
| **65** | 64.25 | 65.75 |
| **70** | 69.25 | 70.75 |
| **75** | 74.25 | 75.75 |
| **80** | 79.25 | 80.75 |
| **85** | 84.25 | 85.75 |
| **90** | 89.25 | 90.75 |
| **95** | 94.25 | 95.75 |
| **100** | 99.25 | 100.75 |
| **120** | 119.25 | 120.75 |
| **140** | 139.25 | 140.75 |
| **160** | 159.25 | 160.75 |
| **180** | 179.25 | 180.75 |
| **200** | 199.25 | 200.75 |

[1] Other tolerances. for example a11. c11. f8. as agreed between customer and supplier.

[2] For nominal lengths above 200 mm. steps of 20 mm.

# 4.6 Circlips. external: metric series

Circlip on shaft

Circlip in groove

(Dimensions in millimetres)

| Reference number of circlip | Shaft diameter S | Groove details | | | | | Circlip details | | | | | | | Minimum external clearance | |
|---|---|---|---|---|---|---|---|---|---|---|---|---|---|---|---|
| | | Diameter G | Tolerance | Width W | Tolerance | Edge margin (min.) n | Diameter D | Tolerance | Thickness T | Tolerance | Beam (approx.) M | Lug depth (max.) L | Lug hole diameter (min.) d | Fitted C | During fitting (C1) |
| S003M | 3 | 2.8 | 0 / −0.06 | 0.5 | | 0.33 | 2.66 | +0.06 / −0.15 | 0.4 | | 0.8 | 1.9 | 0.8 | 6.6 | 7.2 |
| S004M | 4 | 3.8 | | 0.5 | | 0.33 | 3.64 | | 0.4 | | 0.9 | 2.2 | 1.0 | 8.2 | 8.8 |
| S005M | 5 | 4.8 | 0 / −0.075 | 0.7 | | 0.33 | 4.64 | | 0.6 | 0 / −0.04 | 1.1 | 2.5 | 1.0 | 9.8 | 10.6 |
| S006M | 6 | 5.7 | | 0.8 | | 0.45 | 5.54 | | 0.7 | | 1.3 | 2.7 | 1.15 | 11.1 | 12.1 |
| S007M | 7 | 6.7 | 0 / −0.09 | 0.9 | | 0.45 | 6.45 | +0.09 / −0.18 | 0.8 | | 1.4 | 3.1 | 1.2 | 12.9 | 14.0 |
| S008M | 8 | 7.6 | | 0.9 | | 0.60 | 7.35 | | 0.8 | | 1.5 | 3.2 | 1.2 | 14.0 | 15.2 |
| S009M | 9 | 8.6 | | 1.1 | | 0.60 | 8.35 | | 1.0 | | 1.7 | 3.3 | 1.2 | 15.2 | 16.6 |
| S010M | 10 | 9.6 | | 1.1 | +0.14 / 0 | 0.60 | 9.25 | | 1.0 | | 1.8 | 3.3 | 1.5 | 16.2 | 17.6 |
| S011M | 11 | 10.5 | | 1.1 | | 0.75 | 10.20 | +0.18 / −0.36 | 1.0 | 0 / −0.06 | 1.8 | 3.3 | 1.5 | 17.1 | 18.6 |
| S012M | 12 | 11.5 | | 1.1 | | 0.75 | 11.0 | | 1.0 | | 1.8 | 3.3 | 1.7 | 18.1 | 19.6 |
| S013M | 13 | 12.4 | | 1.1 | | 0.90 | 11.9 | | 1.0 | | 2.0 | 3.4 | 1.7 | 19.2 | 20.8 |
| S014M | 14 | 13.4 | 0 / −0.11 | 1.1 | | 0.90 | 12.9 | | 1.0 | | 2.1 | 3.5 | 1.7 | 20.4 | 22.0 |
| S015M | 15 | 14.3 | | 1.1 | | 1.10 | 13.8 | | 1.0 | | 2.2 | 3.6 | 1.7 | 21.5 | 23.2 |
| S016M | 16 | 15.2 | | 1.1 | | 1.20 | 14.7 | | 1.0 | | 2.2 | 3.7 | 1.7 | 22.6 | 24.4 |
| S017M | 17 | 16.2 | | 1.1 | | 1.20 | 15.7 | | 1.0 | | 2.3 | 3.8 | 1.7 | 23.8 | 25.6 |
| S018M | 18 | 17.0 | | 1.3 | | 1.50 | 16.5 | | 1.2 | | 2.4 | 3.9 | 2.0 | 24.8 | 26.8 |

(continued)

(Dimensions in millimetres)

| Reference number of circlip | Shaft diameter S | Groove details | | | | | Circlip details | | | | | | | Minimum external clearance | |
|---|---|---|---|---|---|---|---|---|---|---|---|---|---|---|---|
| | | Diameter G | Tolerance | Width W | Tolerance | Edge margin (min.) n | Diameter D | Tolerance | Thickness T | Tolerance | Beam (approx.) M | Lug depth (max.) L | Lug hole diameter (min.) d | Fitted C | During fitting (C₁) |
| S019M | 19 | 18.0 | | 1.3 | | 1.50 | 17.5 | | 1.2 | | 2.5 | 3.9 | 2.0 | 25.8 | 27.8 |
| S020M | 20 | 19.0 | | 1.3 | | 1.50 | 18.5 | | 1.2 | | 2.6 | 4.0 | 2.0 | 27.0 | 29.0 |
| S021M | 21 | 20.0 | | 1.3 | | 1.50 | 19.5 | | 1.2 | | 2.7 | 4.1 | 2.0 | 28.2 | 30.2 |
| S022M | 22 | 21.0 | 0 −0.21 | 1.3 | | 1.50 | 20.5 | | 1.2 | | 2.8 | 4.2 | 2.0 | 29.4 | 31.4 |
| S023M | 23 | 22.0 | | 1.3 | | 1.50 | 21.5 | +0.21 −0.42 | 1.2 | | 2.9 | 4.3 | 2.0 | 30.6 | 32.6 |
| S024M | 24 | 22.9 | | 1.3 | | 1.70 | 22.2 | | 1.2 | | 3.0 | 4.4 | 2.0 | 31.7 | 33.8 |
| S025M | 25 | 23.9 | | 1.3 | | 1.70 | 23.2 | | 1.2 | | 3.0 | 4.4 | 2.0 | 32.7 | 34.8 |
| S026M | 26 | 24.9 | | 1.3 | +0.14 0 | 1.70 | 24.2 | | 1.2 | 0 | 3.1 | 4.5 | 2.0 | 33.9 | 36.0 |
| S027M | 27 | 25.6 | | 1.3 | | 2.10 | 24.9 | | 1.2 | | 3.1 | 4.6 | 2.0 | 34.8 | 37.2 |
| S028M | 28 | 26.6 | | 1.6 | | 2.10 | 25.9 | | 1.5 | | 3.2 | 4.7 | 2.0 | 36.0 | 38.4 |
| S029M | 29 | 27.6 | | 1.6 | | 2.10 | 26.9 | | 1.5 | | 3.4 | 4.8 | 2.0 | 37.2 | 39.6 |
| S030M | 30 | 28.6 | | 1.6 | | 2.10 | 27.9 | | 1.5 | | 3.5 | 5.0 | 2.0 | 38.6 | 41.0 |
| S031M | 31 | 29.3 | | 1.6 | | 2.60 | 28.6 | | 1.5 | | 3.5 | 5.1 | 2.5 | 39.5 | 42.2 |
| S032M | 32 | 30.3 | | 1.6 | | 2.60 | 29.6 | | 1.5 | | 3.6 | 5.2 | 2.5 | 40.7 | 43.4 |
| S033M | 33 | 31.3 | | 1.6 | | 2.60 | 30.5 | | 1.5 | | 3.7 | 5.3 | 2.5 | 41.9 | 44.4 |

| Ref | Size | | | | | | | | | | | | | |
|-----|------|------|-------|------|------|------|-------|------|-------|-----|-----|-----|------|------|
| S034M | 34 | 32.3 | | 1.6 | 2.60 | 31.5 | | 1.5 | | 3.8 | 5.4 | 2.5 | 43.1 | 45.8 |
| S035M | 35 | 33.0 | | 1.6 | 3.00 | 32.2 | | 1.5 | | 3.9 | 5.6 | 2.5 | 44.2 | 47.2 |
| S036M | 36 | 34.0 | | 1.85 | 3.00 | 33.2 | | 1.75 | | 4.0 | 5.6 | 2.5 | 45.2 | 48.2 |
| S037M | 37 | 35.0 | 0 | 1.85 | 3.00 | 34.2 | | 1.75 | | 4.1 | 5.7 | 2.5 | 46.4 | 49.4 |
| S038M | 38 | 36.0 | −0.25 | 1.85 | 3.00 | 35.2 | +0.25 | 1.75 | −0.06 | 4.2 | 5.8 | 2.5 | 47.6 | 50.6 |
| S039M | 39 | 37.0 | | 1.85 | 3.00 | 36.0 | −0.50 | 1.75 | | 4.3 | 5.9 | 2.5 | 48.8 | 51.8 |
| S040M | 40 | 37.5 | | 1.85 | 3.80 | 36.5 | | 1.75 | | 4.4 | 6.0 | 2.5 | 49.5 | 53.0 |
| S041M | 41 | 38.5 | | 1.85 | 3.80 | 37.5 | | 1.75 | | 4.5 | 6.1 | 2.5 | 50.7 | 54.2 |
| S042M | 42 | 39.5 | | 1.85 | 3.80 | 38.5 | | 1.75 | | 4.5 | 6.5 | 2.5 | 52.5 | 56.0 |
| S043M | 43 | 40.5 | | 1.85 | 3.80 | 39.5 | | 1.75 | | 4.6 | 6.6 | 2.5 | 3.7 | 57.2 |
| S044M | 44 | 41.5 | | 1.85 | 3.80 | 40.5 | | 1.75 | | 4.6 | 6.6 | 2.5 | 54.7 | 58.2 |
| S045M | 45 | 42.5 | | 1.85 | 3.80 | 41.5 | +0.39 | 1.75 | | 4.7 | 6.7 | 2.5 | 55.9 | 59.4 |
| S046M | 46 | 43.5 | | 1.85 | 3.80 | 42.5 | −0.78 | 1.75 | | 4.8 | 6.7 | 2.5 | 56.9 | 60.4 |
| S047M | 47 | 44.5 | | 1.85 | 3.80 | 43.5 | | 1.75 | | 4.9 | 6.8 | 2.5 | 58.1 | 61.6 |
| S048M | 48 | 45.5 | | 1.85 | 3.80 | 44.5 | | 1.75 | | 5.0 | 6.9 | 2.5 | 59.3 | 62.8 |
| S049M | 49 | 46.5 | | 1.85 | 3.80 | 44.8 | | 1.75 | | 5.0 | 6.9 | 2.5 | 60.3 | 63.8 |
| S050M | 50 | 47.0 | | 2.15 | 4.50 | 45.8 | | 2.00 | | 5.1 | 6.9 | 2.5 | 60.8 | 64.8 |
| S052M | 52 | 49.0 | | 2.15 | 4.50 | 47.8 | | 2.00 | | 5.2 | 7.0 | 2.5 | 63.0 | 67.0 |

For full range of sizes and types and for full information see BS 3673: Pt 4: 1977.

# 4.7 Circlips, internal: metric series

Circlip in groove

Circlip in bore

| Reference number of circlip | Shaft diameter B | Groove details | | | | | Circlip details | | | | | | | Minimum internal clearance | |
|---|---|---|---|---|---|---|---|---|---|---|---|---|---|---|---|
| | | Diameter G | Tolerance | Width W | Tolerance | Edge margin (min.) n | Diameter D | Tolerance | Thickness T | Tolerance | Beam (approx.) M | Lug depth (max.) L | Lug hole diameter (min.) d | Fitted C | During fitting $C_1$ |
| B008M | 8 | 8.4 | +0.09 / 0 | 0.9 | | 0.6 | 8.7 | | 0.8 | | 1.1 | 2.4 | 1.0 | 3.6 | 2.8 |
| B009M | 9 | 9.4 | | 0.9 | | 0.6 | 9.8 | | 0.8 | | 1.3 | 2.5 | 1.0 | 4.0 | 3.1 |
| B010M | 10 | 10.4 | +0.11 / 0 | 1.1 | +0.14 / 0 | 0.6 | 10.8 | +0.36 / −0.18 | 1.0 | 0 / −0.06 | 1.4 | 3.2 | 1.2 | 4.4 | 3.6 |
| B011M | 11 | 11.4 | | 1.1 | | 0.6 | 11.8 | | 1.0 | | 1.5 | 3.3 | 1.2 | 4.8 | 3.9 |
| B012M | 12 | 12.5 | | 1.1 | | 0.75 | 13.0 | | 1.0 | | 1.7 | 3.4 | 1.5 | 5.7 | 4.7 |
| B013M | 13 | 13.6 | | 1.1 | | 0.9 | 14.1 | | 1.0 | | 1.8 | 3.6 | 1.5 | 6.4 | 5.3 |
| B014M | 14 | 14.6 | | 1.1 | | 0.9 | 15.1 | | 1.0 | | 1.9 | 3.7 | 1.7 | 7.2 | 6.1 |
| B015M | 15 | 15.7 | | 1.1 | | 1.1 | 16.2 | | 1.0 | | 2.0 | 3.7 | 1.7 | 8.3 | 7.1 |
| B016M | 16 | 16.8 | | 1.1 | | 1.2 | 17.3 | | 1.0 | | 2.0 | 3.8 | 1.7 | 9.2 | 7.9 |
| B017M | 17 | 17.8 | | 1.1 | | 1.2 | 18.3 | +0.42 | 1.0 | | 2.1 | 3.9 | 1.7 | 10.0 | 8.7 |
| B018M | 18 | 19.0 | | 1.1 | | 1.5 | 19.5 | | 1.0 | | 2.2 | 4.1 | 2.0 | 10.8 | 9.3 |
| B019M | 19 | 20.0 | | 1.1 | | 1.5 | 20.5 | | 1.0 | | 2.2 | 4.1 | 2.0 | 11.8 | 9.8 |
| B020M | 20 | 21.0 | | 1.1 | | 1.5 | 21.5 | | 1.0 | | 2.3 | 4.2 | 2.0 | 12.6 | 10.6 |

(continued)

**4.7** (continued)

(Dimensions in millimetres)

| Reference number of circlip | Shaft diameter B | Groove details | | | | | Circlip details | | | | | | | Minimum internal clearance | |
|---|---|---|---|---|---|---|---|---|---|---|---|---|---|---|---|
| | | Diameter G | Tolerance | Width W | Tolerance | Edge margin (min.) n | Diameter D | Tolerance | Thickness T | Tolerance | Beam (approx.) M | Lug depth (max.) L | Lug hole diameter (min.) d | Fitted C | During fitting $C_1$ |
| B021M | 21 | 22.0 | +0.21 0 | 1.1 | | 1.5 | 22.5 | −0.21 | 1.0 | | 2.4 | 4.2 | 2.0 | 13.6 | 11.6 |
| B022M | 22 | 23.0 | | 1.1 | | 1.5 | 23.5 | | 1.0 | | 2.5 | 4.2 | 2.0 | 14.6 | 12.6 |
| B023M | 23 | 24.1 | | 1.1 | | 1.5 | 24.6 | | 1.0 | | 2.5 | 4.2 | 2.0 | 15.7 | 13.6 |
| B024M | 24 | 25.2 | | 1.1 | | 1.8 | 25.9 | | 1.0 | | 2.6 | 4.4 | 2.0 | 16.4 | 14.2 |
| B025M | 25 | 26.2 | | 1.1 | | 1.8 | 26.9 | | 1.0 | | 2.7 | 4.5 | 2.0 | 17.2 | 15.0 |
| B026M | 26 | 27.2 | | 1.1 | | 1.8 | 27.9 | | 1.0 | | 2.8 | 4.7 | 2.0 | 17.8 | 15.6 |
| B027M | 27 | 28.4 | | 1.1 | | 2.1 | 29.1 | +0.92 −0.46 | 1.0 | | 2.8 | 4.7 | 2.0 | 19.0 | 16.6 |
| B028M | 28 | 29.4 | | 1.3 | | 2.1 | 30.1 | | 1.2 | | 2.9 | 4.8 | 2.0 | 19.8 | 17.4 |
| B029M | 29 | 30.4 | +0.25 0 | 1.3 | | 2.1 | 31.1 | | 1.2 | | 3.0 | 4.8 | 2.0 | 20.8 | 18.4 |
| B030M | 30 | 31.4 | | 1.3 | | 2.1 | 32.1 | | 1.2 | | 3.0 | 4.8 | 2.0 | 21.8 | 19.4 |
| B031M | 31 | 32.7 | | 1.3 | | 2.6 | 33.4 | +0.50 −0.25 | 1.2 | | 3.2 | 5.2 | 2.5 | 22.3 | 19.6 |
| B032M | 32 | 33.7 | | 1.3 | +0.14 0 | 2.6 | 34.4 | | 1.2 | | 3.2 | 5.4 | 2.5 | 22.9 | 20.2 |
| B033M | 33 | 34.7 | | 1.3 | | 2.6 | 35.5 | | 1.2 | | 3.3 | 5.4 | 2.5 | 23.9 | 21.2 |
| B034M | 34 | 35.7 | | 1.6 | | 2.6 | 36.5 | | 1.5 | | 3.3 | 5.4 | 2.5 | 24.9 | 22.2 |

| Code | Size | | | | | | | | | | | |
|---|---|---|---|---|---|---|---|---|---|---|---|---|
| B035M | 35 | 37.0 | 1.6 | 3.0 | 37.8 | 1.5 | | 3.4 | 5.4 | 2.5 | 26.2 | 23.2 |
| B036M | 36 | 38.0 | 1.6 | 3.0 | 38.8 | 1.5 | | 3.5 | 5.4 | 2.5 | 27.2 | 24.2 |
| B037M | 37 | 39.0 | 1.6 | 3.0 | 39.8 | 1.5 | | 3.6 | 5.5 | 2.5 | 28.0 | 25.0 |
| B038M | 38 | 40.0 | 1.6 | 3.0 | 40.8 | 1.5 | | 3.7 | 5.5 | 2.5 | 29.0 | 26.0 |
| B039M | 39 | 41.0 | 1.6 | 3.0 | 42.0 | 1.5 | | 3.8 | 5.6 | 2.5 | 29.8 | 26.8 |
| | | +0.25 / 0 | | | +0.78 / −0.39 | | 0 / −0.06 | | | | | |
| B040M | 40 | 42.5 | 1.85 | 3.8 | 43.5 | 1.75 | | 3.9 | 5.8 | 2.5 | 30.9 | 27.4 |
| B041M | 41 | 43.5 | 1.85 | 3.8 | 44.5 | 1.75 | | 4.0 | 5.9 | 2.5 | 31.7 | 28.2 |
| B042M | 42 | 44.5 | 1.85 | 3.8 | 45.5 | 1.75 | | 4.1 | 5.9 | 2.5 | 32.7 | 29.2 |
| B043M | 43 | 45.5 | 1.85 | 3.8 | 46.5 | 1.75 | | 4.2 | 6.0 | 2.5 | 33.5 | 30.0 |
| B044M | 44 | 46.5 | 1.85 | 3.8 | 47.5 | 1.75 | | 4.3 | 6.0 | 2.5 | 34.5 | 31.0 |
| B045M | 45 | 47.5 | 1.85 | 3.8 | 48.5 | 1.75 | | 4.3 | 6.2 | 2.5 | 35.1 | 31.6 |
| B046M | 46 | 48.5 | 1.85 | 3.8 | 49.5 | 1.75 | | 4.4 | 6.3 | 2.5 | 35.9 | 32.4 |
| B047M | 47 | 49.5 | 1.85 | 3.8 | 50.5 | 1.75 | | 4.4 | 6.4 | 2.5 | 36.7 | 33.2 |
| | | +0.30 / 0 | | | +0.92 / −0.46 | | | | | | | |
| B048M | 48 | 50.5 | 1.85 | 3.8 | 51.5 | 1.75 | | 4.5 | 6.4 | 2.5 | 37.7 | 34.2 |
| B049M | 49 | 51.5 | 1.85 | 3.8 | 52.5 | 1.75 | | 4.5 | 6.4 | 2.5 | 38.7 | 35.2 |
| B050M | 50 | 53.0 | 2.15 | 4.5 | 54.2 | 2.0 | | 4.6 | 6.5 | 2.5 | 40.0 | 36.0 |
| B051M | 51 | 54.0 | 2.15 | 4.5 | 55.2 | 2.0 | | 4.7 | 6.5 | 2.5 | 41.0 | 37.0 |
| B052M | 52 | 555.0 | 2.15 | 4.5 | 56.2 | 2.0 | | 4.7 | 6.7 | 2.5 | 41.6 | 37.6 |

For full range of sizes and types and for full information see BS 3673: Pt 4: 1977.

## 4.8 Toroidal sealing rings (O-rings) and their housings (inch series)

The following notes are intended only as an introduction. For full information, essential for satisfactory performance, see BS 1806: 1989 and Amendment 1: 1992.

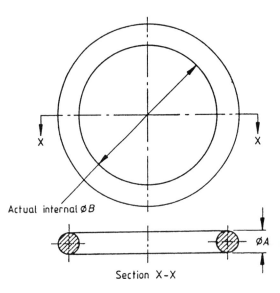

**Fig. 1** *'O'-ring*

### 4.8.1 'O'-ring sizes

(Dimensions in inches

| 'O'-ring size number range | Cross-section (A) (see Fig. 1) | Internal diameter (B) (see Fig. 1) | Nominal diameters $(C)^{\ddagger}$ | $(D)^{\ddagger}$ |
|---|---|---|---|---|
| 001 | 0.040 | 0.029 | § | § |
| 002 | 0.050 | 0.042 | § | § |
| 003 | 0.060 | 0.056 | § | § |
| 004 – 050 | 0.070 ($1/16$ nom.) | 0.070 – 5.239 | $5/64$ – $5\,1/4$ | $13/64$ – $5\,3$ |
| 102 – 178 | 0.103 ($3/32$ nom.) | 0.049 – 9.737 | $1/16$ – $9\,3/4$ | $1/4$ – $9\,15/$ |
| 201 – 281 | 0.139 ($1/8$ nom.) | 0.171 – 14.984 | $3/16$ – 15 | $7/16$ – 15 |

| 309–395 | 0.210<br>(³/₁₆ nom.) | 0.412–25.940 | ⁷/₁₆–26 | ¹³/₁₆–26³/₈ |
| 425–475 | 0.275<br>(¹/₄ nom.) | 4.475–25.940 | 4¹/₂–26 | 5–26¹/₂ |

---

† See Figs 2(b), 3 and 4.

‡ See Figs 2(a) and 3.

§ Since 'O' ring sizes 001, 002 and 003 are extremely small, special care should be taken in the selection of housing sizes to achieve an adequate squeeze.

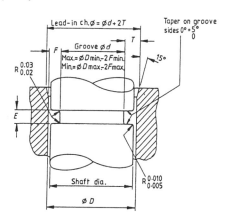

(a) Piston groove

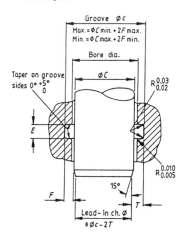

(b) Gland groove

All linear dimensions are in inches.

**g. 2** *Housing grooves (amendment 1: 1992)*

**4.8.2 Dimensions of housings for diametral sealing**

| 'O' ring Size no. | Cross-section diameter, A | Diameter, C † | Diameter, D † | Groove width, E ± 0.005 | Radial depth, F Max. | Radial depth, F Min. | Minimum cross-sectional squeeze of 'O'-ring* | Maximum diametral clearance, G | 2T |
|---|---|---|---|---|---|---|---|---|---|
| | in | | | in | in | in | in | in | in |
| 004 to 050 | 0.070 | Equal to nominal diameter C | Equal to nominal diameter D | 0.094 | 0.062 | 0.060 | 0.005 | 0.005 | 0.170 |
| 102 to 178 | 0.103 | | | 0.141 | 0.094 | 0.091 | 0.006 | 0.005 | 0.240 |
| 201 to 281 | 0.139 | | | 0.188 | 0.125 | 0.122 | 0.010 | 0.006 | 0.320 |
| 309 to 395 | 0.210 | | | 0.281 | 0.188 | 0.184 | 0.017 | 0.007 | 0.460 |
| 425 to 475 | 0.275 | | | 0.375 | 0.250 | 0.245 | 0.019 | 0.008 | 0.600 |

*This is included for reference purposes only.
† See Tables 1 to 5 in BS 1806: 1989.

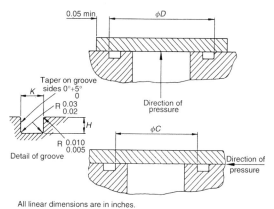

All linear dimensions are in inches.

**Fig. 3**  *Face sealing*

## 4.8.3 Dimensions of housings for static face sealing

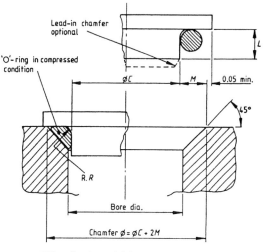

All linear dimensions are in inches.

**Fig. 4**  *Triangular housings*

| 'O-ring Size no. | Cross-section diameter, A | Diameter, C ±0.010 | Diameter, D ±0.010 | Groove depth, H ±0.005 | Minimum recess width, K | Minimum cross-sectional squeeze of 'O-ring* |
|---|---|---|---|---|---|---|
| | in | | | in | in | in |
| | | Equal to nominal diameter C† | Equal to nominal diameter D† | | | |
| 004 to 050 | 0.070 | | | 0.056 | 0.095 | 0.006 |
| 102 to 178 | 0.103 | | | 0.086 | 0.140 | 0.009 |
| 201 to 281 | 0.139 | | | 0.115 | 0.190 | 0.015 |
| 309 to 395 | 0.210 | | | 0.175 | 0.280 | 0.025 |
| 425 to 475 | 0.275 | | | 0.236 | 0.370 | 0.028 |

*This is included for reference purposes only.
† See Tables 1 to 5 in BS 1806: 1989.

**4.8.4 Dimensions of triangular housings for static face sealings**

| Size no. | 'O'-ring Cross-section diameter, A | Diameter, C | Maximum diametral clearance, G | Chamfer, M +0.005 −0 | Maximum radius on spigot, R | Minimum spigot length, L |
|---|---|---|---|---|---|---|
| | in | | in | in | in | in |
| 004 to 050 | 0.070 | Equal to | 0.005 | 0.095 | 0.030 | $\frac{3}{16}$ |
| 102 to 178 | 0.103 | nominal | 0.005 | 0.145 | 0.040 | $\frac{1}{4}$ |
| 201 to 281 | 0.139 | diameter | 0.006 | 0.195 | 0.060 | $\frac{5}{16}$ |
| 309 to 395 | 0.210 | $C^{\dagger}$ | 0.007 | 0.295 | 0.090 | $\frac{7}{16}$ |
| 425 to 475 | 0.275 | | 0.008 | 0.395 | 0.100 | $\frac{9}{16}$ |

†See Tables 1 to 5 in BS 1806: 1989.

## 4.9 Toroidal sealing rings (O-rings) and their housings (metric series)

The following notes are intended only as an intro-duction. For full information, essential for satisfactory performance, see BS 4518: 1982.

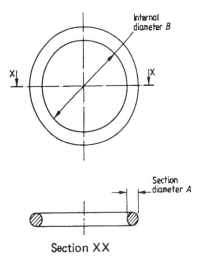

**Fig. 1** *Toroidal sealing ring ('O'-ring)*

## 4.9.1 'O'-ring sizes

### Table 1

(Dimensions in millimetres)

| 'O'-ring size number range | Cross-section (A) (see Fig. 1) | Internal diameter (B) (see Fig. 1) | Nominal housing dimensions† shaft diameter $d_1$ | Cylinder diameters $D_1$ |
|---|---|---|---|---|
| 0031-16 to 00371-16 | } 1.6 | 3.1–37.1 | 3.5–37.5 | 6.0–40.0 |
| 0036-24 to 0696-24 | } 2.4 | 3.6–69.6 | 41–70 | 8–74 |
| 0195-30 to 2495-30 | } 3.0 | 19.5–249.5 | 20–250 | 25–255 |
| 0443-57 to 4993-57 | } 5.7 | 44.3–499.3 | 45–500 | 55–510 |
| 1441-84 to 2941-84 | } 8.4 | 144.1–249.1 | 145–250 | 160–265 |

† See Figs 2 and (BS 4518).

*Notes:*

(1) For 'O'-ring selection charts see BS 4518 Appendix A Tables 7 and 8.

(2) For obsolete series of 'O'-rings see BS 4518 Appendix B Tables B1 to B5 inclusive.

## 4.9.2 'O'-ring housings (piston and cylinder)

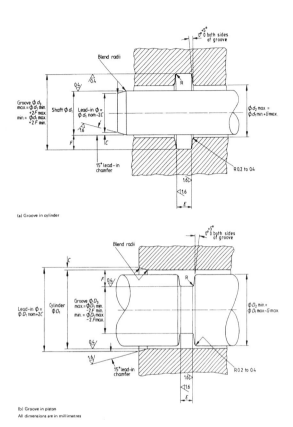

(a) Groove in cylinder

(b) Groove in piston

All dimensions are in millimetres

**Fig. 2**  *Groove for diametral sealing*

**Table 2** Groove dimensions for static diametral sealing (see Fig. 2)

(All dimensions in millimetres)

| 'O'-ring ref. no. | Cross-section diameter A | Radial depth F Max. | Radial depth F Min. | Groove width $E_0^{+0.2}$ | Total diametral clearance G (max.) | Lead-in chamfer C | Max. radius R |
|---|---|---|---|---|---|---|---|
| 0031-16 to 0371-16 | 1.6 | 1.25 | 1.18 | 2.3 | 0.12 | 0.6 | 0.5 |
| 0036-24 to 0696-24 | 2.4 | 1.97 | 1.84 | 3.1 | 0.14 | 0.7 | 0.5 |
| 0195-30 to 2495-30 | 3.0 | 2.50 | 2.35 | 3.7 | 0.15 | 0.8 | 1.0 |
| 0443-57 to 4993-57 | 5.7 | 4.95 | 4.70 | 6.4 | 0.18 | 1.2 | 1.0 |
| 1441-84 to 2491-84 | 8.4 | 7.50 | 7.20 | 9.0 | 0.20 | 1.5 | 1.0 |

**Table 3** *Groove dimensions for dynamic diametral sealing in hydraulic applications (see Fig. 2)*

(All dimensions in millimetres)

| 'O'-ring ref. no. | Cross-section diameter A | Radial depth F | | Groove width $E_0^{+0.2}$ | Total diametral clearance G (max.) | Lead-in chamfer C | Max. radius R |
|---|---|---|---|---|---|---|---|
| | | Max. | Min. | | | | |
| 0036-24 to 0176-24 | 2.4 | 2.09 | 1.97 | 3.2 | 0.14 | 0.6 | 0.5 |
| 0195-30 to 0445-30 | 3.0 | 2.65 | 2.50 | 4.0 | 0.15 | 0.7 | 1.0 |
| 0443-57 to 1443-57 | 5.7 | 5.18 | 4.95 | 7.5 | 0.18 | 1.0 | 1.0 |
| 1441-84 to 2491-84 | 8.4 | 7.75 | 7.50 | 11.0 | 0.20 | 1.2 | 1.0 |

**Table 4** *Groove dimensions for dynamic diametral sealing in pneumatic applications (see Fig. 2)*

(All dimensions in millimetres)

| 'O'-ring ref. no. | Cross-section diameter *A* | Radial depth *F* Max. | Radial depth *F* Min. | Groove width $E_0^{+0.2}$ | Total diametral clearance *G* (max.) | Lead-in chamfer *C* | Max. radius *R* |
|---|---|---|---|---|---|---|---|
| 0036-24 to 0176-24 | 2.4 | 2.20 | 2.13 | 3.2 | 0.14 | 0.6 | 0.5 |
| 0195-30 to 0445-30 | 3.0 | 2.77 | 2.70 | 4.0 | 0.15 | 0.7 | 1.0 |
| 0443-57 to 1443-57 | 5.7 | 5.38 | 5.22 | 7.5 | 0.18 | 1.0 | 1.0 |
| 1441-84 to 2491-84 | 8.4 | 7.96 | 7.75 | 11.0 | 0.20 | 1.2 | 1.0 |

## 4.9.3 Static face sealing

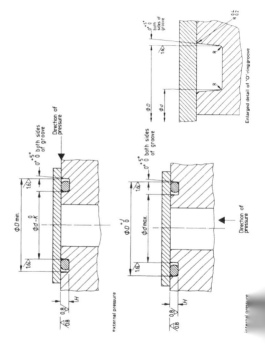

external pressure

internal pressure

Enlarged detail of 'O'-ring groove

**Table 5** *Groove dimensions for static face sealing (see Fig. 3)*

(All dimensions in millimetres)

| 'O'-ring ref. no. | Internal pressure | | | External pressure | | | $H$ | $R$ (max.) |
|---|---|---|---|---|---|---|---|---|
| | $d$ (max.) | $D$ | $J$ | $D$ (min.) | $d$ | $K$ | | |
| 0031-16 | 1.0 | 6.3 | | 7.5 | 3.5 | | | |
| 0041-16 | 2.3 | 7.3 | 0.09 | 8.5 | 4.5 | 0.075 | | |
| 0051-16 | 3.3 | 8.3 | | 9.5 | 5.5 | | | |
| 0061-16 | 4.3 | 9.3 | | 10.5 | 6.5 | | | |
| 0071-16 | 5.8 | 10.3 | | 11.5 | 7.5 | 0.09 | | |
| 0081-16 | 6.8 | 11.3 | | 12.5 | 8.5 | | | |
| 0091-16 | 7.8 | 12.3 | | 13.5 | 9.5 | | | |
| 0101-16 | 8.8 | 13.3 | | 14.5 | 10.5 | | | |
| 0111-16 | 9.8 | 14.3 | 0.11 | 15.5 | 11.5 | | | 0.2 |
| 0121-16 | 10.8 | 15.3 | | 16.5 | 12.5 | | $1.2^{+0.1}_{0}$ | |
| 0131-16 | 11.8 | 16.3 | | 17.5 | 13.5 | 0.11 | | |
| 0141-16 | 12.8 | 17.3 | | 18.5 | 14.5 | | | |
| 0151-16 | 14.0 | 18.3 | | 19.5 | 15.5 | | | |
| 0161-16 | 15 | 19.3 | 0.13 | 20.5 | 16.5 | | | |
| 0171-16 | 16 | 20.3 | | 21.5 | 17.5 | | | |
| 0181-16 | 17 | 21.3 | | 22.5 | 18.5 | 0.13 | | |

*(continued)*

**Table 5** (*continued*)

(All dimensions in millimetres)

| 'O'-ring ref. no. | Internal pressure | | | External pressure | | | | |
|---|---|---|---|---|---|---|---|---|
| | $d$ (max.) | $D$ | $J$ | $D$ (min.) | $d$ | $K$ | $H$ | $R$ (max.) |
| 0191-16 | 18 | 22.3 | | 23.5 | 19.5 | | | |
| 0221-16 | 21 | 25.3 | 0.13 | 26.5 | 22.5 | | | |
| 0251-16 | 24 | 28.3 | | 29.5 | 25.5 | 0.13 | | |
| 0271-16 | 26 | 30.3 | — | 31.5 | 27.5 | | $1.2^{+0.1}_{0}$ | 0.2 |
| 0291-16 | 28 | 32.3 | | 33.5 | 29.5 | | | |
| 0321-16 | 31 | 35.3 | 0.16 | 36.5 | 32.5 | | | |
| 0351-16 | 34 | 38.3 | | 39.5 | 35.5 | 0.16 | | |
| 0371-16 | 36 | 40.3 | | 41.5 | 37.5 | | | |
| 0036-24 | – | 8.4 | | 10 | 4 | | | |
| 0046-24 | 1.0 | 9.4 | 0.09 | 11 | 5 | 0.075 | | |
| 0056-24 | 2.5 | 10.4 | — | 12 | 6 | | | |
| 0066-24 | 4.0 | 11.4 | | 13 | 7 | | | |
| 0076-24 | 5.0 | 12.4 | 0.11 | 14 | 8 | 0.09 | | |
| 0086-24 | 6.4 | 13.4 | | 15 | 9 | | | |
| 0096-24 | 7.4 | 14.4 | | 16 | 10 | | | |
| 0106-24 | 8.4 | 15.4 | | 17 | 11 | 0.11 | | |

| | | | | | | | |
|---|---|---|---|---|---|---|---|
| 0116-24 | 9.5 | 16.4 | | 18 | 12 | | |
| 0126-24 | 10.5 | 17.4 | | 19 | 13 | | |
| 0136-24 | 11.5 | 18.4 | 0.11 | 20 | 14 | 0.11 | |
| 0156-24 | 12.5 | 19.4 | | 21 | 15 | | |
| 0166-24 | 13.5 | 20.4 | | 22 | 16 | | |
| 0176-24 | 14.5 | 21.4 | | 23 | 17 | | |
| 0186-24 | 15.5 | 22.4 | 0.13 | 24 | 18 | | |
| 0196-24 | 16.5 | 23.4 | | 25 | 19 | | |
| 0206-24 | 17.5 | 24.4 | | 26 | 20 | | |
| 0216-24 | 18.5 | 25.4 | | 27 | 21 | 0.13 | $1.7^{+0.1}_{0}$ |
| 0246-24 | 19.5 | 26.4 | 0.16 | 28 | 22 | | |
| 0276-24 | 22.5 | 29.4 | | 31 | 25 | | |
| 0296-24 | 25.5 | 32.4 | | 34 | 28 | | |
| 0316-24 | 27.5 | 34.4 | | 36 | 30 | 0.16 | |
| 0346-24 | 29.5 | 36.4 | | 38 | 32 | | |
| 0356-24 | 32.5 | 39.4 | | 41 | 35 | | |
| | 33.5 | 40.4 | | 42 | 36 | | 0.5 |

(continued)

**Table 5** (*continued*)

(All dimensions in millimetres)

| 'O'-ring ref. no. | Internal pressure | | | External pressure | | K | H | R (max.) |
|---|---|---|---|---|---|---|---|---|
| | d (max.) | D | J | D (min.) | d | | | |
| 0376-24 | 35.5 | 42.4 | | 44 | 38 | | | |
| 0396-24 | 37.5 | 44.4 | | 46 | 40 | | | |
| 0416-24 | 39.5 | 46.4 | 0.16 | 48 | 42 | 0.16 | | |
| 0446-24 | 42.5 | 49.4 | | 51 | 45 | | | |
| 0456-24 | 43.5 | 50.4 | | 52 | 46 | | | |
| 0476-24 | 45.5 | 52.4 | | 54 | 48 | | | |
| 0496-24 | 47.5 | 54.4 | | 56 | 50 | | $1.7^{+0.1}_{0}$ | 0.5 |
| 0516-24 | 49.5 | 56.4 | | 58 | 52 | | | |
| 0546-24 | 52.5 | 59.4 | | 61 | 55 | | | |
| 0556-24 | 53.5 | 60.4 | | 62 | 56 | | | |
| 0576-24 | 55.5 | 62.4 | 0.19 | 64 | 58 | 0.19 | | |
| 0586-24 | 56.5 | 63.4 | | 65 | 59 | | | |
| 0596-24 | 57.5 | 64.4 | | 66 | 60 | | | |
| 0616-24 | 59.5 | 66.4 | | 68 | 62 | | | |
| 0626-24 | 60.5 | 67.4 | | 69 | 63 | | | |
| 0646-24 | 62.5 | 69.4 | | 71 | 65 | | | |

| | 65.5 / 67.5 | 72.4 / 74.4 | 0.19 | 74 / 76 | 68 / 70 | 0.19 | 1.7+0.1/0 | 0.5 |
|---|---|---|---|---|---|---|---|---|
| 0676-24 | 65.5 | 72.4 | | 74 | 68 | | 1.7$^{+0.1}_{0}$ | 0.5 |
| 0696-24 | 67.5 | 74.4 | | 76 | 70 | | | |
| 0195-30 | 17 | 25 | | 28 | 20 | 0.19 | | |
| 0215-30 | 19 | 27 | 0.13 | 30 | 22 | 0.13 | | |
| 0225-30 | 20 | 28 | | 31 | 23 | | | |
| 0245-30 | 22 | 30 | — | 33 | 25 | — | | |
| 0255-30 | 23 | 31 | | 34 | 26 | | | |
| 0266-30 | 24 | 32 | | 35 | 27 | | | |
| 0275-30 | 25 | 33 | | 36 | 28 | | | |
| 0295-30 | 27 | 35 | | 38 | 30 | | | |
| 0315-30 | 29 | 37 | | 40 | 32 | | | |
| 0325-30 | 30 | 38 | | 41 | 33 | | 2.2$^{+0.1}_{0}$ | 1.0 |
| 0345-30 | 32 | 40 | 0.16 | 43 | 35 | 0.16 | | |
| 0355-30 | 33 | 41 | | 44 | 36 | | | |
| 0365-30 | 34 | 42 | | 45 | 37 | | | |
| 0375-30 | 35 | 43 | | 46 | 38 | | | |
| 0395-30 | 37 | 45 | | 48 | 40 | | | |
| 0415-30 | 39 | 47 | | 50 | 42 | | | |
| 0425-30 | 40 | 48 | | 51 | 43 | | | |
| 0445-30 | 42 | 50 | — | 53 | 45 | — | | |

(continued)

**Table 5** (continued)

(All dimensions in millimetres)

| 'O'-ring ref. no. | Internal pressure | | | External pressure | | | H | R (max.) |
|---|---|---|---|---|---|---|---|---|
| | d (max.) | D | J | D (min.) | d | K | | |
| 0495-30 | 47 | 55 | | 58 | 50 | | | |
| 0545-30 | 52 | 60 | | 63 | 55 | | | |
| 0555-30 | 53 | 61 | | 64 | 56 | | | |
| 0575-30 | 55 | 63 | | 66 | 58 | | | |
| 0595-30 | 57 | 65 | 0.19 | 68 | 60 | | | |
| 0625-30 | 60 | 68 | | 71 | 63 | 0.19 | | |
| 0645-30 | 62 | 70 | | 73 | 65 | | | |
| 0695-30 | 67 | 75 | | 78 | 70 | | | |
| 0745-30 | 72 | 80 | | 83 | 75 | | | |
| 0795-30 | 77 | 85 | — | 88 | 80 | — | $2.2^{+0.1}_{0}$ | 1.0 |
| 0845-30 | 82 | 90 | | 93 | 85 | | | |
| 0895-30 | 87 | 95 | | 98 | 90 | | | |
| 0945-30 | 92 | 100 | 0.22 | 103 | 95 | | | |
| 0995-30 | 97 | 105 | | 108 | 100 | 0.22 | | |
| 1045-30 | 102 | 110 | | 113 | 105 | | | |
| 1095-30 | 107 | 115 | | 118 | 110 | | | |
| 1145-30 | 112 | 120 | | 123 | 115 | | | |

| | | | | | | |
|---|---|---|---|---|---|---|
| 1245-30 | 122 | 130 | | 133 | 125 | |
| 1295-30 | 127 | 135 | | 138 | 130 | |
| 1345-30 | 132 | 140 | | 143 | 135 | |
| 1395-30 | 137 | 145 | | 148 | 140 | |
| 1445-30 | ?42 | 150 | | 153 | 145 | |
| 1495-30 | 147 | 155 | 0.25 | 158 | 150 | 0.25 |
| 1545-30 | 152 | 160 | | 163 | 155 | |
| 1595-30 | 157 | 165 | | 168 | 160 | |
| 1645-30 | 162 | 170 | | 173 | 165 | |
| 1695-30 | 167 | 175 | | 178 | 170 | |
| 1745-30 | 172 | 180 | | 183 | 175 | |
| 1795-30 | 177 | 185 | — | 188 | 180 | — |
| 1845-30 | 182 | 190 | | 193 | 185 | $2.2^{+0.1}_{0}$ |
| 1895-30 | 187 | 195 | | 198 | 190 | |
| 1945-30 | 192 | 200 | | 203 | 195 | |
| 1995-30 | 197 | 205 | 0.29 | 208 | 200 | 0.29 |
| 2095-30 | 207 | 215 | | 218 | 210 | |
| 2195-30 | 217 | 225 | | 228 | 220 | |
| 2295-30 | 227 | 235 | | 238 | 230 | 1.0 |
| 2395-30 | 237 | 245 | | 248 | 240 | |
| 2445-30 | 242 | 250 | | 253 | 245 | |
| 2495-30 | 247 | 255 | | 258 | 250 | |

(continued)

327

**Table 5** (continued)

(All dimensions in millimetres)

| 'O'-ring ref. no. | Internal pressure | | | External pressure | | | H | R (max.) |
|---|---|---|---|---|---|---|---|---|
| | d (max.) | D | J | D (min.) | d | K | | |
| 0443-57 | 41 | 55 | | 59 | 45 | | | |
| 0453-57 | 42 | 56 | | 60 | 46 | 0.16 | | |
| 0493-57 | 46 | 60 | | 64 | 50 | | | |
| 0523-57 | 49 | 63 | | 67 | 53 | | | |
| 0543-57 | 51 | 65 | | 69 | 55 | | | |
| 0553-57 | 52 | 66 | 0.19 | 70 | 56 | | | |
| 0593-57 | 56 | 70 | | 74 | 60 | | | |
| 0623-57 | 59 | 73 | | 77 | 63 | 0.19 | | |
| 0643-57 | 61 | 75 | | 79 | 65 | | | 1.0 |
| 0693-57 | 66 | 80 | | 84 | 70 | | | |
| 0743-57 | 71 | 85 | | 89 | 75 | | $4.4^{+0.1}_{0}$ | |
| 0793-57 | 76 | 90 | 0.22 | 94 | 80 | | | |
| 0843-57 | 81 | 95 | | 99 | 85 | | | |
| 0893-57 | 86 | 100 | | 104 | 90 | | | |
| 0943-57 | 91 | 105 | | 109 | 95 | 0.22 | | |
| 0993-57 | 96 | 110 | | 114 | 100 | | | |
| 1043-57 | 101 | 115 | | 119 | 105 | | | |
| | | | | 121 | 110 | | | |

328

| | | | | | | | |
|---|---|---|---|---|---|---|---|
| 1193-57 | 116 | | 130 | 134 | | 120 | 0.22 |
| 1243-57 | 121 | | 135 | 139 | | 125 | |
| 1293-57 | 126 | | 140 | 144 | | 130 | |
| 1343-57 | 131 | | 145 | 149 | | 135 | |
| 1393-57 | 136 | | 150 | 154 | | 140 | |
| 1443-57 | 141 | | 155 | 159 | | 145 | |
| 1493-57 | 146 | 0.25 | 160 | 164 | 0.25 | 150 | |
| 1543-57 | 151 | | 165 | 169 | | 155 | 0.25 |
| 1593-57 | 156 | | 170 | 174 | | 160 | |
| 1643-57 | 161 | | 175 | 179 | | 165 | |
| 1693-57 | 166 | | 180 | 184 | | 170 | $4.4^{+0.1}_{0}$   1.0 |
| 1743-57 | 171 | | 185 | 189 | | 175 | |
| 1793-57 | 176 | | 190 | 194 | | 180 | |
| 1843-57 | 181 | | 195 | 199 | | 185 | |
| 1893-57 | 185 | | 199 | 204 | | 190 | |
| 1943-57 | 190 | 0.29 | 204 | 209 | 0.29 | 195 | 0.29 |
| 1993-57 | 195 | | 209 | 214 | | 200 | |
| 2093-57 | 205 | | 219 | 224 | | 210 | |
| 2193-57 | 215 | | 229 | 234 | | 220 | |
| 2293-57 | 225 | | 239 | 244 | | 230 | |
| 2393-57 | 235 | | 249 | 254 | | 240 | |

*(continued)*

**Table 5** (continued)

(All dimensions in millimetres)

| 'O'-ring ref. no. | Internal pressure | | | External pressure | | | | |
|---|---|---|---|---|---|---|---|---|
| | $d$ (max.) | $D$ | $J$ | $D$ (min.) | $d$ | $K$ | $H$ | $R$ (max.) |
| 2493-57 | 245 | 259 | | 264 | 250 | 0.29 | | |
| 2593-57 | 255 | 269 | | 275 | 261 | | | |
| 2693-57 | 265 | 279 | 0.32 | 285 | 271 | 0.32 | | |
| 2793-57 | 275 | 289 | | 295 | 281 | | | |
| 2893-57 | 285 | 299 | | 305 | 291 | | | |
| 2993-57 | 295 | 309 | | 315 | 301 | | | |
| 3093-57 | 305 | 319 | | 325 | 311 | | | |
| 3193-57 | 315 | 329 | | 335 | 321 | 0.36 | | |
| 3393-57 | 335 | 349 | 0.36 | 355 | 341 | | | |
| 3593-57 | 355 | 369 | | 375 | 361 | | $4.4^{+0.1}_{0}$ | 1.0 |
| 3793-57 | 375 | 389 | | 395 | 381 | | | |
| 3893-57 | 385 | 399 | | 405 | 391 | | | |
| 3993-57 | 395 | 409 | | 415 | 401 | 0.40 | | |
| 4193-57 | 415 | 429 | 0.40 | 436 | 422 | | | |
| 4393-57 | 435 | 449 | | 456 | 442 | | | |
| 4593-57 | 455 | 469 | | 476 | 462 | | | |
| 4793-57 | 475 | 489 | | 496 | 482 | | | |

| | | | | | |
|---|---|---|---|---|---|
| 1441-84 | 140 | 160 | | 145 | 165 |
| 1491-84 | 145 | 165 | | 150 | 170 |
| 1541-84 | 150 | 170 | | 155 | 175 |
| 1591-84 | 155 | 175 | 0.25 | 160 | 180 |
| 1641-84 | 160 | 180 | | 165 | 185 |
| 1691-84 | 165 | 185 | | 170 | 190 |
| 1741-84 | 170 | 190 | | 175 | 195 |
| 1791-84 | 175 | 195 | | 180 | 200 |
| 1841-84 | 180 | 200 | | 185 | 205 |
| 1891-84 | 185 | 205 | | 190 | 210 |
| 1941-84 | 190 | 210 | | 195 | 215 |
| 1991-84 | 195 | 215 | 0.29 | 200 | 220 |
| 2041-84 | 200 | 220 | | 205 | 225 |
| 2091-84 | 205 | 225 | | 210 | 230 |
| 2191-84 | 215 | 235 | | 220 | 240 |
| 2291-84 | 225 | 245 | | 230 | 250 |
| 2341-84 | 230 | 250 | 0.32 | 235 | 255 |
| 2391-84 | 235 | 255 | | 240 | 260 |
| 2491-84 | 245 | 265 | | 250 | 270 |

0.25   0.29   $6.6^{+0.1}_{0}$   1.0

## 4.9.4 Triangular housings for static seals

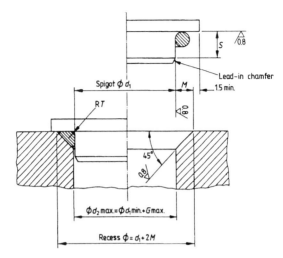

**Fig. 4**   *Triangular housing profile for static sealing*

(All dimensions in millimetres)

...sions of triangular housing for static sealing (see Fig. 4)

| Ref. no. | 'O'-ring Cross-section diameter A | Spigot diameter $d_1$ | Total diametral clearance G (max.) | Chamfer $M^{+0.12}_{0}$ | Maximum radius on spigot T | Spigot length S (min.) |
|---|---|---|---|---|---|---|
| 0031-16 to 0371-16 | 1.6 | | 0.12 | 2.20 | 0.8 | 4.0 |
| 0036-24 to 0696-24 | 2.4 | | 0.14 | 3.30 | 1.3 | 5.0 |
| 0195-30 to 2495-30 | 3.0 | As in Table 1, column 6 | 0.15 | 4.20 | 2.0 | 6.0 |
| 0443-57 to 4993-57 | 5.7 | | 0.18 | 7.80 | 3.0 | 10.0 |
| 1441-84 to 2491-84 | 8.4 | | 0.20 | 11.50 | 4.0 | 14.0 |

## 4.10 Riveted Joints

### 4.10.1 Typical rivet heads and shanks

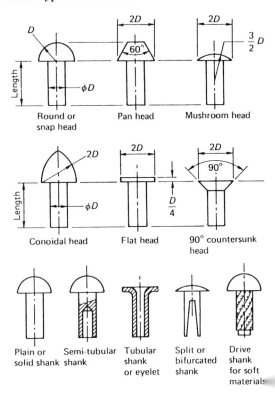

| | | |
|---|---|---|
| Round or snap head | Pan head | Mushroom head |
| Conoidal head | Flat head | 90° countersunk head |

Plain or solid shank    Semi-tubular shank    Tubular shank or eyelet    Split or bifurcated shank    Drive shank for soft materials

### 4.10.2 Typical riveted lap joints

**Single row lap joint**

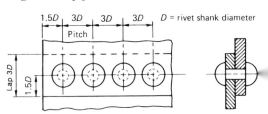

$D$ = rivet shank diameter

## Double row (chain) lap joint

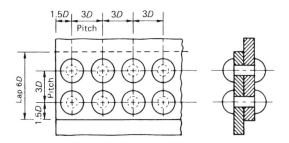

## Double row (zigzag) lap joint

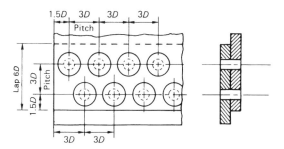

### 4.10.3 Typical riveted butt joints

## Single strap chain riveted butt joint (single row)

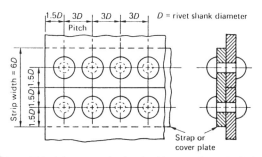

*Note:* This joint may also be double row riveted, chain or zigzag. The strap width = 12D when double riveted (pitch between rows = 3D).

## Double strap chain riveted butt joint (double row)

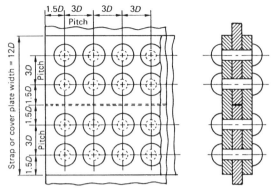

*Note:* This joint may also be double row zigzag riveted (see Section 4.10.2) or it may be single riveted as above.

### 4.10.4 Proportions for hole diameter and rivet length

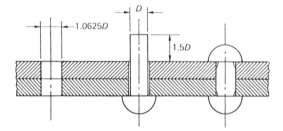

## 4.10.5 Cold forged snap head rivets

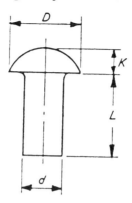

with *d* 16 mm or smaller

$D = 1.75d$
$K = 0.6d$
$L$ = Length

(Dimensions in millimetres)

| Nominal shank diameter* $d$ | Tolerance on diameter $d$ | Nominal head diameter $D$ | Tolerance on diameter $D$ | Nominal head depth $K$ | Tolerance on head depth $K$ | Tolerance on length $L$ |
|---|---|---|---|---|---|---|
| 1 | ±0.07 | 1.8 | ±0.2 | 0.6 | +0.2 / -0.0 | |
| 1.2 | | 2.1 | | 0.7 | | |
| 1.6 | | 2.8 | | 1.0 | | |
| 2.0 | | 3.5 | ±0.24 | 1.2 | +0.24 / -0.0 | |
| 2.5 | | 4.4 | | 1.5 | | |
| 3.0 | | 5.3 | | 1.8 | | +0.5 / -0.0 |
| (3.5) | ±0.09 | 6.1 | ±0.29 | 2.1 | +0.29 / -0.0 | |
| 4 | | 7.0 | | 2.4 | | |
| 5 | | 8.8 | | 3.0 | | |
| 6 | ±0.11 | 10.5 | ±0.35 | 3.6 | +0.35 / -0.0 | |
| (7) | | 12.3 | | 4.2 | | +0.8 / -0.0 |
| 8 | | 14.0 | | 4.8 | | |
| 10 | | 18.0 | ±0.42 | 6.0 | +0.42 / -0.0 | |
| 12 | ±0.14 | 21.0 | | 7.2 | | +1.0 / -0.0 |
| (14) | | 25.0 | | 8.4 | | |
| 16 | | 28.0 | | 9.6 | | |

\* those are non-preferred.

### 4.10.6 Hot forged snap head rivets

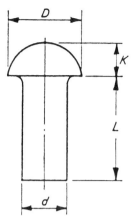

with *d* 14 mm or larger

$D = 1.6d$
$K = 0.65d$
$L$ = Length

(Dimensions in millimetres)

| Nominal shank diameter* d | Tolerance on diameter d | Nominal head diameter D | Tolerance on diameter D | Nominal head depth K | Tolerance on head depth K | Tolerance on length L |
|---|---|---|---|---|---|---|
| (14) | ±0.43 | 22 | ±1.25 | 9 | +1.00 / −0.0 | +10 / −0.0 |
| 16 | | 25 | | 10 | | |
| 18 | | 28 | | 11.5 | | |
| 20 | ±0.52 | 32 | ±1.8 | 13 | +1.5 / −0.0 | +1.6 / −0.0 |
| (22) | | 36 | | 14 | | |
| 24 | | 40 | | 16 | | |
| (27) | ±0.62 | 43 | ±2.5 | 17 | +2.0 / −0.0 | ±3.0 |
| 30 | | 48 | | 19 | | |
| (33) | | 53 | | 21 | | |
| 36 | | 58 | ±3.0 | 23 | +2.5 / −0.0 | |
| 39 | | 62 | | 25 | | |

*Rivet sizes shown in parentheses are non-preferred.
For further information see BS 4620: 1970.

## 4.10.7 Tentative range of nominal lengths associated with shank diameters

(Dimensions in millimetres)

| Nominal shank diameter* d | \ Nominal length* L | | | | | | | | | | | | | | | | | | | | | |
|---|---|---|---|---|---|---|---|---|---|---|---|---|---|---|---|---|---|---|---|---|---|---|
| | 3 | 4 | 5 | 6 | 8 | 10 | 12 | 14 | 16 | (18) | 20 | (22) | 25 | (28) | 30 | (32) | 35 | (38) | 40 | 45 | 50 | 55 |
| 1.0 | × | × | × | × | × | × | × | × | × | | × | | | | | | | | | | | |
| 1.2 | × | × | × | × | × | × | × | × | × | | × | | | | | | | | | | | |
| 1.6 | × | × | × | × | × | × | × | × | × | × | | | | | | | | | | | | |
| 2.0 | × | × | × | × | × | × | × | × | × | | × | × | × | | | | | | | | | |
| 2.5 | × | × | × | × | × | × | × | × | × | × | | × | × | | | | | | | | | |
| 3.0 | × | × | × | × | × | × | × | × | × | × | × | × | × | | | | | | | | | |
| (3.5) | | | | | | | | × | × | × | | | × | | | | | | | | | |
| 4.0 | | | | × | × | × | × | × | × | × | × | × | × | | | | | | | | | |
| 5.0 | | | | × | × | × | × | × | × | × | × | × | × | × | × | × | × | × | | | | |
| 6.0 | | | | × | × | × | × | × | × | × | × | × | × | × | × | × | × | × | | × | | |

*Sizes and lengths shown in parentheses are non-preferred and should be avoided if possible. The inclusion of dimensional data is not intended to imply that all the products described are stock production sizes. The purchaser should consult the manufacturer concerning lists of stock production sizes. For the full range of head types and sizes up to and including 39 mm diameter by 160 mm shank length see BS 4620: 1970.

## 4.11 Self-secured joints

### 4.11.1 Self-secured joints

**Grooved seam**

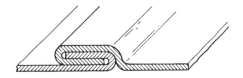

**Double grooved seam**

**Paned down seam**

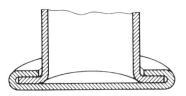

**Knocked up seam**

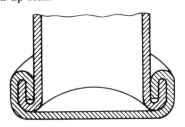

# Making a grooved seam

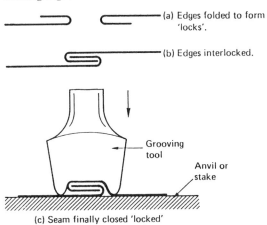

(a) Edges folded to form 'locks'.

(b) Edges interlocked.

Grooving tool

Anvil or stake

(c) Seam finally closed 'locked' using a grooving tool of the correct width.

## 4.11.2 Allowances for self-secured joints

### Grooved seam

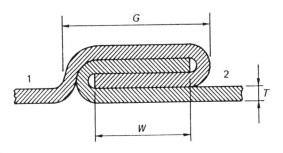

### Double grooved seam

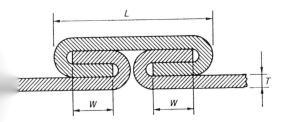

## Paned down seam

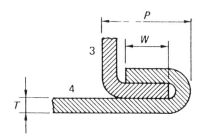

## Knocked up seam

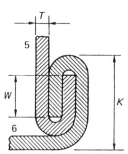

$W$ = width of lock (folded edge)
$G$ = width of grooved seam
$L$ = width locking strip
$P$ = width of paned down seam
$K$ = width of knocked up seam
$T$ = thickness of metal

| Type of joint | Approximate allowance |
|---|---|
| Grooved seam | Total allowance = $3G - 4T$ shared: <br>(a) Equally between limbs 1 and 2; *or* <br>(b) Two-thirds limb 1 and one-third limb 2 where joint centre position is critical. |
| Double grooved seam | Add $W - T$ to the edge of each blank to be joined. <br>Allowance for capping strip = $4W + 4T$, where $L = 2W + 4T$. |
| Paned down seam | Add $W$ to the single edge 3. <br>Add $2W + T$ to the double edge 4. <br>$P = 2W + 2T$. |
| Knocked up joint | Add $W$ to the single edge 5. <br>Add $2W + T$ to the double edge 6. <br>$K = 2W + 3T$. |

## 4.12 The hardening of plain carbon steels

### 4.12.1 Hardening

To quench harden a plain carbon steel, it is cooled rapidly (quenched) from the temperatures shown in the figure below. The degree of hardness the steel achieves is solely dependent upon:

(a) the carbon content;
(b) the rate of cooling.

(a) *Carbon content:* There must be sufficient carbon present to form the hard crystal structures in the steel when it is heated and quenched. The effect of the carbon content on the hardness of the steel after heating and quenching is shown in Table (a).

(b) *Rate of cooling:* The rapid cooling necessary to harden steel is known as **quenching.** The liquid into which the steel is dipped to cause this rapid cooling is called the **quenching bath**.

In the workshop, the quenching bath will contain either:

(a) water;
(b) quenching oil (on no account use lubricating oil).

The more rapidly a plain carbon steel is cooled the harder it becomes.

**Table (a)** Effect of carbon content

| Type of steel | Carbon content (%) | Effect of heating and quenching (*rapid cooling*) |
|---|---|---|
| Mild | Below 0.25 | Negligible |
| Medium carbon | 0.3–0.5 | Becomes tougher |
| | 0.5–0.9 | Becomes hard |
| High carbon | 0.9–1.3 | Becomes very hard |

Unfortunately, rapid cooling also leads to *cracking* and *distortion*. Therefore, the workpiece should not be cooled more rapidly than is required to give the desired degree of hardness. For plain carbon steels, the cooling rates shown in Table (b) are recommended.

**Table (b)**   Rate of cooling

| Carbon content (%) | Quenching bath | Required treatment |
|---|---|---|
| 0.30–0.50 | Oil | Toughening |
| 0.50–0.90 | Oil | Toughening |
| 0.50–0.90 | Water | Hardening |
| 0.90–1.30 | Oil/water | Hardening |

*Notes:*

1. Below 0.5% carbon content, steels are not hardened as cutting tools, so water hardening has not been included.
2. Above 0.9% carbon content, any attempt to harden the steel in water could lead to cracking.

### 4.12.2 Tempering

Hardened plain carbon steel is very brittle and unsuitable for immediate use. A further process known as **tempering** must be carried out to greatly increase the toughness of the steel at the expense of some hardness.

Tempering consists of reheating the steel to a suitable temperature and quenching it in oil or water. The temperature to which the steel is heated depends upon the use to which the component is going to be put. Table (c) gives some suitable temperatures for tempering components made from plain carbon steel.

**Table (c)**   Tempering temperatures

| Component | Temper colour | Temperature (°C) |
|---|---|---|
| Edge tools | Pale straw | 220 |
| Turning tools | Medium straw | 230 |
| Twist drills | Dark straw | 240 |
| Taps | Brown | 250 |
| Press tools | Brownish-purple | 260 |
| Cold chisels | Purple | 280 |
| Springs | Blue | 300 |
| Toughening (crankshafts) | — | 450–600 |

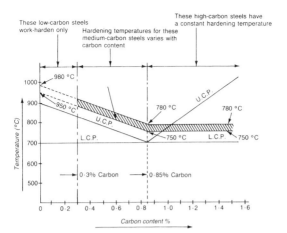

Hardening of plain carbon steels

In the workshop, the tempering temperature is usually judged by the colour of the oxide film that appears on a freshly polished surface of the steel when it is heated. Some tools, such as chisels, only need the cutting edge hardened, the shank being left tough to withstand hammer blows.

### 4.12.3 Overheating carbon steels

It is a common mistake to overheat a steel in the hope that it will become harder. As already stated, the hardness only depends upon the carbon content of the steel and the rate of cooling. Once the correct hardening temperature has been reached, any further increase in temperature only slows up the time taken to cool the workpiece and this tends to reduce the final hardness. Further, overheating also causes *crystal growth*, resulting in a weak and defective component. If the overheating is excessive then 'burning' occurs. This is oxidation of the crystal boundaries of the metal resulting in great weakness and, unlike overheating, the condition cannot be corrected so the workpiece is useless and can only be melted down as scrap.

On the other hand, failure to reach the hardening temperature results in the component not becoming hard no matter how quickly it is quenched.

## 4.12.4 Softening (annealing) plain carbon steels

Plain carbon steels can be softened by heating them to the same temperatures as those shown in the figure above but this time the hot steel is cooled **very slowly**. For large components heated in a furnace, the furnace is turned off, the flue dampers are closed and the furnace and the work slowly cool down together. Small components, heated by a gas torch, can be buried in ground lime stone or fine ashes so that they can cool down slowly.

## 4.12.5 Radiant temperatures

The steel glows at the temperatures required for quench hardening, forging and forge welding. Suitable temperatures and their radiation colours are given in Table (d).

**Table (d)**   Radiant heat colour temperatures

| Colour | Celcius | Fahrenheit |
|---|---|---|
| Just visible red | 500°–600° | 932°–1112° |
| Dull cherry red | 700°–750° | 1300°–1385° |
| Cherry red | 750°–825° | 1385°–1517° |
| Bright cherry red | 825°–875° | 1517°–1600° |
| Brightest red | 900°–950° | 1652°–1750° |
| Orange | 950°–1000° | 1750°–1835° |
| Light Orange | 1000°–1050° | 1835°–1925° |
| Lemon | 1100°–1200° | 2012°–2200° |
| White | 1200°–1300° | 2200°–2372° |

*Note:* The above colours and temperatures are, of course, only roughly approximate.

## Example

Cherry red is suitable for quench hardening ar annealing silver steel.

Bright cherry red is suitable for quench hardeni and annealing medium carbon steels and ground f stock (gauge plate).

Brightest red and orange would be used for forgi (blacksmithing).

Lemon/white heat would be used for forge weldir

(Courtesy of Addison Wesley Longman.)

# 4.13 Types of soft solder and soldering fluxes

| BS Solder | Composition % Tin | Lead | Antimony | Melting range (°C) | Remarks |
|---|---|---|---|---|---|
| A | 65 | 34.4 | 0.6 | 183 – 185 | Free-running solder ideal for soldering electronic and instrument assemblies. Commonly referred to as **electrician's solder.** |
| K | 60 | 39.5 | 0.5 | 183 – 188 | Used for high-class tinsmith's work, and is known as **tinman's solder.** |
| F | 50 | 49.5 | 0.5 | 183 – 212 | Used for general soldering work in coppersmithing and sheet-metal work. |
| G | 40 | 59.6 | 0.4 | 183 – 234 | **Blow-pipe solder.** This is supplied in strip form with a D cross-section 0.3 mm wide. |
| J | 30 | 69.7 | 0.3 | 183 – 255 | **Plumber's solder.** Because of its wide melting range this solder becomes 'pasty' and can be moulded and wiped. |

Non-corrosive types

| Flux | Remarks |
|---|---|
| esin | In its natural form is the gum extracted from the bark of pine trees. It is an amber-coloured substance which is solid at room temperature and does not cause corrosion, but it reacts mildly at soldering temperatures. It is used mainly for electrical work. |
| llow | This is a product of animal fat. It is virtually inactive at room temperature, and like resin is only slightly active at soldering temperatures. This flux is used extensively with 'plumber's solder' for jointing lead sheets and pipes, and with 'body solder' on previously tinned steel for motor-vehicle repair work. |

| Olive oil | This is a natural vegetable oil. It forms a weak vegetable acid at soldering temperatures. A useful flux when soldering pewter. Being non-toxic it is widely used on food containers (canning). |
|---|---|

Corrosive types

| Flux | Remarks |
|---|---|
| Zinc chloride | Commonly called 'Killed Spirits of Salts' and forms the base for most commercially produced fluxes. A good general flux suitable for mild steel, brass, copper, ternplate and tinplate. A proprietary brand is called Baker's fluid. |
| Ammonium chloride | As a soldering flux it is generally used in liquid form when tinning cast iron, brass or copper. |
| Hydrochloric acid | This is known as 'Raw Spirits of Salts' and is extremely corrosive. It is used in dilute form when soldering zinc and galvanised iron or steel. *Zinc chloride (killed spirits of salts is produced by the chemical action of dilute hydrochloric acid on zinc.* |

(Courtesy of Addison Wesley Longman.)

## 4.14 Silver solders and fluxes

| BS 1845 No. | Ag % | Cu % | Zn % | Cd % | Sn % | Mn % | Ni % | Approximate Solidus °C | Liquidus °C | Fry's Metals Ltd | Johnson Matthey Ltd. | Thessco Ltd | Remarks |
|---|---|---|---|---|---|---|---|---|---|---|---|---|---|
| Ag1 | 60 | 15 | 16 | 19 | — | — | — | 620 | 640 | FSB No. 3 | Easyflo | MX20 | For all general work. Fine fillets. |
| Ag2 | 42 | 17 | 16 | 25 | — | — | — | 610 | 620 | FSB No. 2 | Easyflo No. 2 | MX12 | For all general work. Cheaper than Ag1. |
| Ag3 | 38 | 20 | 22 | 20 | — | — | — | 605 | 650 | FSB No. 1 | Argoflo | AG3 | For wider joint gaps than Ag1 and 2. Moderate fillets. |
| Ag11 | 34 | 25 | 20 | 21 | — | — | — | 612 | 688 | FSB No. 15 | Mattibraze 34 | MX4 ⎫ | Cheaper grades, for wider gaps. Larger fillets and wider melting range. |
| Ag12 | 30 | 28 | 21 | 21 | — | — | — | 600 | 690 | FSB No. 16 | Argoswift | MX0 ⎭ | |
| Ag9 | 50 | 15½ | 15½ | 16 | — | — | 3 | 635 | 655 | FSB No. 19 | Easyflo No. 3 | MX20N | Nickel bearing, rather sluggish. For brazing tool-tips. Forms substantial fillets. |
| **CADMIUM FREE ALLOYS** | | | | | | | | | | | | | |
| Ag14 | 55 | 21 | 22 | — | 2 | — | — | 630 | 660 | FSB No. 29 | Silverflo 55 | M25T ⎫ | Cadmium free substitute for Ag1 and Ag2. Very fluid. ⎫ Tin bearing |
| Ag20 | 40 | 30 | 28 | — | 2 | — | — | 650 | 710 | — | Silverflo 40 | M10T ⎬ | |
| Ag21 | 30 | 36 | 32 | — | 2 | — | — | 665 | 755 | FSB No. 33 | Silverflo 302 | M0T ⎭ | |

351

# 4.14 (continued)

## CADMIUM FREE ALLOYS (continued)

| | | | | | | | | | | | | | |
|---|---|---|---|---|---|---|---|---|---|---|---|---|---|
| Ag13 | 60 | 26 | 14 | — | — | — | — | 695 | 730 | FSB No. 4 | Silverflo 60 | H0 | Low zinc. Recommended for nickel-bearing alloys. |
| Ag5 | 43 | 37 | 20 | — | — | — | — | 690 | 770 | FSB No. 5 | Silverflo 43 | — | Useful for step brazing. |
| Ag7 | 72 | 28 | — | — | — | — | — | MP | 780 | FSB No. 17 | AgCu Eutectic | H12 | Very fluid indeed; for vacuum brazing. |
| Ag18 | 49 | 16 | 23 | — | 7.5 | 4.5 | — | 680 | 705 | FSB No. 37 | Argobraze 49H 15 Mn–Ag | M19MN | For brazing carbide tool-tips. |
| Ag19 | 85 | — | — | — | 15 | — | — | 960 | 970 | — | — | — | Very costly; the silver–manganese eutectic. |

### From BS.1845/1977, but still available

| | | | | | | | | | | | | | |
|---|---|---|---|---|---|---|---|---|---|---|---|---|---|
| Ag10* | 40 | 19 | 21 | 20 | — | — | — | 595 | 630 | FSB No. 10 | DIN Argoflo | MX10/DIN | Cheaper than Ag1 and 2 but slightly longer melting range. Fluid. |
| Ag15* | 44 | 30 | 26 | — | — | — | — | 675 | 735 | FSB No. 39 | Silverflo 44 | M14 | Ag15–17 form a series of alloys suitable for successive step brazing. |
| Ag16 | 30 | 38 | 32 | — | — | — | — | 680 | 770 | FSB No. 25 | Silverflo 30 | M0 | |
| Ag17* | 25 | 41 | 34 | — | — | — | — | 700 | 800 | FSB No. 23 | Silverflo 25 | L18 | |

*These alloys also conform to the DIN specifications. Note that the alloy contents shown lie at the midrange of the specifications, and slight differences may be expected between makers. The same applies to the solidus and liquidus figures, which may be a few degrees up or down.

(Reproduced by permission of the British Standards Institution.)

## 4.15 Sparking plug threads 60° SAE standard

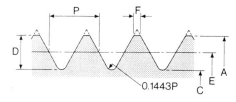

| Dia | Pitch P | Core dia. C | Depth D | Flat F | Tapping drill |
|-----|---------|-------------|---------|--------|---------------|
| 10 | 1.0 | 8.75 | 0.6134 | 0.1250 | 9.10 |
| 12 | 1.25 | 10.44 | 0.7668 | 0.1563 | 10.90 |
| 14 | 1.25 | 12.44 | 0.7668 | 0.1563 | 12.90 |
| 18 | 1.5 | 15.75 | 0.9202 | 0.1875 | 16.50 |

## 4.16 Wood screw hole sizes

| Gauge of screw | Clearance hole | | Pilot hole | |
|----------------|----------------|--------|------------|--------|
| | Imp | Metric | Imp | Metric |
| 0 | 5/64 | 2.00 | 3/64 | 1.00 |
| 2 | 3/32 | 2.50 | 1/16 | 1.50 |
| 4 | 1/8 | 3.00 | 5/64 | 2.00 |
| 6 | 5/32 | 4.00 | 3/32 | 2.50 |
| 8 | 3/16 | 4.50 | 7/64 | 2.50 |
| 10 | 7/32 | 5.00 | 1/8 | 3.00 |
| 12 | 1/4 | 6.00 | 9/64 | 3.50 |
| 14 | 17/64 | 6.50 | 5/32 | 4.00 |

(Courtesy A.J. Reeves (B'ham) Ltd.)

# Appendix 1

## BSI Standards — sales order and enquiry contacts

BSI Standards provides a variety of products and services to help standards users to manage their collection and to make standards work for their organisation. The following is a brief summary of the services available and appropriate contact numbers.

### Customer Services
Tel: (020) 8996 7000
Fax: (020) 8996 7001

— *for identifying, price quotations and ordering British and Foreign standards and other publications*
— *'PLUS' — Private List Updating Service*

### Information Centre
Tel: (020) 8996 7111
Fax: (020) 8996 7048

— *for detailed information and searches on British and overseas standards*
— *Technical Help to Exporters*
— *certification and testing requirements overseas*
— *EC standardisation developments*

### Membership Services
Tel: (020) 8996 7002
Fax: (020) 8996 7001

— *members' help desk*
— *membership administration*

### Translations and Language Services
Tel: (020) 8996 7222
Fax: (020) 8996 7047

— *for technical, standards and commercial translations*

**Copyright**
Tel: (020) 8996 7070
Fax: (020) 8996 7400

— *copyright licences and enquires*

**Library Services**
Tel: (020) 8996 7004
Fax: (020) 8996 7005

— *library services and enquiries*

**Electronic Products Help Desk**
Tel: (020) 8996 7333
Fax: (020) 8996 7047

— *Perinorm*
— *electronic product development*

**Ordering BSI publications**

Orders can be placed by post, phone, fax or telex through BSI Customer Services

Post: BSI Customer Services
BSI Standards
389 Chiswick High Rd
London W4 4AL

Tel: (020) 8996 7000
Fax: (020) 8996 7001

The hours of opening are 8.30 am to 5.30 pm Mondays to Fridays exc. public holidays.

PLUS
Private List Updating Service

Contact: Tel: (020) 8996 7398
Fax: (020) 8996 7001

Electronic media and databases — Help desk

Contact: Fax: (020) 8996 7047

Copyright licences
Contact: Pamela Danvers
Tel: (020) 8996 7070
Fax: (020) 8996 7001

### Non-members

Non-members should send remittance with order, based on the prices given, or pay directly by credit card

### Members

Members will be invoiced in the usual way and will receive the appropriate discounts

### Prices

The Group (Gr) number of each entry, in conjunction with the key below, indicates the UK price of the document. Postage and packing is included.

| Group No. | Non-members | Members |
|---|---|---|
| 0 | 2.60 | 1.30 |
| 1 | 2.60 | 1.30 |
| 2 | 6.70 | 3.35 |
| 3 | 12.00 | 6.00 |
| 4 | 17.50 | 8.75 |
| 5 | 26.00 | 13.00 |
| 6 | 32.60 | 16.30 |
| 7 | 47.00 | 23.50 |
| 8 | 61.50 | 30.75 |
| 9 | 72.50 | 36.25 |
| 10 | 86.50 | 43.25 |
| 11 | 92.00 | 46.00 |
| 12 | 104.00 | 52.00 |
| 13 | 116.50 | 58.25 |
| 14 | 127.50 | 63.75 |
| 15 | 130.00 | 65.00 |

The above prices are for guidance and may be subject to variation

### Members' order hotline

Tel: (020) 8996 7003

For members who know exactly what they want to order.

Speeds up the placing of telephone orders

**Priority Service**

Urgent orders received before 12.00 hours, by phone, fax or telex can be sent by Priority Service and will be despatched on the same day by first class mail (*orders for the Priority Service should be clearly marked*)

The charge for this service is 10% of invoice value, with a minimum charge of £1.00 and a maximum charge of £50.00

**Sales outlets**

See the BSI Catalogue for a full list of BSI sales outlets

**Information centre**

For detailed information and searches on British and overseas standards:

— *certification and testing requirements overseas*
— *Technical help to Exporters*
— *EC development re standardisation*

Tel: (020) 8996 7021
    Electrical
Tel: (020) 8996 7022
    Consumer Products
Tel: (020) 8996 7023
    Construction
Tel: (020) 8996 7024
    Mechanical
Fax: (020) 8996 7048

— *library loans — international and foreign standards and related technical documents are available for loan to BSI members. The current price of tokens is £35.00 per book for orders of 10 or more books (please quote token numbers when requesting items for loan) Library contact*:

Tel: (020) 8996 7004
Fax: (020) 8996 7005

---

***BSI News Update***
**complied and edited by**
**Kay Westlake**
**Tel: (020) 8996 7060**
**Fax: (020) 8996 7089**

---

## Membership services

Tel: (020) 8996 7002
Fax: (020) 8996 7001

## Translations & language services

Tel: (020) 8996 7222
Fax: (020) 8996 7047

— *for technical, standards and commercial translations*

## Copyright

Copyright subsists in all BSI publications. **BSI also holds the copyright, in the UK**, of the publications of the international standardisation bodies. Except as permitted under the Copyright, Designs and Patents Act 1988 no extract may be reproduced, stored in a retrieval system or transmitted in any form or by any means — electronic, photocopying, recording or otherwise — without prior written permission from BSI. If permission is granted, the terms may include royalty payments or a licensing agreement.

Details and advice can be obtained from the:

Copyright Manager,
BSI, 389 Chiswick High Road,
London W4 4AL

Tel: (020) 8996 7070

## BSI print on demand policy

- BSI has revised its production processes following intensive investment and process redesign enabling the rapid production of products on a variety of media from paper to electronic books
- Increasingly, all British, European and International standards you order will be printed on demand from images stored in electronic files
  By providing standards as looseleaf, hole-punched documents, amendments can be integrated more easily as replacement pages to provide improved, up-to-date and complete working documentation.

Eventually the messy and time-consuming cut-and-paste methods will no longer be necessary to update your standards

- During the transition phase from existing processes the product may arrive in various formats. We apologise for this and also for the quality of original material we have to use until the changeover is complete. We hope to start receiving live files from our international colleagues over the next 12 months which will significantly improve the print quality of BS ISOs and BS EN ISOs
- All of the new-format standards are printed on special watermarked paper with the words 'licensed copy' inside the paper, so that you can show that you own the official standard supplied by BSI

**BSI Standards**
389 Chiswick High Road
London W4 4AL

Tel: (020) 8996 9000
Fax: (020) 8996 7400

**Customer Services**
Tel: (020) 8996 7000
Fax: (020) 8996 7001

**Membership Administration**
Tel: (020) 8996 7002
Fax: (020) 8996 7001

**Information Centre**
Tel: (020) 8996 7111
Fax: (020) 8996 7048

**BSI Quality Assurance**
Tel: 01908 220908
Fax: 01908 220671

**BSI Testing**
Tel: 01442 230442
Fax: 01442 321442

**BSI Product Certification**
Tel: 01908 312636
Fax: 01908 695157

**BSI Training Services**
Tel: (020) 8996 7055
Fax: (020) 8996 7364

— Materials and chemicals
— Health and environment
— Consumer products
— Engineering
— Electrotechnical
— Management systems
— Information technology
— Building and civil engineering

• British Standards
• Corresponding International Standards
• European Standards
• Handbooks and other publications

# Appendix 2

---

## Library sets of British Standards in the UK

The following libraries hold full sets of British Standards in either paper, CD-ROM or microfiche formats. The names of public libraries are shown in italics. For other libraries, it is advisable to make prior written application in order to ascertain the hours and conditions for access. **These sets are for reference only and attention is drawn to Copyright Law**.

**ENGLAND**
**Avon**
Bath                    *Central Library*
                        University of Bath
Bristol                 *Commercial Library*

**Bedfordshire**
Bedford                 *Central Library*
Cranfield               Cranfield University
Leighton Buzzard        *Public Library*
Luton                   *Central Library*
                        University of Luton

**Berkshire**
Reading                 *Central Library*
                        University of Reading
                        Reading College
Slough                  *Central Library*

**Buckinghamshire**
Aylesbury               *County Hall*
High Wycombe            College of H & E
Milton Keynes           *Central Library*

**Cambridgeshire**
Cambridge               *Central Library*
Peterborough            *Central Library*

**Cheshire**
Crewe                   *Central Library*
South Wirral            Ellesmere Port Library
Stockport               *Central Library*
Warrington              *Central Library*

**Cleveland**
Cleveland               *County Library*

362

**Hertfordshire**

| | |
|---|---|
| Hatfield | *Central Library* |
| | University of Hertfordshire |
| Stevenage | *Central Library* |
| Watford | *Central Library* |

**Humberside**

| | |
|---|---|
| Grimsby | *Central Library* |
| Hull | *Central Library* |
| Scunthorpe | *Central Library* |

**Jersey**

| | |
|---|---|
| St Helier | *The Jersey Library* |

**Kent**

| | |
|---|---|
| Bexley Heath | *Central Library* |
| Bromley | *Central Library* |
| Chatham | *Central Library* |
| Maidstone | *County Library* |
| Margate | *Public Library* |
| Tonbridge | *Central Library* |

**Lancashire**

| | |
|---|---|
| Blackburn | *Central Library* |
| Bolton | *Central Library* |
| | Institute of Higher Education |
| Preston | *Central Library* |
| | Lancashire Polytechnic |

**Leicestershire**

| | |
|---|---|
| Leicester | De Montfort University |
| | *Information Centre* |
| | University of Leicester |
| Loughborough | University of Technology |

**Lincolnshire**

| | |
|---|---|
| Lincoln | *Central Library* |

**Greater London**

| | |
|---|---|
| Battersea | *Reference Library* |
| Chiswick | BSI Library |
| City University | Reference Library |
| Gower Street | University College London |
| Hammersmith | *Central Library* |
| Haringey | Middx University, Bounds Green |
| | *Central Library* |
| Hendon | *The Burroughs* |
| Holborn | *The British Library* |
| Islington | *Central Library* |
| Kensington | Imperial College of Science |
| | *Educational Libraries* |
| Kingston | Kingston University |
| Palmers Green | *Reference Library* |
| Southwark | South Bank University |

| Stratford | Newham Community College |
| | *Reference Library* |
| Swiss Cottage | *Reference Library* |
| Waltham Forest | Waltham College |
| Westminster | University of Westminster |
| | *Westminster Libraries* |
| Woolwich | *Woolwich Central Library* |

**Greater Manchester**

| Ashton-under-Lyne | *Public Library* |
| Manchester | John Rylands |
| | University Library |
| | Metro University Library |
| | *Public Library* |
| | UMIST Library |
| | University of Manchester |
| Oldham | *Reference Library* |
| Salford | College of Technology |
| | University of Salford |
| Wigan | Wigan & Leigh College |

**Merseyside**

| Birkenhead | *Central Library* |
| Liverpool | *Central Reference Library* |
| | John Moores University |
| | University, Harold Cohen Library |
| St Helens | Gamble Institute |

**Middlesex**

| Uxbridge | Brunel University |

**W Midlands**

| Birmingham | Aston University |
| | *Chamberlain Square* |
| | University of Birmingham |
| | University of Central England |
| Coventry | *Central Library* |
| | Coventry University |
| | Lanchester |
| | University Library |
| | University of Warwick |
| Dudley | *Reference Library* |
| Solihull | *Central Library* |
| Walsall | *Central Library* |
| West Bromwich | *Central Library* |
| Wolverhampton | *Central Library* |

**Norfolk**

| Norwich | *County Hall* |

**Northamptonshire**

| Northampton | *Central Library* |

**Northumberland**
Ashington — College of Arts & Technology

**Nottinghamshire**
Nottingham — *County Library*
Trent University Library
University of Nottingham

**Oxfordshire**
Didcot — Rutherford Appleton Laboratory
Oxford — *Central Library*
Oxford Brookes University

**Shropshire**
Telford — *St Quentin Gate*

**Somerset**
Bridgewater — *County Library*

**Staffordshire**
Hanley — *Library*

**Suffolk**
Ipswich — *County Library*
Lowestoft — *Central Library*

**Surrey**
Croydon — *Central Library*
Guildford — College of Technology
Sutton — *Central Library*
Woking — *Public Library*

**E Sussex**
Brighton — *Reference Library*

**W Sussex**
Brighton — University of Brighton
Crawley — *Public Library*

**Tyne and Wear**
Gateshead — *Central Library*
Newcastle — *Central Library*
Polytechnic University of Newcastle
North Shields — *Central Library*
South Shields — *Central Library*
South Tyneside College
Sunderland — *County Library*
University of Sunderland
Washington — *Central Library*

**Warwickshire**
Rugby — *Central Library*

**Wiltshire**
Trowbridge — *Public Library*

**N Yorkshire**
Northallerton      *County Library*
York      Askham Bryan College
     *Central Library*

**S Yorkshire**
Barnsley      *Central Library*
Doncaster      *Central Library*
Rotherham      *Central Library*
Sheffield      *Central Library*
     City Polytechnic Library
     University of Sheffield

**W Yorkshire**
Bradford      *Public Library*
     University Library
Huddersfield      *Central Library*
     University of Huddersfield
Leeds      *Central Library*
     Metropolitan University
     University, Edward
     Boyle Library
Wakefield      *Library*
     *Headquarters*

**N IRELAND**
**Antrim**
Ballymena      *Area Library*
Belfast      *Central Library*
     College of Technology
     Science Library,
     Queens University
Newtonabbey      University of Ulster

**Armagh**
Armagh      *Southern Education & Library Board*

**Down**
Ballynahinch      *South Eastern Education & Library Board*

**Tyrone**
Omagh      *County Library*

**SCOTLAND**
**Grampian**
Aberdeen      *Central Library*

**Lanarkshire**
Hamilton      Bell College of Technology

**Lothian**
Edinburgh      *Central Library*
     Heriot-Watt University
     Napier Polytechnic University,
     Engineering Library

**Strathclyde**
East Kilbride        *Central Library*
Glasgow              Glasgow University Library
                     *The Mitchell Library*
                     Strathclyde University Library

**Tayside**
Dundee               *Central Library*
                     Institute of Technology
Forfar               *Forfar Central Library*

**WALES**
**Clwyd**
Mold                 *County Civic Centre*

**Dyfed**
Llanelli             *Public Library*

**S Glamorgan**
Cardiff              *Central Library*

**W Glamorgan**
Swansea              *Central Library*
                     University Library

# Appendix 3

## Contributing companies

A. J. Reeves (Birmingham) Limited
Holly Lane
Marston Green
Birmingham B37 7AW
Tel: 0121 779 6831
Fax: 0121 779 5205

Suppliers to Model Engineering.

Addison Wesley Longman
Edinburgh Gate
Harlow
Essex
CM 20 2JE
Tel: +44 (0) 1279 623623
Fax: +44 (0) 1279 431059
Website http://www.pearsoned-ema.com

# INDEX